Organic Light Emitting Diode (OLED) Toward Smart Lighting and Displays Technologies

The book *Organic Light Emitting Diode (OLED) Toward Smart Lighting and Displays Technologies,* edited by Laxman Singh, Rituraj Dubey, and Prof. R. N. Rai, strives to address the multiple aspects of OLEDs and their applications in developing smart lightings and displays. OLEDs have been used in almost all kinds of digital displays like those of mobile phones, laptops, tablets, phablets, TVs, etc., due to their outstanding features, including superior color quality, low cost, wide viewing angle, easy fabrication, mercury-free manufacture, tenability, stretchability, flexibility, etc. Investigations related to the synthesis of new organic materials and fabrication techniques have inspired us to write this book, which will fulfil the desire and thirst of OLEDs-based researchers.

Features

- Nanolithographic techniques used and the challenges involved.
- Printing technology for fabrication.
- Designing of hybrid perovskites.
- Stretchable and flexible materials used.
- Metal–dielectric composites and efficiency of organic semiconductor via molecular doping for OLEDs applications.
- Organic small molecule materials and display technologies involved.
- New generation of organic materials with respect to photophysical approach.
- Mixed valence π-conjugated coordination polymers used.
- Electroluminescent polymer used.
- Blue fluorescent and phosphorescent organic materials used.

In comparison to other books available related to similar topics, this book aims at those audiences who are looking for a single source for a comprehensive understanding of strategies and their challenges with respect to material fabrication of OLEDs. This book covers the pace and productivity at a uniform level in each chapter with respect to the audiences, from doctoral student to postdoctoral researchers or from postdoctoral researchers to multidisciplinary field researchers with a background in physics, chemistry, materials science, and engineering, who are already working with organic materials and their applications.

Organic Light Emitting Diode (OLED) Toward Smart Lighting and Displays Technologies

Material Design Strategies, Challenges and Future Perspectives

Edited by

Laxman Singh
Department of Chemistry, Siddharth University, UP, India

Rituraj Dubey
Ramadhin College Sheikhpura, Munger University, Bihar, India

R. N. Rai
Banaras Hindu University (BHU), Varanasi, UP, India

CRC Press
Taylor & Francis Group
Boca Raton London New York

CRC Press is an imprint of the
Taylor & Francis Group, an **informa** business

First edition published 2024
by CRC Press
2385 NW Executive Center Drive, Suite 320, Boca Raton FL 33431

and by CRC Press
4 Park Square, Milton Park, Abingdon, Oxon, OX14 4RN

CRC Press is an imprint of Taylor & Francis Group, LLC

© 2024 selection and editorial matter, Dr Laxman Singh, Dr Rituraj Dubey and Prof. R. N. Rai; individual chapters, the contributors

Library of Congress Cataloging-in-Publication Data
Names: Singh, Laxman (Chemistry professor), editor. | Dubey, Rituraj, editor. | Rai, R. N. (Rama Nand) editor.
Title: Organic light emitting diode (OLED) toward smart lighting and displays technologies : material design strategies, challenges and future perspectives / edited by Dr. Laxman Singh, RRS College, India, Dr. Rituraj Dubey, B. N. M. College, Munger University, India, Prof. R. N. Rai, Banaras Hindu University (BHU), Varanasi.
Other titles: OLED toward smart lighting and displays technologies
Description: First edition. | Boca Raton : CRC Press, [2024] | Includes bibliographical references and index. | Contents: Nanolithographic techniques for OLEDs / Rituraj Dubey and Laxman Singh—Printing technology for fabrication of advanced OLEDs materials / Pragati Kushwaha and Rituraj Dubey—Design of hybrid Perovskites for OLEDs / Ashish Kumar Srivastava and Arvind Kumar Singh.
Identifiers: LCCN 2023012511 (print) | LCCN 2023012512 (ebook) | ISBN 9781032197036 (hbk) | ISBN 9781032197043 (pbk) | ISBN 9781003260417 (ebk)
Subjects: LCSH: Organic light emitting diodes. | Light emitting diodes. | Organic semiconductors.
Classification: LCC TK7871.89.O74 O745 2024 (print) | LCC TK7871.89.O74 (ebook) | DDC 621.3815/22—dc23/eng/20230821
LC record available at https://lccn.loc.gov/2023012511
LC ebook record available at https://lccn.loc.gov/2023012512

ISBN: 978-1-032-19703-6 (hbk)
ISBN: 978-1-032-19704-3 (pbk)
ISBN: 978-1-003-26041-7 (ebk)

DOI: 10.1201/9781003260417

Typeset in Times
by Apex CoVantage, LLC

Contents

Chapter 10 Improvement in the Efficiency of Organic Semiconductors via
Molecular Doping for OLEDs Applications 203

Sunil Kumar, Pankaj Kumar Chaurasia, and Sandeep K. S. Patel

Chapter 11 Recent Development of Blue Fluorescent Organic Materials for
OLEDs 221

*Pawan Kumar Sada, Pranshu K. Gupta, Tanu Gupta, Abhishek
Rai, and Alok Kumar Singh*

Foreword

There is a continuous increase in the use of organic light emitting diode (OLED) displays and some of the reports indicate that the global market may reach up to more than $150 billion by 2030. The market for display panels is driven by a number of factors, including increasing demands for large liquid crystal display (LCD) televisions, smart phone-sized screens, and display screens for all types of vehicles. The driving force for the display panels includes a variety of applications, such as the introduction of ultrathin LCD televisions, slim smart phone designs, and improved user interfaces such as touch screens for car displays. The developments in the area of flexible substrates have enabled foldable display panels for tablets, smart phones, and notebooks. These flexible substrates made of polymers, plastics, metals, and flexible glass enable bending, as well as size and weight control. Some of these are sturdy and nearly shatter-proof. This area of flexible display technology is already in production by several companies to meet customer demands for the above-mentioned applications.

The book *Organic Light Emitting Diode (OLED) Toward Smart Lighting and Displays Technologies,* edited by Dr. Laxman Singh, Dr. Rituraj Dubey, and Prof. R. N. Rai, strives to address the multiple aspects of OLEDs and their applications in developing smart lightings and displays. They have put very good efforts in bringing together the researchers and the scholars from different parts of the world to contribute a timely and comprehensive picture of the organic light emitting diodes (OLEDs), especially, toward smart lighting and display technologies. Their expertise and good academic and research credentials in the diverse fields of soft materials for energy, biomedical, and environment applications have helped significantly in compiling and editing the chapters and preparing the book. The book has the potential to serve the community interested in OLEDs and their advancement toward smart lighting and display technologies.

I want to wish all the best to the publisher, editors, and contributors for great success. I hope the readers of the book and the policy makers and other stakeholders will find it valuable for further research and commercialization aspects of OLEDs-based smart lightings and displays.

N. B. Singh, PhD
Professor, University of Maryland, Baltimore (USA)

Preface

Organic light emitting diodes (OLEDs) have become an indispensable part of our daily life. They are being used in almost all kinds of digital displays, such as those of mobile phones, laptops, tablets, phablets, TVs, etc., due to their outstanding features, including superior color quality, low cost, wide viewing angle, easy fabrication, mercury-free manufacture, tenability, stretchability, flexibility, etc. Still, intensive research is being carried out nowadays to yield more light-weight, flexible, stretchable, and energy-efficient OLEDs. Investigations related to the synthesis of new organic materials and fabrication techniques have inspired us to write this book, which will fulfill the desire and thirst of OLEDs-based researchers.

This book aims at those audiences who are looking for a single source of comprehensive understanding of strategies and their challenges with respect to OLEDs material fabrication. It has never been an easy task for an author to cover the pace and productivity at a uniform level in each chapter of such a book with respect to the audiences, from doctoral student to postdoctoral researchers or from postdoctoral researchers to multidisciplinary field researchers, and so on. Still, we have tried our level best to deliver sufficient accessibility to all those. In the past couple of decades, several books on OLEDs have been published but none is able to serve as comprehensive guide to this field. Hence, we may say that this book should be a useful desk reference book for both academic and industrial researchers with backgrounds in physics, chemistry, materials science, and engineering, who are already working with organic materials and their applications. However, the subject matter of each chapter has been supported by extensive reference citations so that interested readers may go further to enhance their knowledge.

A total of 12 chapters have been authored by academic research and development experts in their respective fields and included herein to explore the topic of our book.

Chapter 1 includes the extensive details of the nanolithographic techniques used and their challenges involved in OLEDs. Chapter 2 provides an overview of printing technology for fabrication of advanced OLEDs materials. Furthermore, the design of hybrid perovskites for OLEDs (in Chapter 3), stretchable and flexible materials for OLEDs (in Chapter 4), metal–dielectric composites for OLEDs applications (in Chapter 5), organic small molecule materials and display technologies for OLEDs (in Chapter 6), a new generation of organic materials with respect to photophysical approach and OLEDs applications (in Chapter 7), mixed valence π-conjugated coordination polymers for OLEDs (in Chapter 8), synthesis of electroluminescent polymer for OLEDs (in Chapter 9), improvement in the efficiency of organic semiconductors via molecular doping for OLEDs applications (in Chapter 10), recent development of blue fluorescent organic materials for OLEDs (in Chapter 11), and a fundamental perspective of phosphorescent organic materials for OLEDs (in Chapter 12) have also been discussed with respect to the strategies and challenges involved. We would like to express our gratitude to all the contributing authors for investing their time and effort into writing as accurate, complete, and up-to-date chapters as possible and for keeping on schedule as much as possible.

Dr. Laxman Singh
Dr. Rituraj Dubey
Prof. R. N. Rai

About the Editors

Laxman Singh is currently working as Head and Associate Professor in the Department of Chemistry, Siddharth University, UP. Dr. Singh has more than a decade of research and teaching experience (<5 of them are from abroad) in the field of materials chemistry. He has visited several countries, including France, China, and Japan. He has been selected as National Research Fellow, South Korea. He has published more than 75 research articles in well-reputed international journals listed in the Science Citation Index which include books, book chapters, research papers, short commentary, and editorials. He has been granted five international patents from the Korean Intellectual Property Office of South Korea. Dr. Singh is Editor and Editorial Board members of more than 20 international well reputed journals. He is Editor-in-Chief of *Journal of Materials, Processing and Design* (International, Clausius Scientific Press, Canada); One of the Editors of *International Journal of Chemistry and Applications* (JCA, United States) and *Current Nano-Materials* (Guest Editor and Editorial Member, Bentham Science, United States); Associate Editor of *Frontiers in Sustainability* (Switzerland); *Asian Journal of Materials Chemistry* (Asian Publication Corporation, India); Editorial Board Member of *International Journal of Materials Science and Applications* (IJMSA, Science Publishing Group, United States), and *Advanced Energy Conversion Materials* (Universal Wiser Publisher, Singapore); International Advisory Board Member or Editor of *Materials Chemistry, Physical Sciences* (Cambridge Scholars Publishing, UK) and *Progress in Materials Chemistry and Physics* (Clausius Scientific Press, Canada), Review Editor of *Frontiers in Chemistry: Electrochemistry Section* and *Frontiers in Materials* (Frontiers Press, Switzerland); Reviewer of Scientific Proposals, Chilean National Science and Technology Commission. Dr. Singh is well recognised reviewers of more than 30 international well reputed journals: *Journal of Materials Chemistry A, Progress in Crystal Growth and Characterization of Materials, Materials and Design, Journal of Applied Physics, Materials Characterizations, Material Science in Semiconductor Processing, Emerging Materials Research, Journal of Materials Science: Materials in Electronics, International Journal of Materials Research, Superlattices and Microstructures, Tribology International, Advances in Chemical Science, Journal of Alloys and Compounds, Innovative Food Science and Emerging Technologies, Journal of Materiomics, Progress in Natural Science: Materials International, Superlattices and Microstructures, Advanced Powder Technology*, and many more. Dr. Singh has extensive experience in the synthesis and characterizations of nanomaterials, metals, mixed metal oxide (perovskite oxides, spinels), ceramics energy related perovskite materials for Li ion battery, fuel cell, capacitor, supercapacitor, etc.

Rituraj Dubey is currently working as an Assistant Professor and Head of the Department of Chemistry, Ramadhin College Sheikhpura, Munger University, Munger, Bihar, India. Previously, he worked as National Science Council Postdoctoral Researcher in the Department of Chemistry, National Cheng Kung University, Taiwan, under the supervision of Distinguished Prof. Shu-Hui Chen, in the field of bioanalytical chemistry. He obtained his PhD (organoanalytical chemistry) from the Indian Institute of Technology, Roorkee, India, under the supervision of Prof. Ravi Bhushan. He has more than 20 publications, which include books, book chapters, research papers, short commentary, and editorials of international reputation. He has expertise in the fields of fabrication of soft-material-based biochips, aptasensors, biosensors, liquid chromatography-mass spectrum–based proteomics, enantioresolution, high-performance liquid chromatography, thin-layer chromatography, etc.

R. N. Rai currently serves as Professor in the Department of Chemistry, Institute of Science, Banaras Hindu University, Varanasi, India. He joined the department in 2005 as Reader. He served the Birla Institute of Science and Technology, Pilani, Rajasthan, India as a lecturer and Assistant Professor from 2002 to 2005. He also served the Department of Atomic Energy, RRCAT, Indore, India, as a visiting scientist. He completed his postdoctoral research at National Taiwan University, Taipai, Taiwan, ROC, and he has also served as Postdoctoral Fellow at the Indian Institute of Science, Bangalore, India. He was awarded his PhD degree from the Department of Chemistry, Banaras Hindu University, Varanasi, India, and he is also a recipient of the Young Scientist award from the Department of Science and Technology, New Delhi. Prof. Rai has been actively involved in materials science, taking an interest inorganic material: their synthesis, characterization, single crystal growth of pure and binary materials by studying the phase diagram, thermal, optical, nonlinear optical, fluorescence, and medicinal properties.

Acknowledgments

I would like to take this opportunity to thank those who had a significant contribution in shaping this book. I thank the entire team at CRC Press for collaborating with us in preparing this book and providing all the necessary assistance. My special thanks go to Ms. Jyotsna Jangra, the Editorial Assistant at CRC Press, Taylor and Francis Books India Pvt. Ltd., who was there to promptly respond and address all my queries and provide the necessary assistance from the commencement of the book until its completion. I acknowledge and thank all the authors who contributed their works to the book. The authors submitted the work very timely, even during the period of difficulty due to the prevailing COVID-19 pandemic. The authors were very kind to address the comments that enhanced the quality of the chapters.

Very special thanks are due to my coeditors, Dr. Rituraj Dubey and Prof. R. N. Rai, as none of this would have been possible without their relentless efforts and support. I am grateful to Prof. K. D. Mandal and Prof. U. S. Rai for inspiring in me the zeal to do work on the contemporary issues related to the environment and energy. I also express my thanks to Prof. Prakriti Rai, Dean, Faculty of Science and my colleagues at the Department of Chemistry, Siddharth University, Kapilvastu, for supporting me and always being there to encourage me in my achievements and accomplishments. I would like to acknowledge the support and love of my parents (the Late Shri Dr. Indrapal Singh and Smt. Shanti Devi), my older brother (the late Shri Veerendra Singh), who always supported and encouraged me in my whole academic career, my loving and beautiful wife Radha, and other members of my family. I acknowledge my lovely daughters, Mansi Singh, Trisha Singh, Himanshi Rajput and my son, Pulkit Singh, who bring joy and happiness to my life that rejuvenates me in my work.

(Laxman Singh)

Editing a book on *Organic Light Emitting Diode (OLED) Toward Smart Lighting and Displays Technologies* has been an impressive journey in the past 18 months. I am very much grateful to my coeditors, especially, Dr. Laxman Singh. Without him, we could not have imagined this book. His hard work, guidance, affection for me, and tireless efforts are next to impossible to beat, and it helps me in my work. I am also thankful to my other coeditor Prof. R. N. Rai for his motivation and guidance in this work. I thank all the contributing authors who have worked hard to make this book a success. I also especially thank the publisher CRC Press for giving us the opportunity to contribute through this book information for the entire scientific community.

I am eternally grateful to my parents, Smt. Prem Vasi Dubey and Shri Gyaneshwar Dubey, for everything they have done to help me succeed in life. I express my heartfelt thanks to Dr. Diwakar Kumar (Principal, Ramadhin (RD) College Sheikhpura), Dr. Ashutosh K. Mishra (Assistant Professor, Indian Institute of Technology (IIT) Hydrabad), Dr Rajendra Kumar (Associate Professor, Central University (CU), Himachal Pradesh),

Dr. Tarun K. Sharma (Associate Professor, Gujarat Biotechnical University), and Prof. K. R. Justin Thomas (OLEDs expert, Department of Chemistry, IIT Roorkee) for their guidance, motivation, and affection during all of my struggles and successes. I would like to acknowledge the support and love of my loving and beautiful wife Nidhi, my daughter, Aadhishree (Saanvi), who brings joy and happiness in my life that rejuvenates me in my work. I also express my thanks to my dear friends Dr Sumant Mani, Dr Shivendra Nath, Dr Mu Tao Hsieh (National Cheng-Kung University Taiwan (NCKU), Taiwan) and Mr. Shailesh Yadav for supporting me and always being there to encourage me in my achievements and accomplishments.

(Rituraj Dubey)

I would like to thankfully acknowledge the Department of Chemistry, Institute of Science, Banaras Hindu University, for the time, support, and infrastructure facility to make this work possible.

(R. N. Rai)

Contributors

Bhawna
Department of Chemistry, University of Delhi, Delhi, India

Shikha Jyoti Borah
Special Centre for Nano Sciences, Jawaharlal Nehru University, Delhi, India

Pankaj Kumar Chaurasia
P. G. Department of Chemistry, L. S. College (a Constituent Unit of B. R. A. Bihar University), Muzaffarpur, India

Neelu Dheer
Department of Chemistry, Acharya Narendra Dev College, University of Delhi, India

Rituraj Dubey
Department of Chemistry, National Cheng-Kung University, Taiwan (R. O. C.); Department of Chemistry, Indian Institute of Technology Roorkee, UK, India; Department of Chemistry, Ramadhin College Sheikhpura, Munger University, Bihar, India

Bhanu Pratap Singh Gautam
Department of Chemistry, Laxman Singh Mahar Campus, Pithoragarh (Soban Singh Jeena University, Almora), UK, India

Manjul Gondwal
Department of Chemistry, Laxman Singh Mahar Campus, Pithoragarh (Soban Singh Jeena University, Almora), UK, India

Akanksha Gupta
Department of Chemistry, Sri Venkateswara College, University of Delhi, India

Pranshu K. Gupta
Department of Chemistry, Center of Advanced Study, Institute of Science, Banaras Hindu University, UP, India

Tanu Gupta
Department of Chemistry, Rajendra College, Jai Prakash University, Chapra, Bihar, India

Mushraf Hussain
NUIST-Reading Academy, Nanjing University of Information Science and Technology, 219 Ningliu Road, Nanjing, Jiangsu, P. R. China

Barnali Jana
Department of Chemistry, Haldia Government College, Debhog (Haldia), Purba Medinipur, West Bengal, India

Dhananjay Kumar
Department of Chemistry, A. M. College, Gaya, Magadh University, Bodh Gaya, Bihar, India

Rajeev Kumar
Chemistry Industry Research Institute, University of Ulsan, South Korea; Materials Research Centre, I. I. Sc. Bangalore, India

Sanjeev Kumar
Department of Chemistry, University of Delhi, Delhi, India; Department of Chemistry, Kirori Mal College, University of Delhi, Delhi, India

Sumit Kumar
Department of Chemistry, Magadh
   University, Bodh Gaya,
   Bihar, India

Sunil Kumar
Department of Chemistry, L. N.
   T. College, and Department
   of Chemistry, R. B. B. M.
   College (Constituent Units of
   B. R. A. Bihar University),
   Muzaffarpur, India

Vinod Kumar
Special Centre for Nano Sciences,
   Jawaharlal Nehru University,
   Delhi, India

Virendra Kumar
Department of Chemistry, Guru Nanak
   Dev University, Amritsar, Punjab,
   India

Juhi Kumari
Department of Chemistry, A. M.
   College, Gaya, Magadh University,
   Bodh Gaya, Bihar, India

Pragati Kushwaha
Medicinal and Process Chemistry
   Division, Central Drug Research
   Institute, Lucknow, UP, and
   Department of Chemistry,
   Lucknow University, Lucknow,
   UP, India

Sushovan Paladhi
Department of Chemistry, Thakur
   Prasad Singh (T.P.S.) College,
   Patliputra University, Patna,
   Bihar, India

Himanshu Pandey
Department of Chemistry, Ben-Gurion
   University of the Negev, Beer
   Sheva, Israel

Sandeep K. S. Patel
Department of Chemistry, M. M.
   V., Banaras Hindu University,
   Varanasi, India

Abhishek Rai
Department of Chemistry, L. N. Mithila
   University, Darbhanga, Bihar, India

Syed S. Razi
Department of Chemistry, Gaya College
   Gaya, Magadh University, Bodh
   Gaya, Bihar, India; Maria Zambrano
   Fellow, Department of Chemistry,
   University of Murcia, Spain

Subhasis Roy
Department of Chemistry, K. S.
   S. College Lakhisarai, Munger
   University, Bihar, India

Pawan Kumar Sada
Department of Chemistry, L. N. Mithila
   University, Darbhanga, Bihar, India

Nidhi Sharma
School of Applied and Life Sciences,
   Uttaranchal University, Dehradun,
   Uttarakhand, India

Ritika Sharma
Department of Biochemistry, University
   of Delhi, India

Alok Kumar Singh
Department of Chemistry, Deen Dayal
   Upadhyaya Gorakhpur University,
   UP, India

Arvind Kumar Singh
Department of Chemistry, S.B. College,
   V. K. S. University, Ara, Bihar, India

Laxman Singh
Department of Chemistry, Siddharth
   University, UP, India.

Ashish Kumar Srivastava
Department of Basic Sciences and
  Humanities, Pranveer Singh Institute
  of Technology, Kanpur, UP, India

Girijesh Kumar Verma
Department of Chemistry, Deen Dayal
  Upadhyaya Gorakhpur University,
  UP, India

1 Nanolithographic Techniques for OLEDs

Rituraj Dubey and Laxman Singh

1.1 INTRODUCTION

No one can even imagine the existence, innovation and development of nanopatterned structures without the availability of approaches to fabricate nano-sized patterns on a specified surface or substrate.[1] These approaches are called fabrication (nanolithographic) approaches in which new techniques and tools are used to produce and modify the nanostructures on the 1–100 nm length scale as on that scale new phenomena and functionality in materials can emerge.[2] A wide variety of fabrication (nanolithographic) approaches have been developed and are being used nowadays to manipulate surface structures on a nanoscale, such as microcontact printing (μCP), nano-imprint lithography (NIL), capillary force lithography (CFL), beam pen lithography (BPL) and dip-pen nanolithography (DPN). These approaches have their own advantages, disadvantages and applications. For example, μCP uses an elastomer-based stamp form printing of desired surface, and NIL involves the transferring of a pattern from a mould to the substrate. NIL is based upon maskless lithographic methods and hence involves less experimental setup, such as a vacuum chamber, electron sources, ultraviolet (UV) light and x-ray sources. NIL eventually became a cost effective and less time-consuming technique in comparison to other lithographic techniques. CFL is a combination of the essential features of nanoimprint lithography (the moulding of a polymer melt) and that of microcontact printing (the use of an elastomeric mould).[3] In BPL, pen arrays made of coated elastomeric pyramids with an opaque gold layer are used to direct the light to a surface in the near field (below the diffraction limit) to pattern features.[4] NEL is used to fabricate micro- and nanostructures on conductive substrates (metallic or semiconductor surfaces). It is performed in the ambient environment and doesn't require photoresist processing steps, which is required in conventional lithography; hence, NEL has been found to be useful for patterning proteins. It forms an oxide film in touched confinement which is formed between a conductive stamp and the specimen surface due to applied voltage and water bridge formed.[5] DPL uses direct patterning of molecules on the substrates via transport molecular and material "inks" (e.g., alkane thiols, biological molecules such as DNA, viruses, and proteins, polymers and nanoparticles) to many surfaces in a high-throughput fashion with high resolution and eventually enables the synthesis and study of complex chemical and biological structures.[6]

DOI: 10.1201/9781003260417-1

The soft lithographic techniques have several advantages over other lithographic techniques based upon photolithography and electron beam lithography as these can be performed on any surface, such as planar, flexible, curved and soft surfaces of OLEDs.[7] NIL and μCP involve mould creation, the concept of self-assembled monolayers (SAMs) and replica moulding. SAM fabrication is basically the creation of chemisorptions on the substrate surface through either ligand solution or reactive substance vapour, while replica moulding is used in three-dimensional morphology to replicate the shape, structure and features of a master stamp to the substrate in a single step, which can't be even imagined in common lithographic techniques.[8] In this chapter, we discuss the theory and fabrication process involved for each of the approaches mentioned above.

1.2 MICROCONTACT PRINTING

1.2.1 BACKGROUND

As the name defines, μCP is, basically, transferring (printing) of an "ink" pattern from the surface of a relief elastomeric stamp to a substrate just by contact. It is a kind of soft lithography because it uses elastomeric materials, exclusively; the stamp is usually made of polydimethylsiloxane (PDMS) or its structured forms due to its low Young's modulus and high hydrophobicity (contact angle of ~110° with water).[9] It has gained very high popularity since its inception, due to its highly versatile nature for developing surfaces using SAMs.[10] The key steps involved in μCP are as follows: fabrication of stamps with a predesigned pattern, followed by inking (the "ink" is usually made of molecules such as alkanethiols, or protein/enzyme-type biomolecules) the stamp and transferring (with almost 100% efficiency) the biomolecules onto the desired substrate, made up of a metal (Au, Ag, Cu, Pd and Pt), metal oxide, glass or semiconductor after only a few seconds of contact without losing any kind of biological activity.[8,11] The first step uses photolithography (a microlithographic technique not a nanolithographic one), which is one of the most challenging tasks as it involves the fabrication of a stamp with a precise pattern. We know that in photolithography light-sensitive resist layers are used for transferring the pattern drawn on the mask to a silicon template. But it cannot be applied upon nonpolar and curved surfaces and can use only photosensitive polymers for developing the pattern on the mask.[12]

1.2.2 GENERAL PROCEDURE

First, a master is fabricated using photolithography, micro-machining or a diffraction gratings stamp (using microlithographic techniques) for the preparation of a PDMS-based stamp.[8] The mixture of PDMS prepolymer and curing agent is poured on the surface of a master pattern relief structure, which is dipped in a Petri dish. After some time, when the PDMS gets cured and hardened, it is peeled off to obtain the desired stamp having every minute detail of the imprinted region of the master stamp. PDMS is used, mostly, due to its hydrophobicity (as mentioned above), viscoelasticity, elastomericity, inertness and low surface energy. In the next "inking" process step, the

PDMS stamp is wetted with the solution of inks (as mentioned above) for a few seconds and then dried by a stream of air to remove all the liquid content. Molecules that are capable of forming SAMs, such as alkanethiol (HS(CH$_2$)NX) coupled with a gold surface and siloxanes coupled with a hydroxylated surface, are used as inks. In the next and final step this inked stamp is kept in direct contact with the substrate and gets transferred to the substrate and gives the desired pattern after the release of the stamp (Figure 1.1). The advantages and disadvantages of this technique are summarized in Table 1.1.

1.2.3 RECENT APPLICATIONS

- The μCP technique is commonly used for patterning proteins on substrates, thus enabling exploration of the effect of cellular morphology (shape and size) on their growth and survival.[14]
- To study the differentiation of stem cells.[15]
- To study the contractility of cells.[16]
- Used in biomedical research such as DNA hybridization.[17]
- To study the targeted differentiation into cells with specific properties for engraftment.[18]
- Used in controlling the size of embryoid bodies.[19]

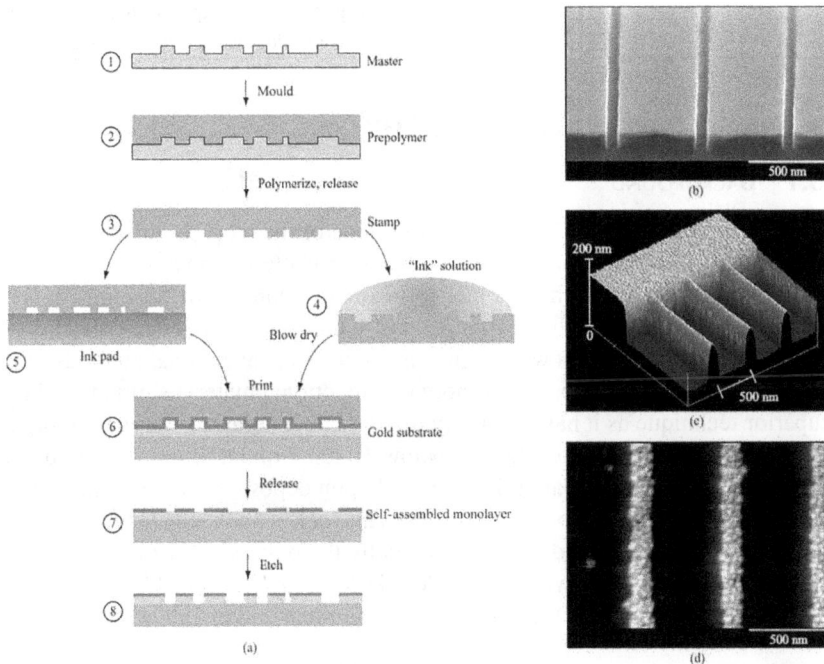

FIGURE 1.1 Schematic diagram of general procedure involved in microcontact printing–based lithographic technique.[13]

TABLE 1.1

Advantages and Disadvantages of Microcontact Printing Lithographic Technique

Advantages	Disadvantages
1. It is a very easy and simple technique of patterning a surface to follow.	1. Ink from polydimethylsiloxane (PDMS) may fade during patterning.
2. It fabricates the surface so easily in a laboratory that it has become a native technique for lab-on-a-chip protocols.	2. Sometimes shrinkage of the stamp may occur, which could lead to a disturbance in the dimensions of the pattern.
3. It can be useful for replication of patterns as multiple stamps can be created from one master stamp.	3. The stamp must be treated with oxygen plasma first; otherwise, protein solutions do not spread evenly.
4. It is a very cost effective and robust technique.	4. Deposition of dirt on the surface of the substrate may occur.
5. It is a very time-saving procedure as the time between the step of drying the stamp (after inking) and the step of printing proteins must be short.	5. Sometimes, lifting off the PDMS from the template may lead to distortion of the stamp and block the master plate, ultimately making it nonreusable.
	6. There is the possibility of deformation in the stamp with time.
	7. Variation in a single parameter such as size or shape of the print requires a new stamp as it cannot be done using the same stamp.

1.3 NANO-IMPRINT LITHOGRAPHY

1.3.1 BACKGROUND

Nano-imprint lithography (NIL) is known as one of the most promising patterning technologies for the fabrication of nano- and micropatterns on various substrates. It is an evolving technique, which comes from an old technique of "casting". It is categorized as a soft lithographic technique due to the simple process involved, with a high throughput yield and low cost. In comparison to conventional methods of fabricating nanopatterns composed of inorganic functional materials, it is found to be a superior technique as it has the ability to fabricate inorganic or organic-inorganic hybrid nanopatterns up to resolutions below 10 nm without issues of light diffraction.[20] The mechanism of patterning is based upon deposition of a polymer film on the substrate with the help of a rigid mould. This means that it consists of a polymer layer, a substrate and a rigid mould. Eventually the polymer resist is mechanically deformed just by pressing with the mould, like "casting" (Figure 1.2).[21] The advantages and disadvantages of this technique are summarized in Table 1.2.

1.3.2 GENERAL PROCEDURES

Based on the various general procedures, there are various types of prime NIL techniques, which are described in the following sections.

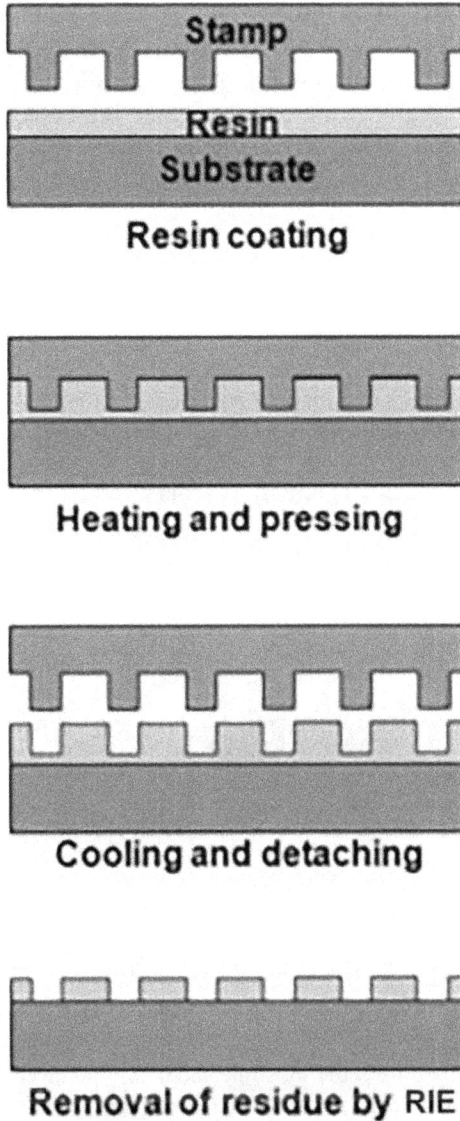

FIGURE 1.2 Schematic diagram of the general procedure involved in the nano-imprint lithography technique.[22]

1.3.2.1 Laser-Assisted Direct Imprint

The laser-assisted direct imprint (LADI) technique was proposed by Chou et al., in which a Si substrate surface is melted using the pulse of an excimer laser and controlled by a quartz mould (a transparent mould).[23] In the first step, a quartz master mould is put on the Si wafer substrate. In the second step, an excimer laser is irradiated onto that surface. As a result, the surface of the Si wafer substrate melts locally and eventually, in the third step, the melted Si is moved into the cavity of the quartz

TABLE 1.2

Advantages and Disadvantages of the Nano-Imprint Lithography Technique

Advantages	Disadvantages
1. It involves a very simple and flexible process.	1. The role of pressure is important: at low pressure, fabrication is not possible and at high pressure, the mould may be damaged; hence, precise pressure is required.
2. It does not require expensive high-resolution resist and costly equipment. Due to its cost effectiveness it is known as a lithographic technique suitable for commercial use.	2. Sometimes adhesion at the mould-resist interface becomes so strong that the possibility of damage in the generated nano-patterns may occur during the releasing of the mould.
3. It can develop nano-patterns at a large range and even in a short span of time, hence it is considered a swift lithographic technique.	3. Since in this technique extensive pressure and temperature are applied between the substrate and mould, there is quite good possibility of rapid damage of mould and hence it requires regular replacement.
4. OLED surfaces fabricated with high throughput and having high resolution can be produced.	4. The scope for further improvement in high throughput is still possible to meet the present market's requirements.
5. Since it doesn't require any kind of wet development process, resist sticking or falling never occurs.	5. Sometimes resolution is restricted by the mould.
6. Many kinds of thermoplastics can be used for fine pattern fabrication.	6. Remains of residual polymer are also observed, in some cases.
7. Designing of cross-sectional profile is possible by the mould.	7. This lithographic technique is not easy for a stepped substrate.
	8. A layer-by-layer alignment system is not possible in this lithographic technique.
	9. For the mould patterning step, thermal compensation correction is required.

mould just by mechanical pressing. After completion of this LADI-NIL process the quartz mould is detached from the substrate and the liquid layer solidifies rapidly and yields the desired Si nanopatterns on the surface of the Si wafer.[24]

1.3.2.2 Room Temperature

As proposed by Chou et al., fabrication of functional materials becomes possible by applying high pressure at room temperature (RT) during simple pressing.[23] In other words, we can say that in RT-NIL technique the imprinted thin film is being deformed by pressurizing with the imprinting mould. Pressure above 1×10^7 Pa is enough for RT-NIL as most of the inorganic thin films have little fluidity in themselves. Besides, the imprinting mould should be more rigid than imprinted thin films in the RT-NIL process.[25] This method is very expensive and harsh conditions may ruin and contaminate the mould and thus eventually limits its use for various substrates for practical industrial applications.

1.3.2.3 Ultraviolet-Photo

Willson's group proposed the fabrication of nanopatterns at low pressure and room temperature.[26] In this process, an ultraviolet (UV)-photo-curable polymer (monomer or oligomer) is used as the imprinting resin on a transparent substrate. The mould material is made up of translucent materials, such as quartz and fused silica. With the applications of very low pressure ($<1 \times 10^5$ Pa) and UV light exposure, this resist can be imprinted in the cavity of the stamp as its viscosity is very low. In this way rapid curing (solidification of patterns) occurs by the cross-linking of the monomer or oligomer for polymerization. This method can be used to fabricate nanopatterns on entire wafers with the assistance of stepper lithographic equipment.

1.3.2.4 Soft-Nano-Imprint Lithography

In this kind of NIL technique, a polymeric mould is used rather than a conventional rigid-material-based mould, such as quartz, Si, nickel, etc. With the help of a replication process such as nano- or micro-moulding, these polymeric moulds can be fabricated.[27] In this case, too, as in the case of µCP, PDMS is used as a soft polymer mould due to its high gas permeability,[28] and the capability of absorbing an organic solvent. Functional resists are made by mixing precursors or nanoparticles in an organic solvent and then categorized as sol-gel solution,[29] nanoparticle-dispersed solution,[29b,30] metal nano-ink,[31] spin-on-glass[32] resist and then further used for fabrication by NIL. Other reasons for using elastomeric PDMS mould are that its surface energy[33] is so small that it may not only uniformly transfer the functional nanopatterns to the substrate but also easily detach from the functional resist and it is so soft that it may easily achieve the conformal contact with the substrate. The removal of organic solvent and the gas products generated from the hydrolysis-condensation reaction[34] are necessary for NIL process; hence, the high permeability of PDMS helps in these steps too.

1.3.2.5 Thermal-Nano-Imprint Lithography

Chou's group proposed a procedure in which a thermoplastic polymer film (resist) is coated on a substrate with the help of a mould just by applying uniform external force (pressure $>5 \times 10^6$ Pa; 80 to 130 atm) but maintaining (via heating) the temperature of both the mould and the polymer film (resist) above the glass transition temperature.[20] Most commercially available materials as a NIL resist are poly(methyl methacrylate) (PMMA) and polystyrene. Before detaching the stamp from the polymer resin/substrate, the stack of the stamp and the substrate with the coated polymer layer must be cooled to the glass transition temperature of the polymer resin. Eventually, a negative phase of the mould is developed over the imprinted polymer pattern on the substrate. It is difficult to release the rigid mould, which adheres to the polymer film after the imprinting process, and for that purpose reactive ion etching is applied. This difficulty is the main reason for using mould materials of low surface energy as it reduces the adhesion between the polymer and the mould. This shortcoming of thermal-NIL should be removed by new discoveries for evolving new ideal resists, which should have the ability to detach neatly from the mould during the demoulding process and also it should not be adhesive with the mould.

1.3.3   RECENT APPLICATIONS

- Production of cheap plasmonic devices and light emitting diodes (LEDs) can be achieved with the help of this technique.
- It is used in cheap mass production of systems of gigabit-scale integration.
- Even generation of 3D nanopatterns can be done with its help at lower cost, less time and fewer steps involved in comparison to conventional 3D structural facbrication techniques.[35]
- Diffractive optical elements, nonlinear optical devices and biochemical analysis systems (such as such as a DNA or protein analysis chip) are fabricated using this lithographic technique.[36]
- Nanoscale polymer structures can be fabricated without using a layer-by-layer process or additional etching processes.
- It is also used in the fabrication of light-emitting devices including LEDs and OLEDs,[37] patterned magnetic media,[38] next-generation random access memory (RAM) devices, including resistive RAM and phase change RAM,[39] and organic and thin film solar cells.[40]

1.4   CAPILLARY FORCE LITHOGRAPHY

1.4.1   BACKGROUND

We know that capillarity does not use any kind of external force as rising of the liquid occurs just by wetting a capillary tube with a liquid due to lowering in the free energy, and hence it is a useful concept for the patterning of polymeric materials. This phenomenon has been used in micromoulding in capillaries (MIMIC).[41] Thus, in other words, we can say that if capillary force is used as the driving force for lithographic fabrication the technique is collectively termed "capillary force lithography" (CFL). This approach has its own advantages over other lithographic techniques and results in geometry-controllable, robust micro- and nanostructures over a large area.[3,42] The work of MIMIC was further improved and advanced up to solvent-assisted microcontact moulding, in which the thin surface layer of the polymer film on a substrate was softened by wetting the polymer surface with a suitable solvent for moulding.[43]

Technically, CFL is a combination of NIL and μCP, that is, it uses moulding of a polymer melt from NIL and an elastomeric mold from μCP or other soft lithographic techniques, and hence eventually delivers the retention of NIL over μCP by keeping the stringent pattern fidelity requirements for the fabrication of integrated circuits on the one hand, and by eliminating the need to use the extremely high pressure that is needed in NIL on the other hand, and that without using the etching steps for the removal of any residual resist.[3] The CFL technique was first introduced in the MIMIC method, in which liquid prepolymers and masks with open ends were utilized.[41a,44] This approach was found suitable for fabricating large area patterns as even the slow capillary filling rate cannot affect the outcome of CFL. Classification of CFL, based upon the patterning methods, is shown in Figure 1.3. The advantages and disadvantages of this technique are summarized in Table 1.3.

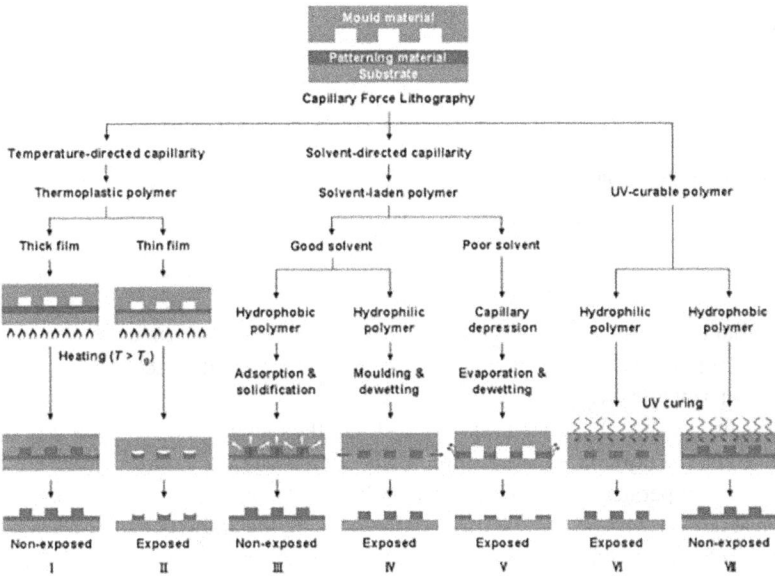

FIGURE 1.3 Classification of capillary force lithography (based upon the patterning methods).[45]

TABLE 1.3
Advantages and Disadvantages of the Capillary Force Lithography Technique

Advantages	Disadvantages
1. It can fabricate geometry-controllable, robust micro- and nanostructures over a large area.	1. It uses polydimethylsiloxane, which gives rise to deformation, buckling or collapse of shallow relief features or rounding of sharp corners when released from the master as possible shortcomings of this technique.
2. It is not affected by the limitation of slow capillary filling rates observed in micromoulding in capillaries.	2. In the experiments, care should be taken to avoid air traps and non-uniform contact, which result in defects and non-uniform distribution of the thickness profile.
3. Designing of cross-sectional profile is possible by the mould.	3. In many cases, rapid pattern formation with the direct exposure of the substrate surface is required for cell microarrays and selective protein adsorption.
4. It has also a wide range of biological and biomedical applications, including cell and tissue engineering.	4. Sometimes dewetting takes place during or after the capillary rise depending on the competition between capillary and dispersion forces due to the use of relatively thin polymer (<100 nm).
5. It has a wide range of material features, biological compatibility, ease of processing and prototyping, and cost effectiveness.	5. In certain designs, if the aspect ratio of the stamp is too high, stresses originating from gravity, adhesion or capillary forces may result in lateral collapse of relief structures and if the aspect ratio is too small, the recessed region will not be able to maintain its structure during stamping.

(Continued)

TABLE 1.3 (Continued)

Advantages	Disadvantages
6. It allows both quick and affordable fabrication.	
7. It does not require clean-room facilities.	

1.4.2 GENERAL PROCEDURE

As the name describes, CFL involves the capillarity phenomenon, defined as the rise of liquid (in the case of CFL it is a polymer melt) in the capillary tube due to lowering of free energy after wetting the capillary tube with the liquid. In this technique, first, a polymer layer is spin-coated onto a substrate, then an elastomeric mould (PDMS) is placed on the polymer layer followed by heating of the whole stack above the glass transition temperature so that polymer melt fills the void space due to capillarity action (Figure 1.4). The stack is then cooled down to the ambient temperature before removal of the mould and eventually fabrication of a negative replica of the mask pattern onto the substrate is completed.[46]

1.4.3 RECENT APPLICATIONS

- For the fabrication of high-resolution (~100 nm), second-generation moulds for soft lithography, CFL is used in combination with other techniques, such as replica moulding, NIL, wet etching, and μCP.[48]
- To fabricate biomimetic structures, the lotus leaf, hierarchical micropillar arrays, spatula-like, slanted nanohairs on setae-like microhairs and PMMA nanohairs on gecko feet CFL are used successfully.[49]
- It has shown its potential application in regulating cell growth, for example, in cell attachment, migration and aggregation.[50] Besides, it is also useful in influencing the stem cell fate.[51] Thus, we may say that CFL can be used for the development of cell-material-based therapeutics.[52]
- It is used in cell and tissue engineering.[45]
- It may provide an alternative to delivering vaccines by fabricating rapidly dissolvable microneedle patches for minimally invasive drug delivery. Hence, it can be used in fabricating drug delivery platforms.[53]
- The CFL approach allowed for the simple and facile fabrication of highly periodic nanostructures with a narrow size range. Due to this existing property of CFL, it can be used to fabricate hybrid core–shell nanostructures.[54] These nanostructures have a diverse range of applications in bio-nanotechnology,[55] optoelectrics,[56] and energy research.[57]
- CFL can be used to fabricate linear assemblies of Pd (palladium) nanocubes, which are further integrated into a circuit as a miniature hydrogen gas sensor with the help of electrostatic interactions.[58]
- CFL may be used for imprinting of nanostructures onto the active layer of a solar cell to enhance light trapping without removing the experimental setup, which requires removal of samples from the nitrogen glove box;

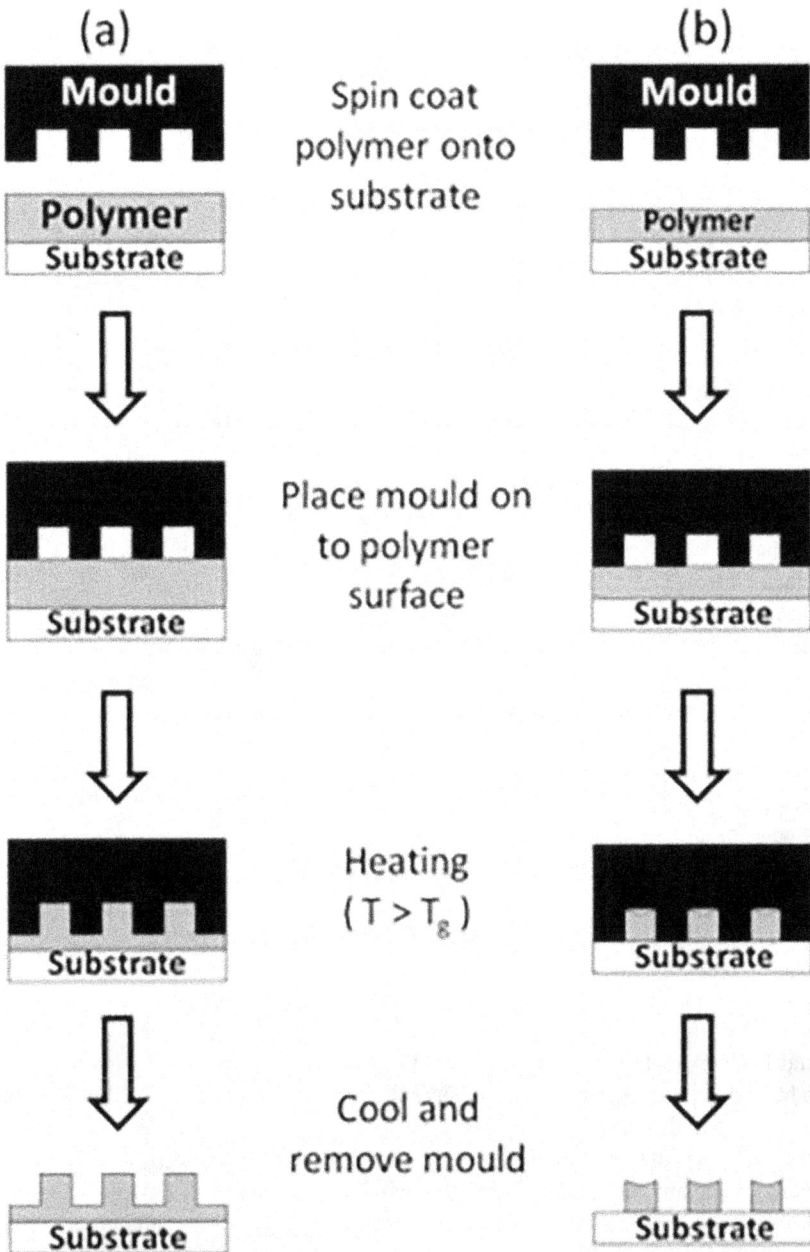

FIGURE 1.4   Schematic of the general procedure involved in the capillary force lithography technique: (a) thick and (b) thin polymer films.[47]

hence integrity of the material is maintained successfully. This protocol also shortens the pathway for charge transfer, leading to an improved collection of charge and an increase in fill factor.[59] Thus, eventually, CFL significantly improves solar cell efficiency, due to relatively greater absorption and electrical enhancement.[60]

- It is used in the manufacture of thin-film transistor liquid crystal displays (TFT-LCDs).[61] Besides, it is also used in patterning optically active materials such as micro- and nanoscale diffractive optical element structures onto a 3D concave lens.[62]

1.5 BEAM PEN LITHOGRAPHY

1.5.1 BACKGROUND

Beam pen lithography (BPL) utilizes near-field apertures to overcome the diffraction limit and is a unique technique in which apertures with sub-wavelength dimensions at the tip of each gold-coated elastomeric pen in a massive array are used to perform scanning near-field optical (NSOM)-based lithography and that eventually combines with polymer-based lithography.[4a] However, it does not allow for the fabrication of sub-diffraction limit features in arbitrary geometries in spite of using transparent PDMS stamps for controlling the light propagation in the context of surface patterning.[63] Thus, in other words, we may say that in BPL one can direct light to a surface in the near field to pattern features below the diffraction limit by coating the elastomeric pyramids also known as pen arrays with an opaque gold layer and subsequently etching apertures at the tip of each pen and eventually utilizing these apertures.[4] Recently, individual actuation of each pen was achieved such that each pen can simultaneously write a distinct pattern.[4b] Still, the effect of a liquid medium surrounding the tip on BPL patterning performance and capabilities is not known.

Later, an advanced version of BPL was introduced in which direct writing takes place by using transparent two-dimensional elastomeric tip arrays for delivering various ink molecules to a surface, known as polymer pen lithography (PPL). PPL is a constructive approach in which high throughput and nanometre registration over square centimetre areas can be achieved easily.[64] It can be further modified and utilized in selectively passing the light to a photoresist-coated surface by taking advantage of the transparent polymer pens in the PPL array by coating with an opaque metal layer except at the point of each tip and thus arbitrary patterns on a surface can be created by combining the movement of the stage and selective illumination of desired tips. In this case, near-field distances could be guaranteed just by bringing the BPL array into contact with the substrate without damaging the substrate or array due to elastomeric and reversibly deformable nature of the pens. A schematic of the steps involved in fabricating a BPL tip array, using gold-coated pens, in contact with an adhesive PMMA surface is shown in Figure 1.5. The advantages and disadvantages of this technique are summarized in Table 1.4.

1.5.2 GENERAL PROCEDURE

As reported earlier,[64a] a PDMS array of pyramid-shaped tips was fabricated in the first step in which tapering to a tip had a diameter of 100 nm and each pyramidal pen

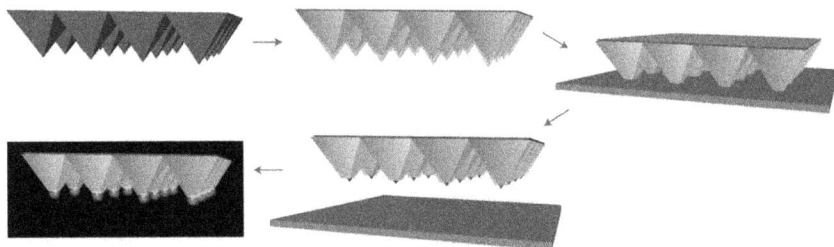

FIGURE 1.5 Schematic of the general procedure involved in the beam pen lithograpy technique.[4a]

TABLE 1.4
Advantages and Disadvantages of the Beam Pen Lithography (BPL) Technique

Advantages	Disadvantages
1. All pens in the array act in unison, making this technique useful for generating replicas of patterns.	1. All pens in the array act in unison, making this technique useful *only* for generating replicas of patterns.
2. Contrary to conventional photolithographic techniques and microcontact printing, BPL provides the flexibility to create arbitrary patterns through piezo-controlled movement of the beam pen array over a substrate.	2. It yields >500 nm pores and hence is not suitable for utilization as a legitimate desktop nanofabrication tool.
3. BPL is a promising platform for performing high-throughput nano- to macro-scale photochemistry.	3. It does not allow for the fabrication of sub-diffraction limit features in arbitrary geometries.
4. It is more cost-effective than photolithographic techniques due to cost-cutting related to photomask fabrication.	4. BPL experiments have mostly been performed in the dry state or with an ink-coated pen array.

had a square base with edges ~10 μm in length. The whole array was then treated with oxygen plasma before deposition of a layer of 80 nm gold and 5 nm titanium (as adhesive) layer with the help of a thermal evaporation procedure. In the next step, a PMMA-coated glass slide was brought into contact with this tip array for the removal of the gold layer from the apex of each tip, to create an aperture exposing the transparent PDMS. The desired average aperture size can be selected by controlling the contact area between the tips and PMMA surface through the application of different external forces on the back side of this BPL tip array coupled with PDMS tip compression. Now, in the proof-of-concept experiment, a positive photoresist coated silicon substrate was brought into contact with the BPL tip array, keeping the tip-to-tip spacing up to 80 mm and subjected to UV light exposure above the beam pen array. For each light exposure, a single dot feature per tip was created through the opaque gold layer on the sidewalls of each tip by strict passage of light through the apertures of the PDMS pyramidal tips. By controlling resist type and resist layer thickness, the diameter of each dot could be modulated easily. However, the tip aperture size is the most important factor contributing to the BPL resolution by controlling the area of resist exposed to light. One more thing should be kept in mind, that the aperture size did not change significantly from tip to tip (less than 10% variation), and just by controlling the contact force (0.002–0.2 N for a 1-cm^2 pen array) it can range from 500 nm to 5 mm in diameter.

1.5.3   RECENT APPLICATIONS

- It is used in DNA array synthesis.[65]
- It is used in protein immobilization.[66]
- It is used in nanoparticle assembly.[67]
- It is used in 3D printing.[68]
- It is used in fabrication of functional devices.[4b]
- It is used in patterning in registry with existing nanostructures, for example, the electrical transport of the nanowire structures exhibited clear semiconducting behaviour, eventually with the help of BPL.[4b]

1.6   DIP-PEN NANOLITHOGRAPHY

1.6.1   BACKGROUND

Dip-pen nanolithography (DPN) is considered a constructive strategy that can directly deliver materials onto a surface. It was introduced by the Mirkin group in 1999 by considering the fabrication of well-defined features, such as lines and dots.[6a] In this lithographic technique, an atomic force microscope (AFM) was used to control chemical reactions locally and to direct the delivery of molecules to a reactive substrate. DPN has been used to transfer a wide variety of "inks", such as octadecanethiol, 16-mercaptohexadecanoic acid, ω-mercaptoundecyl bromoisobutyrate, polymers, DNA, proteins, peptides, colloidal nanoparticles, metal ions and sols to various substrates with sub-50 nm resolution.[6b] The meniscus formed between the AFM tip and the substrate serves as a conduit for ink transport, and arbitrary patterning can be accomplished by moving the tip across the substrate as dictated by computer control. In the last two decades, several researchers have evolved DPN as an advanced lithographic technique with respect to the ink and substrate generality, throughput, and patterning scale and complexity and outcomes have included several advanced commercial applications.[6b,6c,69] DPN has emerged as a tool for the field of nanotechnology that can allow researchers to look towards a new phase of materials discovery just by a simple molecular printing (lithography). A timeline for the evolution of DPN is shown in Figure 1.6. The advantages and disadvantages of this technique are summarized in Table 1.5.

1.6.2   GENERAL PROCEDURE

Based on the various general procedures, there are differing types of prime DPN techniques, which are described in the following sections.

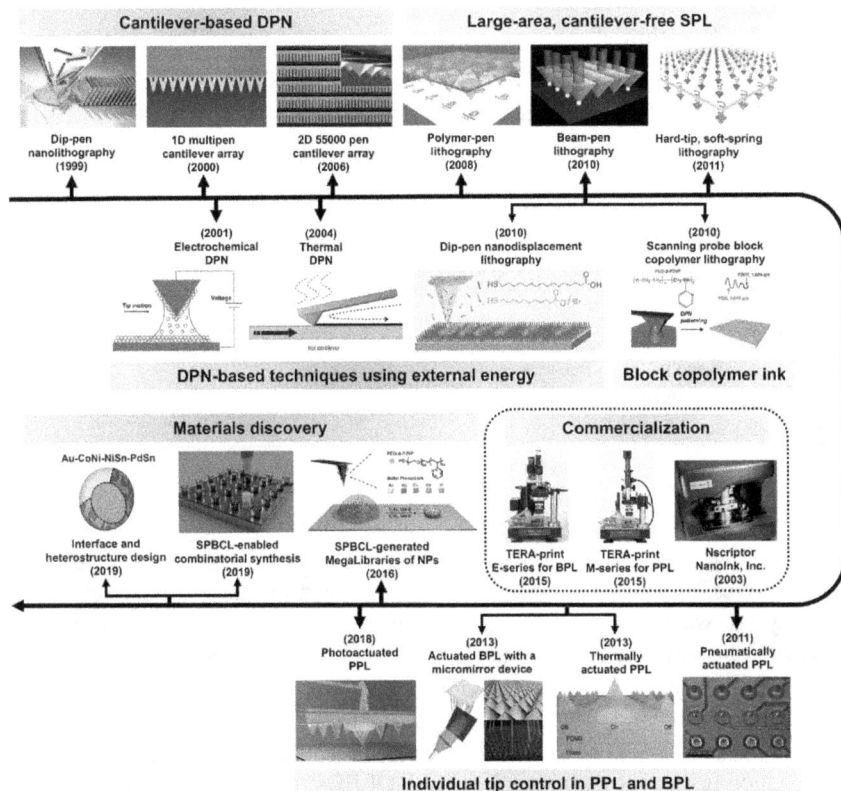

FIGURE 1.6 Schematic of the general procedure involved in the dip-pen nanolithography technique. (Reprinted (adapted) with permission from Liu et al.[70] Copyright (2020) American Chemical Society.)

TABLE 1.5
Advantages and Disadvantages of the Dip-Pen Nanolithography (DPN) Technique

Advantages	Disadvantages
1. It provides high resolution and high registration at low cost.	1. The spatial resolutions of DPN are limited by the radii of curvature of the tips and the size of the meniscus or nanoreactor formed between the tip and the surface.
2. DPN has shown its potential in synthesizing a variety of functional nanostructures essential for electronics, biology, optics and energy storage.	2. Functionalization and design of multicomponent inks, the development of independent control of each tip, the advancement of three-dimensional DPN, and the progression of combinatorial synthesis and screening are required for further advancement.

(Continued)

TABLE 1.5 (Continued)

Advantages	Disadvantages
3. It offers the ability for multiplexed patterning with facile feature and spatial control and high throughput.	
4. It enables rapid development, high-resolution and widespread dissemination.	
5. It gives a materials-general technique that offers good registration and in situ imaging capabilities	
6. It is a maskless and template-free technique; hence, it can be performed without a clean-room facility and under ambient conditions.	

1.6.2.1 Electrochemical Dip-Pen Nanolithograpy

This technique was introduced for direct fabrication of metal and semiconductor nanostructures.[71] In this approach a nanometer-sized electrochemical cell is generated by the water meniscus formed between the tip and the substrate in which the water transports the metal salts to the surface and reduces these precursors to a metallic form. In this way, the researchers can look towards the scope of patterning of materials beyond organic molecules due to this conductive nanofabrication. This DPN approach was used to tailor electronic and electro-optical properties on semiconducting and insulating surfaces like OLEDs (up to sub-100 nm scale),[72] to modify the composition of nanoscale devices[73] and to immobilize biological molecules (histidine-tagged proteins) on a metallic nickel surface by metal chelation (the Ni-His bond).[74] In an advanced stage, a method was developed in which an ink-coated, biased conducting AFM tip was used to oxidize an organic thin film followed by transfer of ink molecules to the oxidized regions in a single step.[75] Electochemical DPN can also be used to fabricate a DPN pattern on organic nanostructures via electrochemical desorption of the molecules from the peripheries of organic nanostructures in an electrochemical cell, with improvement in the feature resolution.[76] In this way, we can say that electrochemical DPN is an important DPN approach that can provide electrochemical synthesis and deposition of ink materials and the selective modification of surfaces like OLEDs during the fabrication process.

1.6.2.2 Thermal Dip-Pen Nanolithography

As the name indicates, in this approach a heated cantilever tip was used to melt and deposit solid organic inks on a surface.[77] In this work octadecylphosphonic acid was transported onto a mica surface for direct-write of a nanoscale line structure via heating above its melting temperature (near 100°C). Interestingly, no significant deformation was observed after cooling. Thermal DPN was used to deposit continuous indium metal lines (up to <80 nm width)[78] and to deposit surface-aligned functional poly(N-isopropylacrylamide) nanostructures on an epoxy-silane SAM-functionalized silicon oxide surface.[79] Thus, we can say that this DPN approach is useful for the deposition of a variety of solid inks, which is not possible by other lithographic techniques like BPL, PPL or other DPN with respect to the context of materials that do not have appreciable water solubility. Interestingly, with the variation in the heating temperature, ink transport and feature size can be controlled easily.[80]

1.6.2.3 Mechanical Dip-Pen Nanolithography

This technique was introduced using the mechanical force of an AFM tip to shave away a pre-assembled SAM followed by transferring the ink onto a diffusion-limited area.[81] Technically, it is a combination of other conventional DPN techniques and nanograting having SAM as the substrate to intentionally suppress the diffusion of ink via confinement and results in a very high resolution. Generally, a mechanical force was applied on the sharp AFM tip for shaving away a preassembled alkanethiol SAM on Au and then the molecules adsorbed on the tip are rapidly grafted on the exposed areas.[82] In the advanced stage, mechanical DPN was used in indirect programming of nanostructures of polymer brushes in which a base was triggered for polymer chain growth via displacement of SAM surface-initiated atom transfer radical polymerization initiators.[83] In this approach, destructive (via SAM damage) and constructive (via ink transport) strategies are performed simultaneously. Thus, we can say that the introduction of mechanical energy to the DPN process leads us to deliver the combination of material deposition with other chemical/physical reactions, and results in the generation of complex nanostructures, mechanical nanofabrication, formation of nanoscale reactors, thermochemical modification, surface functionalization and electrochemical synthesis, which will be very useful in the fields of nanotechnology, material science and interfacial chemistry.

1.6.3 RECENT APPLICATIONS

- DPN-generated patterns can be used not only as resists for wet chemical etching but also as well-ordered chemical templates for the assembly of materials.[84]
- DPN-patterned etch-resists have been used to fabricate arrays of functional nanodevices that play an important role in the miniaturization of microelectrode arrays are used in disease diagnosis, biological analysis and environmental monitoring.[85]
- DPN-defined templates have also been widely used to realize the assembly of nanoparticles.[86]
- It has been used to generate secondary patterns for the indirect assembly of nanomaterials.[87]
- DPN-generated templates have also been used to pattern a special type of polymeric structure, polymer brushes that have shown potential in the fields of electronics, optics and biology.[88]
- DPN-generated chemical templates can be used to locally modify the properties of functional surfaces such as OLEDs.[89]
- It is an efficient technique for nanopatterning a diverse set of biomolecules, including DNA, proteins, peptides, lipids, bacteria and viruses using "direct" and "indirect" strategies.[90]

- DPN-based techniques also show potential for constructing 2D as well as 3D structures.[91]
- It has also been used to pattern other nanostructures (e.g., inorganic nanoparticles, nanotubes and polymeric nanostructures) with diverse utilities in chemical, electrical and optical devices.[92]

1.7 CONCLUSION

In summary, we may say that several advanced fabrication (nanolithograhic) techniques for OLEDs, namely, microcontact printing, nano-imprint lithography, capillary force lithography, beam pen lithography, and dip-pen nanolithography have become indispensible techniques for the fabrication of OLEDs. Each technique has its own merits and demerits and on the basis of that optimum technique could be used for several types of OLEDs. The existing demerits give the researchers a foreseeable task to overcome for better and more advanced OLEDs.

REFERENCES

1. Gates, B. D.; Xu, Q.; Stewart, M.; Ryan, D.; Willson, C. G.; Whitesides, G. M., New Approaches to Nanofabrication: Molding, Printing, and Other Techniques. *Chemical Reviews* **2005,** *105* (4), 1171–1196.
2. (a) Quake, S. R.; Scherer, A., From Micro- to Nanofabrication with Soft Materials. *Science* **2000,** *290* (5496), 1536; (b) Liddle, J. A.; Gallatin, G. M., Nanomanufacturing: A Perspective. *ACS Nano* **2016,** *10* (3), 2995–3014.
3. Suh, K. Y.; Kim, Y. S.; Lee, H. H., Capillary Force Lithography. *Advanced Materials* **2001,** *13* (18), 1386–1389.
4. (a) Huo, F.; Zheng, G.; Liao, X.; Giam, L. R.; Chai, J.; Chen, X.; Shim, W.; Mirkin, C. A., Beam Pen Lithography. *Nature Nanotechnology* **2010,** *5* (9), 637–640; (b) Liao, X.; Brown, K. A.; Schmucker, A. L.; Liu, G.; He, S.; Shim, W.; Mirkin, C. A., Desktop Nanofabrication with Massively Multiplexed Beam Pen Lithography. *Nature Communications* **2013,** *4* (1), 2103.
5. (a) Snow, E. S.; Campbell, P. M., Fabrication of Si Nanostructures with an Atomic Force Microscope. *Applied Physics Letters* **1994,** *64* (15), 1932–1934; (b) Held, R.; Heinzel, T.; Studerus, P.; Ensslin, K.; Holland, M., Semiconductor Quantum Point Contact Fabricated by Lithography with an Atomic Force Microscope. *Applied Physics Letters* **1997,** *71* (18), 2689–2691; (c) Rokhinson, L. P.; Tsui, D. C.; Pfeiffer, L. N.; West, K. W., AFM Local Oxidation Nanopatterning of a High Mobility Shallow 2D Hole Gas. *Superlattices and Microstructures* **2002,** *32* (2), 99–102.
6. (a) Piner, R. D.; Zhu, J.; Xu, F.; Hong, S.; Mirkin, C. A., "Dip-Pen" Nanolithography. *Science* **1999,** *283* (5402), 661–663; (b) Liu, G.; Hirtz, M.; Fuchs, H.; Zheng, Z., Development of Dip-Pen Nanolithography (DPN) and Its Derivatives. *Small* **2019,** *15* (21), 1900564; (c) Ginger, D. S.; Zhang, H.; Mirkin, C. A., The Evolution of Dip-Pen Nanolithography. *Angewandte Chemie International Edition* **2004,** *43* (1), 30–45.
7. Qin, D.; Xia, Y.; Whitesides, G. M., Soft Lithography for Micro- and Nanoscale Patterning. *Nature Protocols* **2010,** *5* (3), 491–502.
8. Xia, Y.; Whitesides, G. M., Soft Lithography. *Annual Review of Materials Science* **1998,** *28* (1), 153–184.
9. Dubey, R.; Bhushan, R., Microcontact Printing in Bioanalysis: Where Are We and Where Shall We Be? *Bioanalysis* **2016,** *8* (20), 2093–2095.

10. (a) Prime, K. L.; Whitesides, G. M., Self-Assembled Organic Monolayers: Model Systems for Studying Adsorption of Proteins at Surfaces. *Science* **1991,** *252* (5009), 1164; (b) Kumar, A.; Whitesides, G. M., Features of Gold Having Micrometer to Centimeter Dimensions Can Be Formed Through a Combination of Stamping with an Elastomeric Stamp and an Alkanethiol "Ink" Followed by Chemical Etching. *Applied Physics Letters* **1993,** *63* (14), 2002–2004.

11. Khadpekar, A. J.; Khan, M.; Sose, A.; Majumder, A., Low Cost and Lithography-Free Stamp Fabrication for Microcontact Printing. *Scientific Reports* **2019,** *9* (1), 1024.

12. Deng, T.; Wu, H.; Brittain, S. T.; Whitesides, G. M., Prototyping of Masks, Masters, and Stamps/Molds for Soft Lithography Using an Office Printer and Photographic Reduction. *Analytical Chemistry* **2000,** *72* (14), 3176–3180.

13. Alom Ruiz, S.; Chen, C. S., Microcontact Printing: A Tool to Pattern. *Soft Matter* **2007,** *3* (2), 168–177.

14. Chen, C. S.; Mrksich, M.; Huang, S.; Whitesides, G. M.; Ingber, D. E., Geometric Control of Cell Life and Death. *Science* **1997,** *276* (5317), 1425.

15. (a) Kilian, K. A.; Bugarija, B.; Lahn, B. T.; Mrksich, M., Geometric Cues for Directing the Differentiation of Mesenchymal Stem Cells. *Proceedings of the National Academy of Sciences* **2010,** *107* (11), 4872; (b) Warmflash, A.; Sorre, B.; Etoc, F.; Siggia, E. D.; Brivanlou, A. H., A Method to Recapitulate Early Embryonic Spatial Patterning in Human Embryonic Stem Cells. *Nature Methods* **2014,** *11* (8), 847–854.

16. Wang, N.; Ostuni, E.; Whitesides, G. M.; Ingber, D. E., Micropatterning Tractional Forces in Living Cells. *Cell Motility* **2002,** *52* (2), 97–106.

17. (a) Lange, S. A.; Benes, V.; Kern, D. P.; Hörber, J. K. H.; Bernard, A., Microcontact Printing of DNA Molecules. *Analytical Chemistry* **2004,** *76* (6), 1641–1647; (b) Thibault, C.; Le Berre, V.; Casimirius, S.; Trévisiol, E.; François, J.; Vieu, C., Direct Microcontact Printing of Oligonucleotides for Biochip Applications. *Journal of Nanobiotechnology* **2005,** *3* (1), 7; (c) Singh, I.; Wendeln, C.; Clark, A. W.; Cooper, J. M.; Ravoo, B. J.; Burley, G. A., Sequence-Selective Detection of Double-Stranded DNA Sequences Using Pyrrole–Imidazole Polyamide Microarrays. *Journal of the American Chemical Society* **2013,** *135* (9), 3449–3457.

18. Van Hoof, D.; Mendelsohn, A. D.; Seerke, R.; Desai, T. A.; German, M. S., Differentiation of Human Embryonic Stem Cells into Pancreatic Endoderm in Patterned Size-Controlled Clusters. *Stem Cell Research* **2011,** *6* (3), 276–285.

19. Karp, J. M.; Yeh, J.; Eng, G.; Fukuda, J.; Blumling, J.; Suh, K.-Y.; Cheng, J.; Mahdavi, A.; Borenstein, J.; Langer, R.; Khademhosseini, A., Controlling Size, Shape and Homogeneity of Embryoid Bodies Using Poly(Ethylene Glycol) Microwells. *Lab on a Chip* **2007,** *7* (6), 786–794.

20. Chou, S. Y.; Krauss, P. R.; Renstrom, P. J., Imprint of Sub-25 nm Vias and Trenches in Polymers. *Applied Physics Letters* **1995,** *67* (21), 3114–3116.

21. Chou, S. Y.; Krauss, P. R.; Renstrom, P. J., Nanoimprint Lithography. *Journal of Vacuum Science & Technology B: Microelectronics and Nanometer Structures Processing, Measurement, and Phenomena* **1996,** *14* (6), 4129–4133.

22. Byeon, K. J.; Lee, H., Recent Progress in Direct Patterning Technologies Based on Nano-Imprint Lithography. *The European Physical Journal Applied Physics* **2012,** *59* (1).

23. Chou, S. Y.; Keimel, C.; Gu, J., Ultrafast and Direct Imprint of Nanostructures in Silicon. *Nature* **2002,** *417* (6891), 835–837.

24. Dongxu, W.; Nitul, S. R.; Xichun, L., Nanoimprint Lithography—The Past, the Present and the Future. *Current Nanoscience* **2016,** *12* (6), 712–724.

25. (a) Oh, S. C.; Bae, B. J.; Yang, K. Y.; Kwon, M. H.; Lee, H., Fabrication of Aluminum Nano-Scale Structures Using Direct-Embossing with a Nickel Template. *Metals and Materials International* **2011,** *17* (5), 771–775; (b) Taniguchi, J.; Tokano, Y.; Miyamoto, I.; Komuro, M.; Hiroshima, H., Diamond Nanoimprint Lithography. *Nanotechnology* **2002,** *13* (5), 592–596.

26. Bailey, T.; Choi, B. J.; Colburn, M.; Meissl, M.; Shaya, S.; Ekerdt, J. G.; Sreenivasan, S. V.; Willson, C. G., Step and Flash Imprint Lithography: Template Surface Treatment and Defect Analysis. *Journal of Vacuum Science & Technology B: Microelectronics and Nanometer Structures Processing, Measurement, and Phenomena* **2000,** *18* (6), 3572–3577.

27. (a) Shin, J.-H.; Lee, S.-H.; Byeon, K.-J.; Han, K.-S.; Lee, H.; Tsunozaki, K., Fabrication of Flexible UV Nanoimprint Mold with Fluorinated Polymer-Coated PET Film. *Nanoscale Research Letters* **2011,** *6* (1), 458; (b) Hong, S. H.; Hwang, J.; Lee, H., Replication of Cicada Wing's Nano-Patterns by Hot Embossing and UV Nanoimprinting. *Nanotechnology* **2009,** *20* (38), 385303.

28. (a) Bender, M.; Plachetka, U.; Ran, J.; Fuchs, A.; Vratzov, B.; Kurz, H.; Glinsner, T.; Lindner, F., High Resolution Lithography with PDMS Molds. *Journal of Vacuum Science & Technology B: Microelectronics and Nanometer Structures Processing, Measurement, and Phenomena* **2004,** *22* (6), 3229–3232; (b) Peng, F.; Liu, J.; Li, J., Analysis of the Gas Transport Performance Through PDMS/PS Composite Membranes Using the Resistances-in-Series Model. *Journal of Membrane Science* **2003,** *222* (1), 225–234.

29. (a) Sun, Y.; Seo, J. H.; Takacs, C. J.; Seifter, J.; Heeger, A. J., Inverted Polymer Solar Cells Integrated with a Low-Temperature-Annealed Sol-Gel-Derived ZnO Film as an Electron Transport Layer. *Advanced Materials* **2011,** *23* (14), 1679–1683; (b) Elfanaoui, A.; Elhamri, E.; Boulkaddat, L.; Ihlal, A.; Bouabid, K.; Laanab, L.; Taleb, A.; Portier, X., Optical and Structural Properties of TiO2 Thin Films Prepared by Sol-Gel Spin Coating. *International Journal of Hydrogen Energy* **2011,** *36* (6), 4130–4133.

30. Metin, C. O.; Bonnecaze, R. T.; Lake, L. W.; Miranda, C. R.; Nguyen, Q. P., Aggregation Kinetics and Shear Rheology of Aqueous Silica Suspensions. *Applied Nanoscience* **2014,** *4* (2), 169–178.

31. (a) Jung, I.; Jo, Y. H.; Kim, I.; Lee, H. M., A Simple Process for Synthesis of Ag Nanoparticlesand Sintering of Conductive Ink for Use in Printed Electronics. *Journal of Electronic Materials* **2012,** *41* (1), 115–121; (b) Kim, J.; Na, S.-I.; Kim, H.-K., Inkjet Printing of Transparent InZnSnO Conducting Electrodes from Nano-Particle Ink for Printable Organic Photovoltaics. *Solar Energy Materials and Solar Cells* **2012,** *98*, 424–432.

32. (a) Yao, J.; Le, A.-P.; Schulmerich, M. V.; Maria, J.; Lee, T.-W.; Gray, S. K.; Bhargava, R.; Rogers, J. A.; Nuzzo, R. G., Soft Embossing of Nanoscale Optical and Plasmonic Structures in Glass. *ACS Nano* **2011,** *5* (7), 5763–5774; (b) Cattoni, A.; Cambril, E.; Decanini, D.; Faini, G.; Haghiri-Gosnet, A.-M., Soft UV-NIL at 20 nm Scale Using Flexible Bi-Layer Stamp Casted on HSQ Master Mold. *Microelectronic Engineering* **2010,** *87*, 1015–1018.

33. (a) Choi, S.-J.; Yoo, P. J.; Baek, S. J.; Kim, T. W.; Lee, H. H., An Ultraviolet-Curable Mold for Sub-100-nm Lithography. *Journal of the American Chemical Society* **2004,** *126* (25), 7744–7745; (b) Kim, E.; Xia, Y.; Whitesides, G. M., Micromolding in Capillaries: Applications in Materials Science. *Journal of the American Chemical Society* **1996,** *118* (24), 5722–5731.

34. (a) Dirè, S.; Tagliazucca, V.; Callone, E.; Quaranta, A., Effect of Functional Groups on Condensation and Properties of Sol-Gel Silica Nanoparticles Prepared by Direct Synthesis from Organoalkoxysilanes. *Materials Chemistry and Physics* **2011,** *126* (3), 909–917; (b) Yu, G.; Zhu, L.; Wang, X.; Liu, J.; Xu, D., Fabrication of Silica-Supported ZrO$_2$ Mesoporous Fibers with High Thermal Stability by Sol-Gel Method Through a Controlled Hydrolysis–Condensation Process. *Microporous and Mesoporous Materials* **2010,** *130* (1), 189–196.

35. Nouri, L.; Posseme, N.; Landis, S.; Milesi, F.; Gaillard, F.; Mariolle, D.; Licitra, C., Silicon Anodization as a New Way to Transfer 3D Nano-Imprinted Pattern into a Substrate. *ECS Transactions* **2017,** *77* (4), 119–125.

36. (a) Li, W.-D.; Chou, S. Y., Solar-Blind Deep-UV Band-Pass Filter (250–350 nm) Consisting of a Metal Nano-Grid Fabricated by Nanoimprint Lithography. *Optics Express* **2010**, *18* (2), 931–937; (b) Yoon, Y.-T.; Lee, H.-S.; Lee, S.-S.; Kim, S. H.; Park, J.-D.; Lee, K.-D., Color Filter Incorporating a Subwavelength Patterned Grating in Poly Silicon. *Optics Express* **2008**, *16* (4), 2374–2380; (c) Su, J.-C.; Lin, T.-M., Polarized White Light Emitting Diodes with a Nano-Wire Grid Polarizer. *Optics Express* **2013**, *21* (1), 840–845.

37. (a) Byeon, K.-J.; Cho, J.-Y.; Kim, J.; Park, H.; Lee, H., Fabrication of SiNx-based Photonic Crystals on GaN-based LED Devices with Patterned Sapphire Substrate by Nanoimprint Lithography. *Optics Express* **2012**, *20* (10), 11423–11432; (b) Song, C. G.; Cha, Y.-J.; Oh, S. K.; Kwak, J. S.; Park, H.-J.; Jeong, T., Optimized Via-Hole Structure in GaN-Based Vertical-Injection Light-Emitting Diodes. *Journal of the Korean Physical Society* **2016**, *68* (1), 159–163.

38. (a) Dong, Q.; Li, G.; Ho, C.-L.; Faisal, M.; Leung, C.-W.; Pong, P. W.-T.; Liu, K.; Tang, B.-Z.; Manners, I.; Wong, W.-Y., A Polyferroplatinyne Precursor for the Rapid Fabrication of L10-FePt-Type Bit Patterned Media by Nanoimprint Lithography. *Advanced Materials* **2012**, *24* (8), 1034–1040; (b) Hellwig, O.; Bosworth, J. K.; Dobisz, E.; Kercher, D.; Hauet, T.; Zeltzer, G.; Risner-Jamtgaard, J. D.; Yaney, D.; Ruiz, R., Bit Patterned Media Based on Block Copolymer Directed Assembly with Narrow Magnetic Switching Field Distribution. *Applied Physics Letters* **2010**, *96* (5), 052511.

39. (a) Meier, M.; Gilles, S.; Rosezin, R.; Schindler, C.; Trellenkamp, S.; Rüdiger, A.; Mayer, D.; Kügeler, C.; Waser, R., Resistively Switching Pt/Spin-on Glass/Ag Nanocells for Non-Volatile Memories Fabricated with UV Nanoimprint Lithography. *Microelectronic Engineering* **2009**, *86* (4), 1060–1062; (b) Jeong, H. Y.; Lee, J. Y.; Choi, S.-Y., Interface-Engineered Amorphous TiO_2-Based Resistive Memory Devices. *Advanced Functional Materials* **2010**, *20* (22), 3912–3917.

40. (a) Ko, D.-H.; Tumbleston, J. R.; Gadisa, A.; Aryal, M.; Liu, Y.; Lopez, R.; Samulski, E. T., Light-Trapping Nano-Structures in Organic Photovoltaic Cells. *Journal of Materials Chemistry* **2011**, *21* (41), 16293–16303; (b) Battaglia, C.; Escarré, J.; Söderström, K.; Erni, L.; Ding, L.; Bugnon, G.; Billet, A.; Boccard, M.; Barraud, L.; De Wolf, S.; Haug, F.-J.; Despeisse, M.; Ballif, C., Nanoimprint Lithography for High-Efficiency Thin-Film Silicon Solar Cells. *Nano Letters* **2011**, *11* (2), 661–665.

41. (a) Kim, E.; Xia, Y.; Whitesides, G. M., Polymer Microstructures Formed by Moulding in Capillaries. *Nature* **1995**, *376* (6541), 581–584; (b) Adamson, A. W.; Gast, A. P., *Physical Chemistry of Surfaces*. Wiley: New York, 1997.

42. Suh, K. Y.; Lee, H. H., Capillary Force Lithography: Large-Area Patterning, Self-Organization, and Anisotropic Dewetting. *Advanced Functional Materials* **2002**, *12* (6–7), 405–413.

43. King, E.; Xia, Y.; Zhao, X.-M.; Whitesides, G. M., Solvent-Assisted Microcontact Molding: A Convenient Method for Fabricating Three-Dimensional Structures on Surfaces of Polymers. *Advanced Materials* **1997**, *9* (8), 651–654.

44. (a) Kim, E.; Xia, Y.; Whitesides, G. M., Two- and Three-Dimensional Crystallization of Polymeric Microspheres by Micromolding in Capillaries. *Advanced Materials* **1996**, *8* (3), 245–247; (b) Zhao, X.-M.; Stoddart, A.; Smith, S. P.; Kim, E.; Xia, Y.; Prentiss, M.; Whitesides, G. M., Fabrication of Single-Mode Polymeric Waveguides Using Micromolding in Capillaries. *Advanced Materials* **1996**, *8* (5), 420–424.

45. Suh, K.-Y.; Park, M. C.; Kim, P., Capillary Force Lithography: A Versatile Tool for Structured Biomaterials Interface Towards Cell and Tissue Engineering. *Advanced Functional Materials* **2009**, *19* (17), 2699–2712.

46. Jo, P. S.; Vailionis, A.; Park, Y. M.; Salleo, A., Scalable Fabrication of Strongly Textured Organic Semiconductor Micropatterns by Capillary Force Lithography. *Advanced Materials* **2012**, *24* (24), 3269–3274.

47. Ho, D.; Zou, J.; Zdyrko, B.; Iyer, K. S.; Luzinov, I., Capillary Force Lithography: The Versatility of This Facile Approach in Developing Nanoscale Applications. *Nanoscale* **2015,** *7* (2), 401–414.

48. (a) Bruinink, C. M.; Péter, M.; de Boer, M.; Kuipers, L.; Huskens, J.; Reinhoudt, D. N., Stamps for Submicrometer Soft Lithography Fabricated by Capillary Force Lithography. *Advanced Materials* **2004,** *16* (13), 1086–1090; (b) Duan, X.; Zhao, Y.; Perl, A.; Berenschot, E.; Reinhoudt, D. N.; Huskens, J., Nanopatterning by an Integrated Process Combining Capillary Force Lithography and Microcontact Printing. *Advanced Functional Materials* **2010,** *20* (4), 663–668.

49. (a) Jeong, H. E.; Lee, S. H.; Kim, P.; Suh, K. Y., Stretched Polymer Nanohairs by Nanodrawing. *Nano Letters* **2006,** *6* (7), 1508–1513; (b) Jeong, H. E.; Lee, J. K.; Kim, H. N.; Moon, S. H.; Suh, K. Y., A Nontransferring Dry Adhesive with Hierarchical Polymer Nanohairs. *Proceedings of the National Academy of Sciences of the United States of America* **2009,** *106* (14), 5639–5644; (c) Zhang, Y.; Lin, C.-T.; Yang, S., Fabrication of Hierarchical Pillar Arrays from Thermoplastic and Photosensitive SU-8. *Small* **2010,** *6* (6), 768–775.

50. Kim, D. H.; Seo, C. H.; Han, K.; Kwon, K. W.; Levchenko, A.; Suh, K. Y., Guided Cell Migration on Microtextured Substrates with Variable Local Density and Anisotropy. *Advanced Functional Materials* **2009,** *19* (10), 1579–1586.

51. (a) Yang, K.; Jung, K.; Ko, E.; Kim, J.; Park, K. I.; Kim, J.; Cho, S. W., Nanotopographical Manipulation of Focal Adhesion Formation for Enhanced Differentiation of Human Neural Stem Cells. *ACS Applied Materials & Interfaces* **2013,** *5* (21), 10529–10540; (b) Ahn, E. H.; Kim, Y.; Kshitiz; An, S. S.; Afzal, J.; Lee, S.; Kwak, M.; Suh, K. Y.; Kim, D. H.; Levchenko, A., Spatial Control of Adult Stem Cell Fate using Nanotopographic Cues. *Biomaterials* **2014,** *35* (8), 2401–2410.

52. Yang, H. S.; Ieronimakis, N.; Tsui, J. H.; Kim, H. N.; Suh, K. Y.; Reyes, M.; Kim, D. H., Nanopatterned Muscle Cell Patches for Enhanced Myogenesis and Dystrophin Expression in a Mouse Model of Muscular Dystrophy. *Biomaterials* **2014,** *35* (5), 1478–1486.

53. Moga, K. A.; Bickford, L. R.; Geil, R. D.; Dunn, S. S.; Pandya, A. A.; Wang, Y.; Fain, J. H.; Archuleta, C. F.; O'Neill, A. T.; Desimone, J. M., Rapidly-Dissolvable Microneedle Patches via a Highly Scalable and Reproducible Soft Lithography Approach. *Advanced Materials* **2013,** *25* (36), 5060–5066.

54. (a) Baek, Y.-K.; Yoo, S. M.; Kang, T.; Jeon, H.-J.; Kim, K.; Lee, J.-S.; Lee, S. Y.; Kim, B.; Jung, H.-T., Large-Scale Highly Ordered Chitosan-Core Au-Shell Nanopatterns with Plasmonic Tunability: A Top-Down Approach to Fabricate Core–Shell Nanostructures. *Advanced Functional Materials* **2010,** *20* (24), 4273–4278; (b) Kim, J.-S.; Jeon, H.-J.; Yoo, H.-W.; Baek, Y.-K.; Kim, K. H.; Kim, D. W.; Jung, H.-T., Generation of Monodisperse, Shape-Controlled Single and Hybrid Core–Shell Nanoparticles via a Simple One-Step Process. *Advanced Functional Materials* **2014,** *24* (6), 841–847.

55. Chithrani, B. D.; Ghazani, A. A.; Chan, W. C. W., Determining the Size and Shape Dependence of Gold Nanoparticle Uptake into Mammalian Cells. *Nano Letters* **2006,** *6* (4), 662–668.

56. Hu, X.; Gong, J.; Zhang, L.; Yu, J. C., Continuous Size Tuning of Monodisperse ZnO Colloidal Nanocrystal Clusters by a Microwave-Polyol Process and Their Application for Humidity Sensing. *Advanced Materials* **2008,** *20* (24), 4845–4850.

57. Zhong, C. J.; Luo, J.; Fang, B.; Wanjala, B. N.; Njoki, P. N.; Loukrakpam, R.; Yin, J., Nanostructured Catalysts in Fuel Cells. *Nanotechnology* **2010,** *21* (6), 062001.

58. Zou, J.; Zdyrko, B.; Luzinov, I.; Raston, C. L.; Swaminathan Iyer, K., Regiospecific Linear Assembly of Pd Nanocubes for Hydrogen Gas Sensing. *Chemical Communications* **2012,** *48* (7), 1033–1035.

59. Cheng, Y.-S.; Gau, C., Efficiency Improvement of Organic Solar Cells with Imprint of Nanostructures by Capillary Force Lithography. *Solar Energy Materials and Solar Cells* **2014**, *120*, 566–571.
60. An, C. J.; Cho, C.; Choi, J. K.; Park, J.-M.; Jin, M. L.; Lee, J.-Y.; Jung, H.-T., Highly Efficient Top-Illuminated Flexible Polymer Solar Cells with a Nanopatterned 3D Microresonant Cavity. *Small* **2014**, *10* (7), 1278–1283.
61. Jeong, J. K., The Status and Perspectives of Metal Oxide Thin-Film Transistors for Active Matrix Flexible Displays. *Semiconductor Science and Technology* **2011**, *26* (3), 034008.
62. Zhang, D.; Yu, W.; Wang, T.; Lu, Z.; Sun, Q., Fabrication of Diffractive Optical Elements on 3-D Curved Surfaces by Capillary Force Lithography. *Optics Express* **2010**, *18* (14), 15009–15016.
63. (a) Qin, D.; Xia, Y.; Black, A. J.; Whitesides, G. M., Photolithography with Transparent Reflective Photomasks. *Journal of Vacuum Science & Technology B: Microelectronics and Nanometer Structures Processing, Measurement, and Phenomena* **1998**, *16* (1), 98–103; (b) Qin, D.; Xia, Y.; Whitesides, G. M., Elastomeric Light Valves. *Advanced Materials* **1997**, *9* (5), 407–410.
64. (a) Huo, F.; Zheng, Z.; Zheng, G.; Giam, L. R.; Zhang, H.; Mirkin, C. A., Polymer Pen Lithography. *Science* **2008**, *321* (5896), 1658; (b) Zheng, Z.; Daniel, W. L.; Giam, L. R.; Huo, F.; Senesi, A. J.; Zheng, G.; Mirkin, C. A., Multiplexed Protein Arrays Enabled by Polymer Pen Lithography: Addressing the Inking Challenge. *Angewandte Chemie International Edition* **2009**, *48* (41), 7626–7629; (c) Huang, L.; Braunschweig, A. B.; Shim, W.; Qin, L.; Lim, J. K.; Hurst, S. J.; Huo, F.; Xue, C.; Jang, J.-W.; Mirkin, C. A., Matrix-Assisted Dip-Pen Nanolithography and Polymer Pen Lithography. *Small* **2010**, *6* (10), 1077–1081.
65. (a) Singh-Gasson, S.; Green, R. D.; Yue, Y.; Nelson, C.; Blattner, F.; Sussman, M. R.; Cerrina, F., Maskless Fabrication of Light-Directed Oligonucleotide Microarrays Using a Digital Micromirror Array. *Nature Biotechnology* **1999**, *17* (10), 974–978; (b) Beier, M.; Hoheisel, J. D., Production by Quantitative Photolithographic Synthesis of Individually Quality Checked DNA Microarrays. *Nucleic Acids Research* **2000**, *28* (4), E11.
66. Naqvi, A.; Nahar, P., Photochemical Immobilization of Proteins on Microwave-Synthesized Photoreactive Polymers. *Analytical Biochemistry* **2004**, *327* (1), 68–73.
67. (a) Zhou, X.; Zhou, Y.; Ku, J. C.; Zhang, C.; Mirkin, C. A., Capillary Force-Driven, Large-Area Alignment of Multi-Segmented Nanowires. *ACS Nano* **2014**, *8* (2), 1511–1516; (b) Lin, Q. Y.; Li, Z.; Brown, K. A.; O'Brien, M. N.; Ross, M. B.; Zhou, Y.; Butun, S.; Chen, P. C.; Schatz, G. C.; Dravid, V. P.; Aydin, K.; Mirkin, C. A., Strong Coupling between Plasmonic Gap Modes and Photonic Lattice Modes in DNA-Assembled Gold Nanocube Arrays. *Nano Letters* **2015**, *15* (7), 4699–4703.
68. Tumbleston, J. R.; Shirvanyants, D.; Ermoshkin, N.; Januszewicz, R.; Johnson, A. R.; Kelly, D.; Chen, K.; Pinschmidt, R.; Rolland, J. P.; Ermoshkin, A.; Samulski, E. T.; DeSimone, J. M., Additive Manufacturing. Continuous Liquid Interface Production of 3D Objects. *Science* **2015**, *347* (6228), 1349–1352.
69. Salaita, K.; Wang, Y.; Mirkin, C. A., Applications of Dip-Pen Nanolithography. *Nature Nanotechnology* **2007**, *2* (3), 145–155.
70. Liu, G.; Petrosko, S. H.; Zheng, Z.; Mirkin, C. A., Evolution of Dip-Pen Nanolithography (DPN): From Molecular Patterning to Materials Discovery. *Chemical Reviews* **2020**, *120* (13), 6009–6047.
71. Li, Y.; Maynor, B. W.; Liu, J., Electrochemical AFM "Dip-Pen" Nanolithography. *Journal of the American Chemical Society* **2001**, *123* (9), 2105–2106.
72. Maynor, B. W.; Filocamo, S. F.; Grinstaff, M. W.; Liu, J., Direct-Writing of Polymer Nanostructures: Poly(thiophene) Nanowires on Semiconducting and Insulating Surfaces. *Journal of the American Chemical Society* **2002**, *124* (4), 522–523.

73. Maynor, B. W.; Li, J.; Lu, C.; Liu, J., Site-Specific Fabrication of Nanoscale Heterostructures: Local Chemical Modification of GaN Nanowires Using Electrochemical Dip-Pen Nanolithography. *Journal of the American Chemical Society* **2004,** *126* (20), 6409–6413.

74. Agarwal, G.; Naik, R. R.; Stone, M. O., Immobilization of Histidine-Tagged Proteins on Nickel by Electrochemical Dip Pen Nanolithography. *Journal of the American Chemical Society* **2003,** *125* (24), 7408–7412.

75. Cai, Y.; Ocko, B. M., Electro Pen Nanolithography. *Journal of the American Chemical Society* **2005,** *127* (46), 16287–16291.

76. (a) Zhang, Y.; Salaita, K.; Lim, J.-H.; Mirkin, C. A., Electrochemical Whittling of Organic Nanostructures. *Nano Letters* **2002,** *2* (12), 1389–1392; (b) Zhang, Y.; Salaita, K.; Lim, J.-H.; Lee, K.-B.; Mirkin, C. A., A Massively Parallel Electrochemical Approach to the Miniaturization of Organic Micro- and Nanostructures on Surfaces. *Langmuir* **2004,** *20* (3), 962–968.

77. Sheehan, P. E.; Whitman, L. J.; King, W. P.; Nelson, B. A., Nanoscale Deposition of Solid Inks via Thermal Dip Pen Nanolithography. *Applied Physics Letters* **2004,** *85* (9), 1589–1591.

78. Nelson, B. A.; King, W. P.; Laracuente, A. R.; Sheehan, P. E.; Whitman, L. J., Direct Deposition of Continuous Metal Nanostructures by Thermal Dip-Pen Nanolithography. *Applied Physics Letters* **2006,** *88* (3), 033104.

79. Lee, W.-K.; Whitman, L. J.; Lee, J.; King, W. P.; Sheehan, P. E., The Nanopatterning of a Stimulus-Responsive Polymer by Thermal Dip-Pen Nanolithography. *Soft Matter* **2008,** *4* (9), 1844–1847.

80. Chung, S.; Felts, J. R.; Wang, D.; King, W. P.; De Yoreo, J. J., Temperature-Dependence of Ink Transport During Thermal Dip-Pen Nanolithography. *Applied Physics Letters* **2011,** *99* (19), 193101.

81. (a) Amro, N. A.; Xu, S.; Liu, G.-Y., Patterning Surfaces Using Tip-Directed Displacement and Self-Assembly. *Langmuir* **2000,** *16* (7), 3006–3009; (b) Liu, G.-Y.; Xu, S.; Qian, Y., Nanofabrication of Self-Assembled Monolayers Using Scanning Probe Lithography. *Accounts of Chemical Research* **2000,** *33* (7), 457–466.

82. (a) Xu, S.; Miller, S.; Laibinis, P. E.; Liu, G.-Y., Fabrication of Nanometer Scale Patterns within Self-Assembled Monolayers by Nanografting. *Langmuir* **1999,** *15* (21), 7244–7251; (b) Xu, S.; Liu, G.-y., Nanometer-Scale Fabrication by Simultaneous Nanoshaving and Molecular Self-Assembly. *Langmuir* **1997,** *13* (2), 127–129.

83. (a) Liu, X.; Li, Y.; Zheng, Z., Programming Nanostructures of Polymer Brushes by Dip-Pen Nanodisplacement Lithography (DNL). *Nanoscale* **2010,** *2* (12), 2614–2618; (b) Zoppe, J. O.; Ataman, N. C.; Mocny, P.; Wang, J.; Moraes, J.; Klok, H.-A., Surface-Initiated Controlled Radical Polymerization: State-of-the-Art, Opportunities, and Challenges in Surface and Interface Engineering with Polymer Brushes. *Chemical Reviews* **2017,** *117* (3), 1105–1318.

84. Basnar, B.; Willner, I., Dip-Pen-Nanolithographic Patterning of Metallic, Semiconductor, and Metal Oxide Nanostructures on Surfaces. *Small* **2009,** *5* (1), 28–44.

85. (a) Huang, X.-J.; O'Mahony, A. M.; Compton, R. G., Microelectrode Arrays for Electrochemistry: Approaches to Fabrication. *Small* **2009,** *5* (7), 776–788; (b) Lin, Z.; Takahashi, Y.; Murata, T.; Takeda, M.; Ino, K.; Shiku, H.; Matsue, T., Electrochemical Gene-Function Analysis for Single Cells with Addressable Microelectrode/Microwell Arrays. *Angewandte Chemie International Edition* **2009,** *48* (11), 2044–2046; (c) Lee, C.-Y.; Tan, Y.-J.; Bond, A. M., Identification of Surface Heterogeneity Effects in Cyclic Voltammograms Derived from Analysis of an Individually Addressable Gold Array Electrode. *Analytical Chemistry* **2008,** *80* (10), 3873–3881.

86. Demers, L. M.; Mirkin, C. A., Combinatorial Templates Generated by Dip-Pen Nanolithography for the Formation of Two-Dimensional Particle Arrays. *Angewandte Chemie International Edition* **2001,** *40* (16), 3069–3071.

87. Li, B.; Lu, G.; Zhou, X.; Cao, X.; Boey, F.; Zhang, H., Controlled Assembly of Gold Nanoparticles and Graphene Oxide Sheets on Dip Pen Nanolithography-Generated Templates. *Langmuir* **2009,** *25* (18), 10455–10458.

88. (a) Chen, T.; Amin, I.; Jordan, R., Patterned Polymer Brushes. *Chemical Society Reviews* **2012,** *41* (8), 3280–3296; (b) Nie, Z.; Kumacheva, E., Patterning Surfaces with Functional Polymers. *Nature Materials* **2008,** *7* (4), 277–290.

89. Zhou, X.; He, S.; Brown, K. A.; Mendez-Arroyo, J.; Boey, F.; Mirkin, C. A., Locally Altering the Electronic Properties of Graphene by Nanoscopically Doping It with Rhodamine 6G. *Nano Letters* **2013,** *13* (4), 1616–1621.

90. (a) Demers, L. M.; Ginger, D. S.; Park, S. J.; Li, Z.; Chung, S. W.; Mirkin, C. A., Direct Patterning of Modified Oligonucleotides on Metals and Insulators by Dip-Pen Nanolithography. *Science* **2002,** *296* (5574), 1836; (b) Hyun, J.; Kim, J.; Craig, S. L.; Chilkoti, A., Enzymatic Nanolithography of a Self-Assembled Oligonucleotide Monolayer on Gold. *Journal of the American Chemical Society* **2004,** *126* (15), 4770–4771; (c) Lee, K.-B.; Park, S.-J.; Mirkin, C. A.; Smith, J. C.; Mrksich, M., Protein Nanoarrays Generated by Dip-Pen Nanolithography. *Science* **2002,** *295* (5560), 1702; (d) Lee, K.-B.; Kim, E.-Y.; Mirkin, C. A.; Wolinsky, S. M., The Use of Nanoarrays for Highly Sensitive and Selective Detection of Human Immunodeficiency Virus Type 1 in Plasma. *Nano Letters* **2004,** *4* (10), 1869–1872.

91. Weinberger, D. A.; Hong, S.; Mirkin, C. A.; Wessels, B. W.; Higgins, T. B., Combinatorial Generation and Analysis of Nanometer- and Micrometer-Scale Silicon Features via "Dip-Pen" Nanolithography and Wet Chemical Etching. *Advanced Materials* **2000,** *12* (21), 1600–1603.

92. (a) Nie, Z.; Petukhova, A.; Kumacheva, E., Properties and Emerging Applications of Self-Assembled Structures Made from Inorganic Nanoparticles. *Nature Nanotechnology* **2010,** *5* (1), 15–25; (b) Auffan, M.; Rose, J.; Bottero, J.-Y.; Lowry, G. V.; Jolivet, J.-P.; Wiesner, M. R., Towards a Definition of Inorganic Nanoparticles from an Environmental, Health and Safety Perspective. *Nature Nanotechnology* **2009,** *4* (10), 634–641; (c) Boles, M. A.; Engel, M.; Talapin, D. V., Self-Assembly of Colloidal Nanocrystals: From Intricate Structures to Functional Materials. *Chemical Reviews* **2016,** *116* (18), 11220–11289.

2 Printing Technology for Fabrication of Advanced OLEDs Materials

Pragati Kushwaha and Rituraj Dubey

2.1 INTRODUCTION

An organic light emitting diode (OLED), also known as an organic electroluminescent (organic EL) diode, is a light emitting diode that functions in response to an electric current. The emissive electroluminescent layer is a film of organic compounds in OLED. OLEDs have appeared as the most commended alternative to liquid crystal displays (LCDs) in portable display devices, such as smartwatches, smartphones, digital cameras, etc. This superiority is because OLEDs offer various disruptive features. The key advantages are their self-emitting property, full colour capability, high luminous efficiency, better contrast, less power consumption, light weight, ultra thin, wide viewing angle, large area colour displays, flexibility and potentially low cost.[1-3]

Although significant development has been made in OLED technology, there are still massive challenges to realize the high efficiency and prolonged life, maintaining high brightness especially for illumination applications.[4,5] To enhance the efficiency of OLEDs, during the past few years there has been a growing interest in printing technology for the fabrication. Here, we will provide an outline of all major printing techniques for the fabrication of OLEDs and discuss their advantages and limitations.

2.2 FUNDAMENTAL STRUCTURE OF OLEDs

The fundamental structure of OLED consists of six different layers (Figure 2.1). The top and bottom layers are called the seal and substrate, respectively, and are made of protective glass or plastic. Between these two layers, there are a negative terminal (cathode) and a positive terminal (anode). Last, between the positive and negative terminals, there are two layers made of organic molecules, called the emissive layer and the conductive layer.

2.2.1 OLEDs MATERIALS

The materials used play a very crucial role in deciding the efficiency and life of OLEDs.[6] A remarkable transformation of OLED efficiency can be found by introducing new materials. Compared to the first generation of fluorescent materials, the

DOI: 10.1201/9781003260417-2

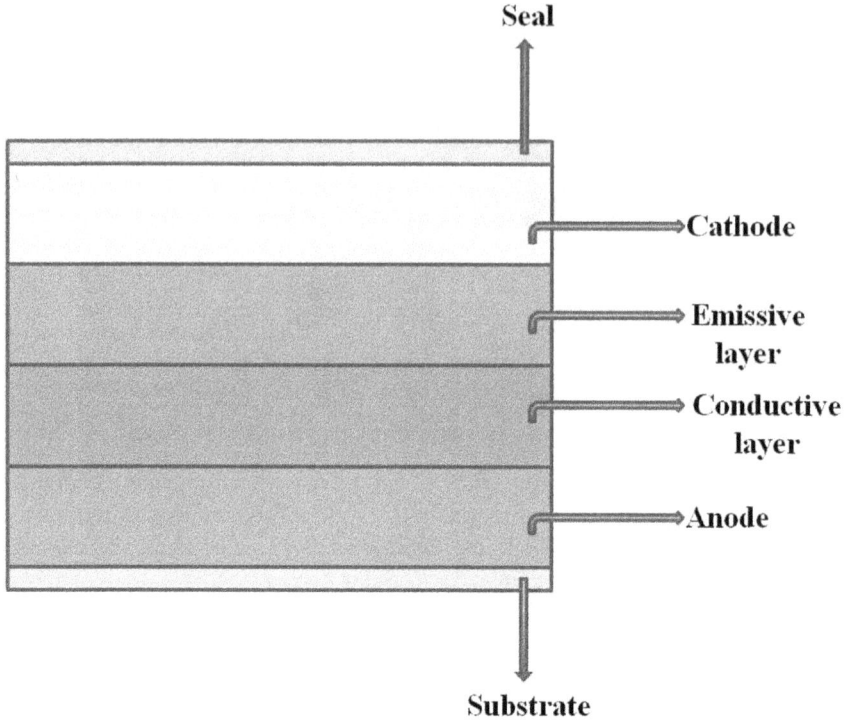

FIGURE 2.1 Basic OLED structure.

efficiency of OLED has multiplied tenfold. Also, the ongoing evolution of OLED materials has enhanced the lifetime of devices significantly.

2.2.1.1 Substrate

The frequently used materials for substrate are foil, plastic and glass. This is used to provide support to the OLED.

2.2.1.2 Anode

Indium tin oxide (ITO) is the most frequently used material for the anode. It is visibly transparent and has a high work function. ITO is deposited on the substrate by vacuum sputtering. Then the substrate is subjected to the surface treatment. Currently, graphene-based anodes are being used which display almost the same properties as ITO.[7]

2.2.1.3 Conductive Layer

The main role of the conductive layer is to transport holes to the emission layers. Commonly, hole transport materials used for OLEDs are aromatic amino compounds, as shown in Figure 2.2. One of the classical hole transport materials was *N,N'*-diphenyl-*N,N'*-bis(m-tolyl)-1,1'-biphenyl-4,4'-diamine, commonly abbreviated

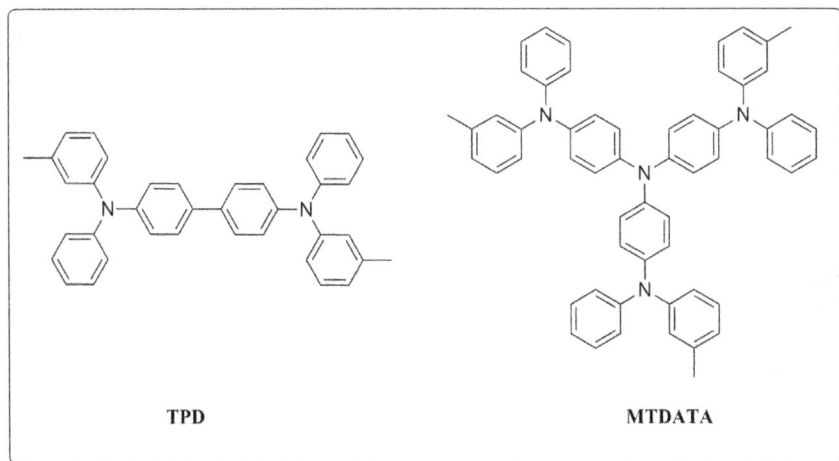

FIGURE 2.2   Commonly used hole transport materials for OLEDs.

TPD. It was first reported by Tang and Van Slyke.[8] TPD has good hole mobility and works through vacuum deposition. Another common hole transport material is *m*-methyl-tris(diphenylamine)triphenylamine) known as MTDATA. Generally, p-type materials are chosen for the conductive layer.

2.2.1.4   Emissive Layer

The important role of materials in the emission layer (EML) is to release highly efficient emissions and requisite colours. Emitting materials strongly determine the emitting efficiency, colours and the life of the EML. The EML is composed of organic plastic molecules that decide the color of light produced. A frequently used material is polyflorene, commonly called PFO. Another widely used material is poly[2-methoxy-5-(2-ethyl-hexyloxy)phenylene vinylene] (MEH-PPV). Materials in emitting layers are generally composed of a guest-host system.

2.2.1.5   Cathode

Conductors having low work function will be selected as the cathode to enhance electron injection. Metals like barium and calcium are generally used as the cathode. These metals are highly reactive, thus deposited with an aluminium layer to avoid degradation. Another benefit of aluminium capping later is that it provides adherence to electrical contacts and back reflection of emitted light out to the ITO layer. The cathode can be transparent or solid metal type depending on the type of OLED used.

2.3   OLEDs FABRICATION TECHNOLOGY

During recent time printing techniques for diode fabrication have gained enhanced recognition because of their great potential and economical manufacturing approaches well suited with both flexible and rigid substrates. In the process of printing, inks are neatly placed over the substrate and then solidified, resulting in a lasting pattern. In printed electronic devices, the inks are spreading widely with specific electrical or opto-electrical properties. Some of the printing techniques are well suited for lab-scale fabrication, such as vacuum deposition, whereas hand roll to roll printing such as inkjet printing or spray coating are more convenient for large-scale fabrication.

2.3.1 PROCESS OF FABRICATION

Generally, an ISO 6 clean room (Class 1000 clean room) is used to fabricate OLEDs. It is a contained space where arrangements are made to lessen the particulate contamination and manage other environmental parameters such as temperature, pressure and humidity. The essential component is a high-efficiency particulate air (HEPA) filtration system that is used to maintain air cleanliness by trapping particles that are 0.3 micron and bigger in size. Particularly in class 1000 clean room air cleanliness is maintained to a maximum of 1000 particles ($\geq$0.5 μm) per cubic meter of inside air. However, OLEDs are tolerant enough to dust that these are insulating and mostly only functioning when the dust particles perch on the surface.[9] In general the process of fabrication involves six steps from the substrate to the devices ready to use, that will be explained here.

2.3.1.1 Substrate Cleaning

To prepare the ITO surface for coating, first the substrates are sonicated in a sodium hydroxide (NaOH) solution following being washed in deionized (DI) water and finally blown dry. First, the substrates are arranged into the cleaning rack so that they all have the same orientation. Then the arranged rack is transferred to a beaker containing a 10% solution of dilute NaOH. The substrates are further sonicated to eliminate the photoresist.

The photoresist may either be dissolved or fractured into layers depending on the work efficiency and temperature of sonicator used. The time required for the process also depends on the ultrasonic bath used, as well as temperature. The substrates are then washed rigorously after sonication to wash out photoresist. The process of sonication is repeated for same amount of time with fresh solution of NaOH to make sure there is no residual layer of photoresist. After this second sonication process, the substrate should again be washed thoroughly with water. To avoid contamination, substrates are kept inside the water until it is ready for blow drying.

2.3.1.2 Applying PEDOT: PSS

PEDOT: PSS is the frequently used hole injection material.[10] The chemical name of PEDOT: PSS is poly(3,4-ethylenedioxythiophene) polystyrene sulfonate which is a polymer mixture of two ionomers. Finest quality PEDOT: PSS is required for efficacious device functioning and is generally the most tedious part of the fabrication process. It has a fresh and hydrophilic surface for coating appropriately. It is also compulsory to make sure that the active areas have not been in contact with any other surfaces because it will affect ITO spinning. PEDOT: PSS is spin coated at 5000 rpm to construct a film having thickness around 40 nm. The time required for the process is about 30 seconds. For restricting material use, the process can be done by pipetting

20 to 30 µL in between the spinning substrate. With the help of a cotton bud dipped in DI water, PEDOT: PSS must be rubbed off after spin coating. Finally, the substrates are placed in the box to keep dust from settling on the devices. Otherwise, it can also be kept in air for a couple of minutes placed directly on a hotplate.

2.3.1.3   Applying Active Layer

Generally, the active layer is applied either in open air or within a glovebox. The performance can be minutely different provided exposure time and minimised light levels. Twenty microlitres of the solution are pipetted out onto the substrate spinning at approximately 2000 rpm. This gives a good uniform coverage with the demanded thickness. The substrate is then rotated for a few more seconds to allow it to dry. After spin coating the samples are heated and then allowed to cool slowly. This process is called annealing. Annealing can be performed with the help of a solvent or in the presence of heat, but in the case of OLED reference solution thermal annealing is preferred at 80°C for 10 minutes. The cathode strip must be wiped off before cathode deposition. At last, the substrates are kept facing down with the cathode strip present at the wide end of the opening.

2.3.1.4   Cathode Evaporation

Conventionally, 100 nm aluminium is evaporated at a rate of pprox. 1.5A/s. Thinner cathodes up to 50 nm have been also utilized without any reduction in initial performance. Working with calcium is comparatively easy as it melts at low temperature. However, the maintenance is crucial as it can be performed only in the glovebox, failing which degradation occurs.

2.3.1.5   Annealing

Further, after cathode deposition, thermal annealing can be done at a temperature of 150°C for about 15 minutes resulting in the desired performance.

2.3.1.6   Encapsulation

Encapsulation is compulsory to protect the devices against degradation in open air once they are taken out of the glovebox. Generally, a glass welding technology is used for encapsulation for long lifetime.

2.4.   PRINTING TECHNIQUES FOR OLEDs FABRICATION

Printing technologies are classified in two main categories:

(a) Dry techniques
(b) Wet techniques

In general, OLEDs are fabricated either by a dry process or a wet process. Dry techniques, for example, vacuum thermal evaporation (VTE) and organic vapour phase deposition (OVPD), have been mostly used for small organic molecules material deposition. In the evolution of fabricated organic light emitting devices, these conventional and digital printing technologies cover various requirements, complementing each other. Conventional printing technologies provide the speed and large-area processing capabilities required for large production, but require higher set-up cost.

Inkjet, as a master-less printing technique, covers instead the different end of the spectrum, being very convenient for the development and functional study of technology demonstrators and for small and specialised production runs. Due to the convenience of depositing large area uniform and homogenous film with dry techniques, and because of the solubility of small organic molecules, it will be too small for solution processing of adequately thick films. In contrast, polymer organic materials are deposited using wet techniques, for example spin coating, ink jet printing and contact stamping. In wet processes, there are procedures for coating without thin patterning and processes for making thin patterns. Typical dry techniques such as vacuum thermal evaporation are not feasible because of their high molecular weight, which causes evaporation temperatures up to a point in excess of their decomposition temperature.

The following section describes the most common methods of various fabrication techniques in detail.

2.4.1   VACUUM EVAPORATION PROCESS

The vacuum evaporation process is the most frequently used technique for fabricating OLED devices. The reason for this is simply that the vacuum thermal evaporation method generates OLEDs having the best performance in terms of efficiency and life duration. Additionally, this method is the most mature fabrication technique. During the process, organic materials set into crucibles are converted to gas phase by enhancing their temperature at low pressures such as $10-5 \sim 10-7$ torr. The materials in gas phase are settled down on substrates, being transformed to solid phase due to the temperature of the substrate being much lower than the crucible.

2.4.1.1   Mask Deposition

Mask deposition techniques are commonly used for fabricating subpixel emitting patterns for perception of full colors.[11] The diagram below (Figure 2.3) represents the mask deposition process.

While this process has an uncomplicated technology in principle, in reality it is not as simple as it looks, precisely for high resolution displays and big screens. This is because of the mask distortion or the possibilities of defects.

2.4.1.2   Evaporation Methods

Evaporation techniques can be classified into three categories on the basis of the correlation between the evaporation source and the substrate, that is, point source evaporation, linear source evaporation and planar source evaporation. The schematic view is shown in Figure 2.4.

Point source evaporation is the most effortless process and commonly used in research and development and the production of small or medium-sized substrates. In this process, an evaporation material is set into the point type of crucible (source) and a substrate is fixed at a distance from the crucible. The substrate is often rotated to obtain better thickness and uniformity of deposited organic layers. Additionally, to

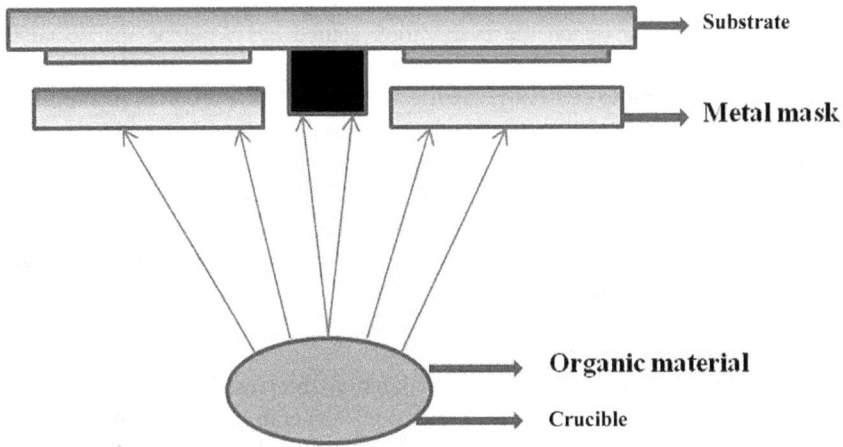

FIGURE 2.3   Overview of the mask deposition process.

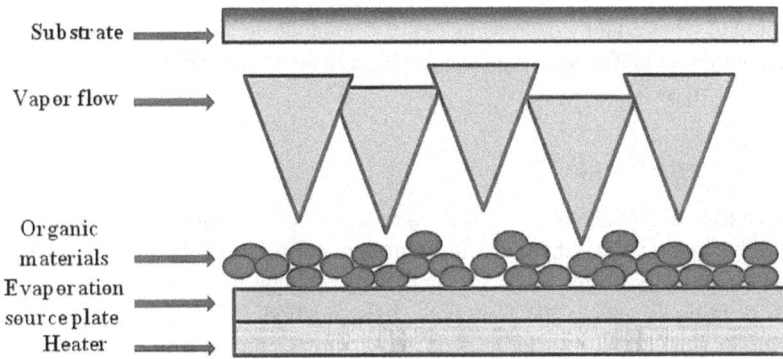

FIGURE 2.4   Schematic for evaporation.

maintain good thickness uniformity of deposited organic layers and to decrease the influence of the radiant rays on the substrate, the distance between the target (source) and the substrate (T/S distance) is very large. Generally, the distance is several tens of centimetres. Due to the large distances, most evaporated materials are deposited on the walls of the vacuum chamber rather than being deposited on the substrate. Thus, the material yield of point source evaporation is generally less than 10%. The evaporation equipment would have to be very large if this process is applied to large substrates. Therefore, the use of point source evaporation is restricted to small and medium-sized substrates. The higher limit of the substrate size seems to be G4 (730 × 920 mm).

In contrast, the linear source method is relevant to large substrates. The difference in this method is that a linear shape evaporation source is used and the substrate moves. The T/S distance can be shorter compared to the point source method.

Therefore, it is possible to secure high material yield. The third method is planar source evaporation, in which the evaporation source has a planar shape and the substrate does not move. The T/S distance can be decreased as it can in the linear source method. Fujimoto et al. of Hitachi Zosen Corporation reported planar source evaporation equipment.[12] They simulated the material yield evaporated by the planar source, taken as the T/S distance is 200 mm and thickness uniformity is better than ±3%. When the substrate size is not bigger, such as G2 or G3, the simulated material yield is about 20%–30%. In contrast, when the substrate size is as large as G6–G8, the material yield is greater than 60%–70%. Indeed, they have fabricated a planar source evaporation system for a G6 substrate.

2.4.1.3 Ultra High Vacuum

The vacuum deposition method can be outlined as the pressure and residual water play a remarkable role in OLED performance.[13] Generally the vacuum pressure in depositions of OLED materials is in the range of 10^{-5} to 10^{-9} torr; thus, the materials do not undergo any collisions while running along with its path towards the substrate, additionally to keep the deposited materials contamination free. Depending on throwing distance, in this technique 70%–99% of the material is deposited on the walls of the vacuum chamber rather than the substrate. Shadow masking allows for patterning of a material by decreasing deposition in the areas where it is not required. Usually, a thin metal foil with a pattern of through apertures across its surface is used as a shadow mask.

2.4.2 SCREEN PRINTING

Although the long-established fabrication process includes spin casting and physical vapor deposition (PVD), there is still great demand for cheaper and more versatile methods, especially for low information content displays as well as general lighting. Screen printing is one of the frequently used methods for fast and cost-effective deposition of dye films over large surfaces. It allows patterning to easily define the substrate area to be deposited. Screen printing is commonly used in industries. The most important factor of the screen printing method consists of a cloth of woven threads. A variety of cloth types is available in which polyester is most commonly used. Nylon cloth and metal cloth can be also used sometimes. Mesh count can be calculated as the number of threads per inch in the cloth. The cloth is extended tightly in a rigid frame to arrange a patterned mask in the frame. Ink is poured on the external surface of the cloth then a substrate is placed below the framed cloth on which the ink is to be printed. Thus, the ink does not reach directly the bottom surface of the cloth. Using a squeegee, the mesh is pushed to the substrate along a line of contact; the resulting ink is passed to the substrate in accordance with the patterns of mesh through the mesh openings.[14] Thus, the ink occupies the pores in the cloth. After that the squeegee pushes the cloth against the substrate underneath from upwards and by slipping horizontally over the surface, squashes out the ink in the open cloth areas on the substrate. This creates the printed pattern. Initially, this pattern remains wet and afterwards is drained. This process is repeated several times by replacing the substrate and printing a new one.[15]

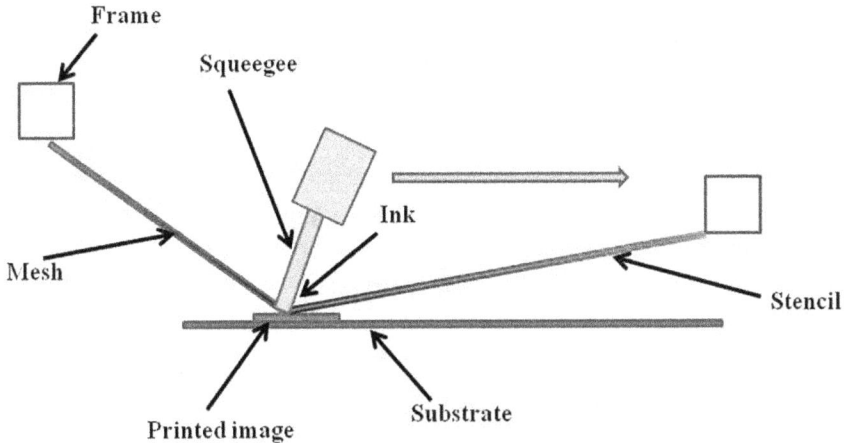

FIGURE 2.5   Schematic representation of the screen printing process.

The tension of the cloth is maintained to prevent it from dangling in the screen. Cloth having a higher mesh count is preferred as it provides both higher print definition and lower theoretical ink volume (i.e., the total volume of ink present in mesh openings per unit area of substrate), but the mesh opening and percent open area are reduced. Generally, the printed layers of light-emitting polymer lamp construction must be the thinnest possible, which can be found using greater mesh count screens with decreased theoretical ink volume values.

Screen printing can be significantly used for solution-processed inorganic and organic materials having various viscosities despite their layer function or substrate flexibility. Also in this method, the material usage is decreased because the materials are only directed to the printed areas. This fabrication process is relatively faster and more versatile than any other techniques. The screen printing method is very easy, affordable, fast and adaptable compared to other fabrication processes. Initially it was thought that screen printing was not relevant to the fabrication of thin films having less than 100 nm thickness, but Pardo et al. proved the utilization of screen printing in the fabrication of an organic active layer having fine thickness, which opened new possibilities as the screen printing process could be used for fabricating OLEDs.[16]

2.4.2.1   Screen

The screen comprises a stencilled fabric mesh. Initially, the first meshes were made of silk, but recently meshes are primarily formed by interwoven stainless steel or polyester threads. The frequency of the threads (number of threads per centimetre) will determine the image quality, while the thickness of the threads will be the deciding factor for defining the thickness of the ink film deposit. A thinner thread will lead to a thinner fabric and hence less ink deposit, and vice versa. Thus, increasing the frequency of the threads will allow a better quality image; also a higher frequency will require thinner threads to leave sufficient openings between the threads.

2.4.2.2 Screen Inks

Screen printing is the printing process with the widest range of applications, hence it requires a greater variety of inks. The viscosity and composition of the ink must complement the fineness of the mesh. Coarser meshes will require the use of higher-viscosity inks to avoid dripping; the mesh opening will limit the maximum size of the pigments to prevent clogging. Screen inks therefore present a variety of viscosities (1 to 100 Pa·s) and solid content can be high as 90%. The solvents being used in screen printing technology generally have low evaporation rates to avoid drying in the mesh opening.

2.4.2.3 Substrates

Screen printing is used on almost every kind of substrate or surface, stiff or flexible. Commonly, screen presses are sheet-fed, printing one sample at a time. By adapting the shape of the mesh and the squeegee, screen printing can be used over complicated products such as ceramic surfaces.

2.4.3 Inkjet Printing

After its establishment in the nineteenth century, inkjet printing has become one of the most popular printing techniques. It is another method to fabricate OLEDs. Inkjet printing presents the superiority of a noncontact technique that shows facile patterning of large-area films with manageable thicknesses and material conservation, which is ideally suited to pattern the organic films for large-area solution-processed OLEDs. There are specific demands for inks during this process, regarding surface tension, viscosity, solvent density, solvent evaporation rate, etc. Although polymers with high molecular weights and amorphous characteristics have been accepted as the suitable choice of materials for inkjet printing, the disadvantages are the complicated purification and the charge unbalance problem that have largely affected the device efficiency and lifetime. In contrast, small molecules are advantageous in providing the high performance needed for practical applications due to their well-defined structures, high chemical purity and controllable reproducibility with no variations.

In this process, a low viscosity ink is deposited in a drop-wise manner with the help of a digitally controlled actuation mechanism, for example, a piezoelectric element, as represented in Figure 2.6. Inkjet printing works as the ejection of droplets of ink from a nozzle onto the substrate. The substrate can be either rigid or flexible. The following methods of drop-on-demand ink ejection exist:

1) The piezoelectric method—piezoelectric ceramic tile is utilized to create pressure to force drops of ink from an ink tank near the nozzle.
2) The thermal method—thermal excitation from a heating element is used to create frequent vapourisation of the ink to form a bubble, causing the sudden pressure enhancement capable of propelling a droplet of ink out through the nozzle.[17]

The chamber filled with liquid is contracted in response to the application of an external voltage. This immediate reduction sets up a shockwave in the liquid,

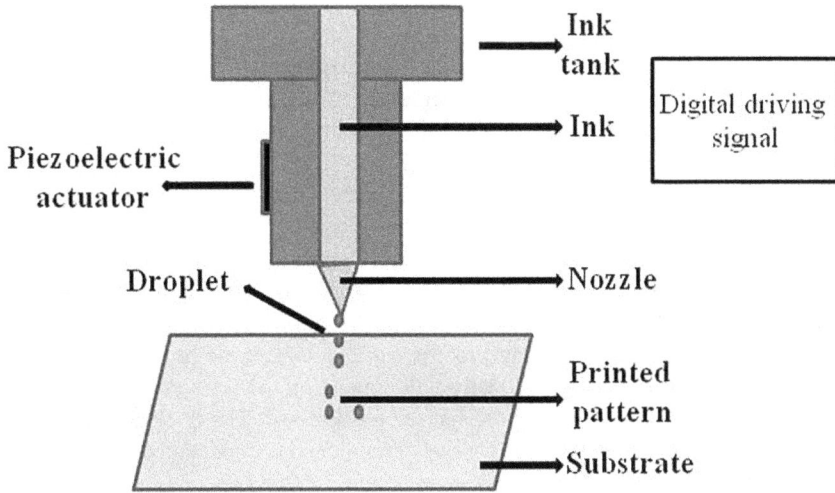

FIGURE 2.6   Basic representation of the inkjet printing process.

which causes a liquid drop to pour out from the nozzle. The ejected drop falls down under the action of gravity and air resistance until it impinges on the substrate. The drop then dries through a solvent evaporation process. Modern developments show that drop spreading and the final printed patterns rely strongly on the viscosity of the inks.

Usually, the inks used in this process are formulated with viscosities of 1–20 centipoise, as of the water, to balance the jettability of the inks. Inkjet printing, a noncontact printing method, can produce fairly intricate patterning without the use of a mask. Initially, inkjet printing was used for graphics printing processes, such as customized packaging, tailored advertising, and office and desktop publishing but recently this process has attracted remarkable interest for patterning functional materials for printed electronics owing to its compatibility with a variety of low-viscosity inks. Contrasting various printing techniques, one of the lowest rates of fabrication waste is produced by inkjet printing, which also has minimal initial startup costs and the ability to perform digital, noncontact and additive patterning. Several challenges of the process in developing materials include time-consuming empirical ink formulation and printing parameter optimization to accommodate fluid property needs and ink-substrate interactions, as well as stringent drop placement accuracy required for complex electronic device layouts. Together, the deposition of a single drop of ink with possible errors in the droplet trajectory controls the application of conventional inkjet printing with respect to miniaturized electronics characterized by a resolution of less than 60 μm.[18] Although inkjet printing offers considerable advantages over traditional fabrication techniques, some associated challenges limit its potential application. Particularly, ink formulation properties, printing parameters and curing methods can all substantially affect the electrical, physical and mechanical properties of the printed structures.

2.4.4 SPIN COATING

In the past few decades, the spin coating process has been utilized for the application of thin films. This process is used to generate a uniform thin film on the desired substrate with source material. This approach often involves applying a small amount of fluid resin to the centre of the substrate. Then the substrate is spun at high speed, generally around 3000 rpm. Figure 2.7 shows the schematic representation of the spin coating process.

Because of high rotation the fluid spins off all the edges of the substrate. The spinning is kept up until the necessary film thickness is achieved. Typically volatile, the applied solvent gradually evaporates. Therefore, the film will be finer the faster the rotating angle is. The spin coating process works on the basis of centrifugal force. The film thickness that is formed lastly will depend on the resin properties, such as viscosity, drying rate, percent solids, surface tension, etc. The thickness of the film also depends on the criteria chosen for the spinning and the concentration of solution and solvent being used in the process. A spin coater, often known as a spinner, is the device used in the spin coating process. The whole process of spin coating can be explained in the following four steps.

2.4.4.1 Dispense Stage

The typical spin coating process of fabrication consists first of a dispense step. In this step, the resin fluid is deposited onto the substrate surface. Two methods of dispension are frequently used: static dispense and dynamic dispense. Static dispense is the deposition of a small puddle of fluid on or near the centre of the substrate. This can range from 1 to 10 cc depending on the fluid's characteristics and the size of the substrate to be printed. High viscosity fluid and large substrate requires more ink to ensure full coverage of the substrate during the fast rotation step. The procedure

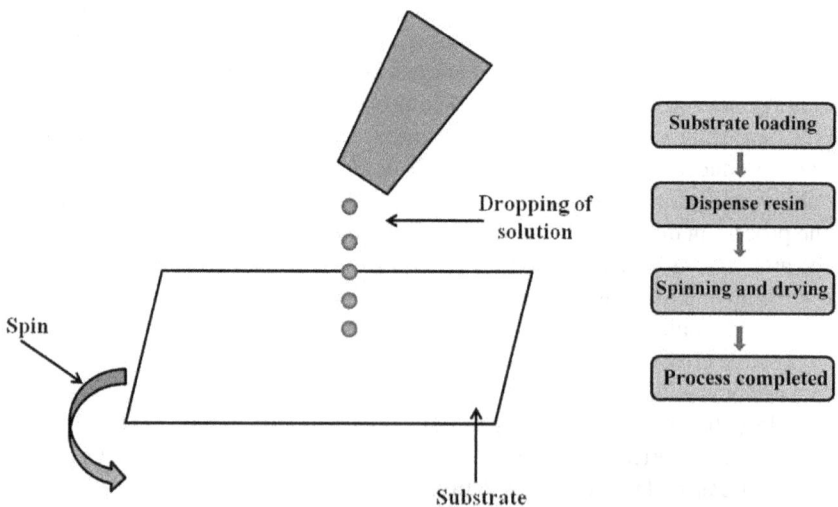

FIGURE 2.7   Overview of the spin coating method.

employed for dispension when the substrate is spinning at a relatively modest speed is known as dynamic dispense. The speed of rotation is about 500 rpm during this step. Because of this fluid is spread all over the substrate. Dynamic dispense causes less waste of resin material because it is usually not required to deposit as much to wet the entire surface of the substrate.

2.4.4.2   Substrate Acceleration Stage

This step is characterized by extensive fluid loss from the wafer surface due to the rotational motion. Eventually, thickness of the fluid should be as fine to be completely co-rotating with the wafer and any possibilities of fluid thickness differences is gone. Finally, the wafer maintains the required speed and the fluid becomes thin enough that the viscous shear drag fairly balances the rotational gain in momentum. This stage involves spinning at speeds between 1500 and 6000 rpm, depending on the substrate and fluid characteristics. It can be completed in 10 seconds to a few minutes. Together the spin speed and the time of this stage will generally define the final film thickness. The spin coating method comprises many variables that tend to cancel and average out during the spin.[19]

2.4.4.3   Substrate Spinning and Fluid Thinning

This stage is characterized by gradual thinning of fluid. Generally fluid thinning is quite even, though in the presence of solutions containing volatile solvents, it is often possible to find interference colours "spinning off", and thus spontaneously coating thickness is reduced. The fluid forms a uniform layer outward, but mostly form droplets at the edge to be flung off; this is called the edge effect. Thus, depending on different parameters, such as surface tension, viscosity, rotation rate, etc., there may be a little bit of coating thickness difference around the rim of the final wafer.

2.4.4.4   Evaporation and Drying

As in the previous steps, the fluid thickness attains a point where the viscosity effects yield only rather minor net fluid flow. After that, the evaporation of any volatile solvent species will become the key factor in the coating process. Thus, a drying step is needed after the high-speed rotation step for drying the film without significantly reducing its thickness. The drying process can be very important for thick films since long drying times may be demanded to enhance the physical stability of the film before being used. Without this step the handling of substrate can be very difficult, as the fluid can spill off the side of the substrate when removing it from the spin coater. In this scenario, drying the film at a moderate spin speed, or roughly 25% of the high spin speed, is usually sufficient without significantly altering the film thickness.

2.4.5   GRAVURE PRINTING

The gravure printing method is a high-volume, high-resolution reel-to-reel printing process. Its basic working mechanism comprises an engraved cylinder, an ink bath, a blade, referred to as the 'doctor blade', and a rubber-covered impression roller. Particularly when compared to inkjet and screen printing, gravure printing has attracted a lot of interest due to its exceptional scalability and competitive resolution.

Compared to printing processes such as offset printing and flexography, gravure printing requires that the desired image be etched or placed onto a cylindrical roller, which transfers the ink to rolls of substrate directly by contact pressure, as shown in Figure 2.8. It allows the printing of small structures in the range of micrometres, as well as the application of large-area functional layers with a size of several square centimetres.

Due to its high throughput, it is also suitable for the large production of printed electronics. The printing process may be divided into four steps: predosing, dosing, transfer and drying phase. During predosing, the fluid is applied on the surface of the printing form cylinder. When the cylinder rotates, the excess ink is removed by the clinging doctor blade and the engraved cells of the gravure cylinder are filled with ink. In the second phase, the substrate carrier, including the substrate itself, is transferred to the printing cylinder. Therefore, the impression cylinder pushes the substrate against the printing cylinder. To decrease slip, the substrate carrier and the printing cylinder are balanced to the same velocity before coming in contact. In the graphic industry, the gravure method is traditionally the printing technique of choice where the quality of the print is critical. Current mass production markets for rotogravure include the production of high-quality packaging and publication printing. Gravure printing is highly suitable for high-volume manufacturing and is regarded as great for industries when combined with a reel to reel (R2R) printing arrangement.[20]

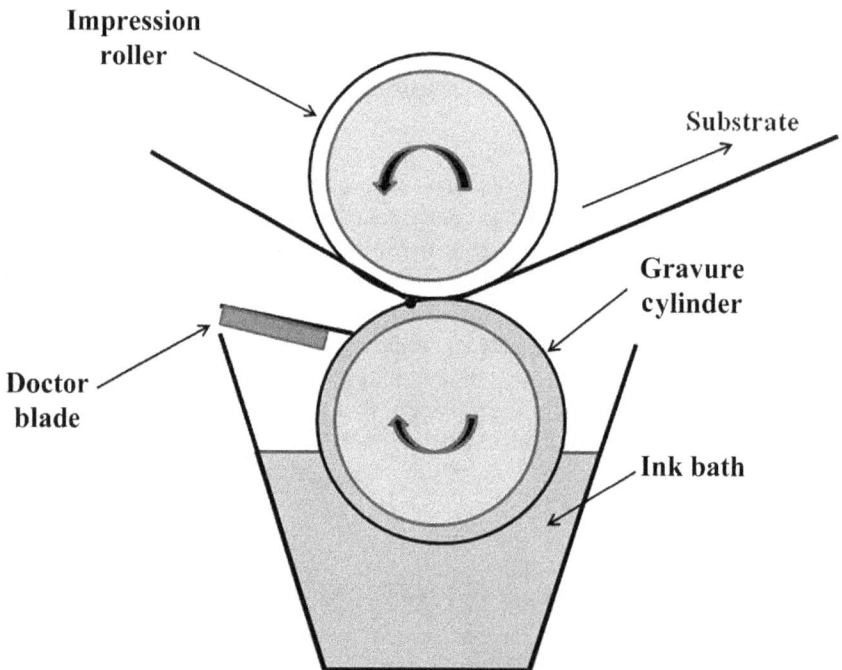

FIGURE 2.8   Representation of the shuttle-based gravure printing process.

2.4.6 Aerosol Jet Printing

A relatively recent printing method called aerosol jet printing uses a contactless direct write approach to produce fine features on a variety of substrates. Originally discovered for the manufacture of electronic circuitry, the technique has been explored for a range of applications, including active and passive electronic components, actuators and sensors, as well as for selective chemical and biological responses. In this printing technique, liquid pigments are first atomized into tiny droplets that are 1 to 5 μm in diameter before being carried to the printing head by a carrier gas. Additionally, an annualer sheath flow is utilised at the printing head to aerodynamically drive the aerosol to be deposited onto the intended substrate.

Additionally, this method works with a variety of materials, including organic and inorganic inks with high viscosities. Another benefit of aerosol jet printing is that it has the unique ability to print patterns on nonflat (3D) surfaces because of the substantial standoff distance between the printing nozzle and the substrate. Wide material compatibility, high resolution and independence of orientation have given novelty in a number of applications when aerosol jet printing is conducted as a digitally driven approach.

2.4.7 In-Line Fabrication

In-line fabrication is a large-scale process technique. It has a vertical in-line tool that utilises a constant substrate flow. This procedure uses linear sources for depositing organic and metallic compounds. Additionally, instead of the point sources that are more frequently employed in OLEDs, the in-line sources are used where material is deposited from a linear tube. This enhances material usage by a factor of 10 and the method is cost effective for mass production. The patterns formed using this technique are stable. In-line fabrication can be used for complex stack structures. Deposition rate and throughput are high.

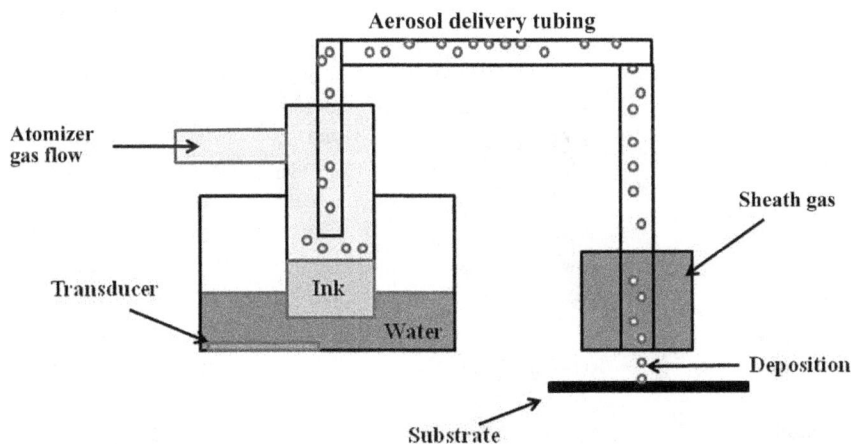

FIGURE 2.9   Diagram illustrating the aerosol jet printing process.

2.5 COMPARISON OF VARIOUS PRINTING TECHNIQUES

After discussing the most popular fabrication techniques for OLED, it is apparent that choosing the best fabrication technique is a really challenging task. The choice of an OLED fabrication method depends on a number of factors, including the need for high-quality growth, the volume of production, the materials used in the process, etc. The expensive physical vapour deposition method is the standard fabrication method. Screen printing is a good alternative for this technique. It is cheaper and a multifaceted fabrication process. But thin film, such as below 100 nm, is difficult to fabricate by screen printing. Ink-jet printing is a noncontact deposition technology that allows for contamination-free development on flexible substrates. In-line fabrication is unquestionably the best method for mass production of OLEDs. Table 2.1 provides an overview of the comparison between different printing techniques focusing on resolution, speed, cost effectiveness, etc.[21,22] The values given in the table are only representations to help the readers for better understanding.

2.6 CONCLUSION

The typical fabrication processes for organic light-emitting diodes are discussed. These methodologies are a key factor for the future development of OLEDs for large-scale, cost-effective production. The simplicity of fabrication techniques and the adoption of reputable manufacturing technology are the main factors, which were not available previously in the electronics industry and have caused OLEDs to gain notable scientific attention. A range of printing techniques are available for the development of OLEDs and other printed electronic products. Each printing and

TABLE 2.1
Comparison of Different Printing Methods

	Vacuum Evaporation	Screen Printing	Inkjet Printing	Spin Coating	Gravure Printing	Aerosol Jet Printing
Max Resolution	High	Medium (50–100 µm)	Variable (20–50 µm)	Variable (5–100 µm)	Very high (<20 µm)	Variable (20–100 µm)
Thickness of layer	Variable	Thick (1–20 µm)	Very thin (<0.5 µm)	Very thin (<0.01 µm)	Variable (<0.1–5 µm)	Thin (0.1–5 µm)
Cost	High	Low	Low	Low	High	Low
Speed	High	Slow	Very slow	High	Very high	High
Ink Viscosity	–	High (1–100 Pa·s)	Low (<0.02 Pa·s)	High (1–20 Pa s)	Low (0.05–0.5 Pa·s)	Low (0.001–2.5 Pa s)
Application	Small volume production	Medium to large area	Small area and volume production	Small volume production	Large volume production	Small volume production

coating method presents its own advantages and limitations. It can be expected that the first fully printed high-volume OLEDs will be produced by integrating various printing and coating techniques. Challenges have still to be overcome, for example, more accurate and homogenous deposition of thin films, depending on the complexity of the pattern to be printed, the area, size and production volume. By conducting attentive study now and in the future, we will become familiar with more accessible, adaptable display and fabrication processes.

REFERENCES

1. Farinola, G.M., Ragni, R. 2011. Electroluminescent materials for white organic light emitting diodes. *Chem. Soc. Rev.* 40:3467–3482.
2. Zhong, C., Duan, C., Huang, F. et al. 2011. Materials and devices toward fully solution processable organic light-emitting diodes. *Chem. Mater. Rev.* 23:326–340.
3. Sasabe, H., Takamatsu, J., Motoyama, T. et al. 2010. High-efficiency blue and white organic light-emitting devices incorporating a blue iridium carbene complex. *Adv. Mater.* 22:5003–5007.
4. Zhou, G.J., Wang, Q., Ho, C.L. et al. 2009. Duplicating sunlight from simple WOLEDs for lighting application. *Chem. Commun.* 24:3574–3576.
5. Zhou, G.J., Wang, Q., Wang, X.Z. et al. 2010. Metallophosphorus of platinum with distinct main-group elements: A versatile approach towards color tuning and white-light emission with superior efficiency/color quality/brightness trade-offs. *J. Mater. Chem.* 20:7472–7484.
6. Karzazi, Y. 2014. Organic light emitting diodes: Devices and applications. *Environ. Sci.* 5:1–12.
7. Wu, J., Agrawal, M., Becerril, H. et al. 2010. Organic light-emitting diodes on solution-processed graphene transparent electrodes. *ACS Nano.* 4:43–48.
8. Tang, C.W., Van Slyke, S.A. 1987. Organic electroluminescent diodes. *Appl. Phys. Lett.* 51:913.
9. Bardsley, J.N. 2004. International OLED technology roadmap, selected topics in quantum electronics. *IEEE J.* 10:3–9.
10. Hwang, J., Amy, F., Kahn, A. 2006. Spectroscopic study on sputtered PEDOT PSS: Role of surface PSS layer. *Org. Electron.* 7:387–396.
11. Kubota, H., Miyaguchi, S., Ishizuka, S. et al. 2000. Organic LED full color passive matrix display. *J Lumin.* 87:56–60.
12. Fujimoto, E., Daiku, H., Kamikawa, K. et al. 2010. OLED manufacturing system equipped by planar evaporation source. *SID Digest.* 10:695–698.
13. Ikeda, T., Murata, H., Kinoshita, Y. et al. 2006. Enhanced stability of organic light-emitting devices fabricated under ultra-high vacuum condition. *Chem. Phys. Lett.* 42:111–114.
14. Khan, S., Lorenzelli, L. Dahiya, R.S. 2015. Technologies for printing sensors and electronics over large flexible substrates: A review. *IEEE Sens. J.* 15:3164–3185.
15. Lee, D.H., Choi, J., Chae, H. et al. 2008. Single-layer organic-light-emitting devices fabricated by screen printing method. *Korean J. Chem. Eng.* 25:176–180.
16. Shaheen, S.E., Radspinner, R., Peyghambarian, N. et al. 2001. Fabrication of bulk heterojunction plastic solar cells by screen printing. *Appl. Phys. Lett.* 79:2996–2998.
17. Wu, W. 2017. Inorganic nanomaterials for printed electronics: A review. *Nanoscale.* 9:7342–7372.
18. Secor, E.B., Prabhumirashi, P.L., Puntambekar, K. et al. 2013. Inkjet printing of high conductivity, flexible graphene patterns. *J. Phys. Chem. Lett.* 4:1347–1351.

19. Mitzi, D.B., Kosbar, L.L., Murray, C.E. et al. 2004. High mobility ultrathin semicon-
    ducting films prepared by spin coating. *Nature*. 428:299–303.
20. Nisato, G., Donald, L., Simone, G. 2016. Organic and Printed Electronics: Fundamentals
    and Applications; CRC Press: Boca Raton, FL.
21. Deganello, D. 2013. Printing techniques for the fabrication of OLEDs. In Organic
    Light-Emitting Diodes (OLEDs) Materials, Devices and Applications (pp. 360–385).
22. Chu, Y., Qian, C., Chahal, P. 2019. Printed diodes: Materials processing, fabrication,
    and applications. *Adv. Sci.* 6:1–29.

3 Design of Hybrid Perovskites for OLEDs

Ashish Kumar Srivastava and Arvind Kumar Singh

3.1 INTRODUCTION: BACKGROUND AND DRIVING FORCES

Lighting accounts for about 20% of worldwide electricity use and emission of 12.3 Gt CO_2 in 2020.[1] Light emitting diodes (LEDs) that emit white light are growing in popularity. Compared to incandescent bulbs and fluorescent tube lights, LEDs are significantly more efficient.[2–4] One of the drawbacks of incandescent bulbs is that they are too hot to touch. The extra heat energy is completely wasted.

In one year, the energy consumption might be decreased by about 1,000 TW h if all common white-light sources were replaced with LED ones. This power is equivalent to 230 coal plants of 500-MW capacity. In turn, it can reduce greenhouse gas emissions up to 200 million tons.[5]

LEDs can be fabricated with the help of luminescent perovskites.[6] The designed LED has various properties, so that it can be used for lighting purposes, for example, successful colour representation, warm white colouration and strong spectral overlap. In addition to this, the gaps in the spectrum should be much less and emitted light should cover the visible range only, otherwise it leads to wastage of energy.[7]

Pervoskite has been known since the 19th century. The hybrid perovskites are emerging as an important LED material owing to high performance, ultra low cost per watt photovoltaic operations, extraordinary degree of structural properties and flexibility.[8,9] In addition, extremely high optical absorption, small effective mass for electron and holes, dominant point defects responsible for generating shallow levels and grain boundaries are advantageous over inorganic and organic LEDs.[10] In contrast to typical inorganic nanomaterials, hybrid perovskite's exceptional colour purity is balanced regardless of crystal size thanks to its inherent quantum-well structure, in contrast with the usual inorganic nanomaterials.[11] To address issues like complex synthesis, high cost, poor colour purity and high ionisation energy, hybrid perovskites can have certain advantages over organic LEDs (OLEDs) and inorganic quantum dot LEDs.[10] To understand the application of hybrid perovskite as LED, it is important to understand its design and changes in physical properties with changing design. In view of this, this chapter focuses on prerovskite, its classification, design and important applications as LEDs.

Perovskites (ABX_3) are a group of compounds that take on the same three-dimensional (3D) crystal structure as the mineral $CaTiO_3$ depicted in Figure 3.1. In the case of organic–inorganic hybrid perovskites, at least one among A, B and X ions is organic.[12–16] Usually cation A is organic($CH_3NH_3^+$, $CH(HN_2)_2^+$, B is metal(Pb^{2+} and Sn^{2+}) and X is halogen(Cl^-, Br^- and I^-).[17]

DOI: 10.1201/9781003260417-3

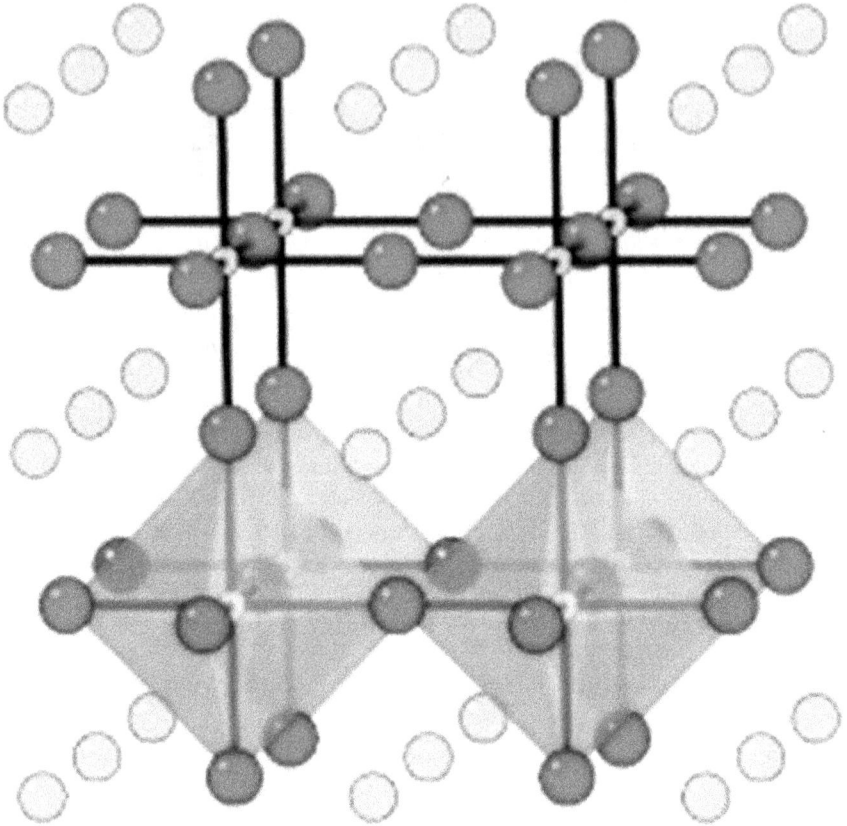

FIGURE 3.1 Combined ball-and-stick model of perovskite. Skeletal (top) and shaded poly-hedral (bottom) representations of the three-dimensional cubic ABX_3 perovskite structure. A, B, and X atoms are shown as gold, cyan, and red spheres, respectively. (Reproduced with permission from Bayrammurad and David.[29] Copyright 2016 American Chemical Society.)

3.2 CLASSIFICATION OF PEROVSKITES

Perovskite can be classified according to dimensionality.

1. Three-dimensional perovskite
2. Lower-dimensional perovskite: (a) <100>-oriented perovskite, (b) <110>-oriented perovskite, (c) <111>-oriented perovskite
3. Exotic framework
4. Molecular perovskite

3.2.1 THREE-DIMENSIONAL PEROVSKITES

While designing 3D perovskite, (a) the concept of tolerance factor, (b) charge balance and (c) bonding and coordination preferences of A, B and X ions are important.[18–20]

The structure of a 3D pervoskite can be easily understood by considering Figure 3.1. The space filling model suggests that A, B and X ions adopt the perovskite framework. A (mostly cation) occupies at 12 fold coordinated holes. The structure is built by sharing the corners of an octahedron (BX_6).

Goldschmidt's tolerance factor (t) for most of the 3D perovskites must lie between 0.8 and 1, and it is given by the following formula:

$$t = \frac{\sqrt{2\left(R_B + R_X\right)}}{\left(R_B + R_X\right)}$$

Here R_A, R_B and Rx are the ionic radii of ions A, B and X, respectively.[18]

In comparison to complex ions and nonspherical organic ions, it is quite easy to assign ionic radii to spherically symmetric inorganic ions. Assumptions are taken for the ionic radii of organic cations.[21] t values have been calculated for a large number of inorganic-organic hybrid pervoskites. More than 500 of them have been found to have consistent t values, while 600 perovskites are completely nonconsistent.[22]

The second important factor for the design of pervoskites is charge balance.[20] In hybrid perovskite ABX_3, cation A is mostly organic and monovalent, such as methylammonium, formamidinium and many more.[23] If A is monovalent and X is a halogen, then cation B must be divalent to make the perovskite a neutral molecule. Cation B can be from the lower side of the p-block {Ge(II), Sn(II), and Pb(II)}, alkine earth metal Ca(II) and lanthanides, which show stable oxidation by donating two electrons, such as Eu(II) and Yb(II).[24]

The third important factor is coordination preferences. In the hybrid perovskite structure, the ABX_3 framework, alternating larger and smaller B-X bonds have been found. The lone pair on the s-orbital of the metal atom B moves away from the metal's octahedral crystallographic centre. This type of shift is observed when the halide ion is replaced by a format anion.[25] Some other examples have also been reported where organic amines are acting as monovalent cation.[26–28]

3.2.2   LOWER-DIMENSIONAL PEROVSKITE

The dimensionality is the number of corners shared in BX_6 octahedra in the crystal structure, as depicted in Figure 3.1.[29] In two-dimensional perovskites the tolerance factor is not a restriction. In addition to this, there are no restrictions on the length of an interlayer cation A. The organic cation A should only have terminal functional groups that can interact ionically with the anionic inorganic substructure, and the rest of the organic molecule shouldn't get in the way of the inorganic parts B and X.[30] The two-dimensional perovskites are derived from 3D perovskite by cutting in different directions, as illustrated in Figure 3.2. They are of the following types:

(a) The group of ⟨100⟩-oriented layered perovskites
(b) The group of ⟨110⟩-oriented perovskites
(c) The group of ⟨111⟩-oriented perovskites

In each of these layered structures, the perovskite framework is separated by a layer of typically larger organic cations.[31–39]

The dimensionality of perovskites is related to their physical properties. An increase in the value of the band gap has been observed with decreasing dimensionality.[40]

3.2.3 EXOTIC FRAMEWORK

Hybrid pervoskites have also been reported that are neither three- nor two-dimensional; they are a combination of $\langle 100 \rangle$- and $\langle 110 \rangle$-oriented layers.[41]

The pervoskite material provides an unending number of structures owing to its structural flexibility and simple handling. One important term, "spectator ion" has also been introduced. Spectatior ions are included in the reaction mixture, although their presence in the final chemical is not very noticeable. The spectator ions might play an important role in the resulting topology. Two different types of structure are found with molecular formula $[NH_3(CH_2)_5NH_3]SnI_4$. The β-phase is formed under regular synthetic condition and on the addition of a spectator CH_3NH_3I, an α-phase is produced.[42] The resulting α-phase can be viewed as a result of the $\langle 100 \rangle$- and $\langle 110 \rangle$-oriented layers. Another possible explanation of the structure of the α-phase is the interconnection of four octahedron long $\langle 100 \rangle$-layers at 45° to form the crisscross morphology, as shown in Figure 3.2. However, the detailed mechanism of formation

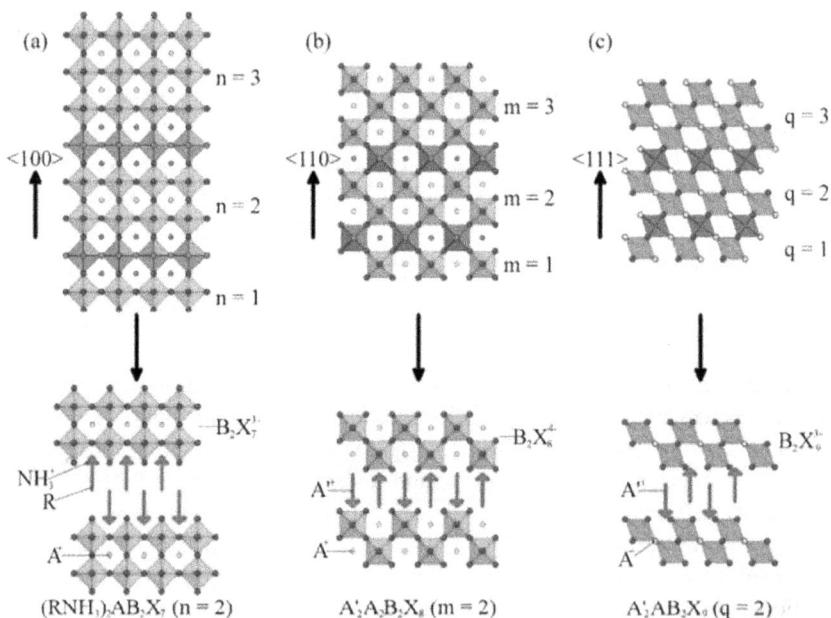

FIGURE 3.2 Schematic representation of the derivation of the lower-dimensional organic–inorganic perovskites (lower sections) from different cuts of the three-dimensional (3D) perovskite structure (top sections). (a) The family of $\langle 100 \rangle$-oriented layered perovskites; (b) cuts along the $\langle 110 \rangle$ direction of the 3D perovskite structure provide the $\langle 110 \rangle$-oriented family; (c) the $\langle 111 \rangle$-oriented family. (Reproduced with permission from Bayrammurad and David.[29] Copyright 2016 American Chemical Society.)

of the α-phase still needs to be understood. The reaction condition and addition of spectator ions are influencing the resulting morphology and fragmentation of three-dimensional hybrid pervoskite.

3.3 EFFECT OF VARIETY OF SUBSTITUENT

3.3.1 EFFECT OF ORGANIC MOLECULES ON HYBRID PEROVSKITES

Organic cations in organic–inorganic hybrid pervoskite provide structural flexibility, as can be seen in the previous examples. Besides providing structural flexibility, organic cations also introduce the functionality because of their conjugated structure. The greater amount of conjugation results in decreased separation of the highest occupied molecular orbital (HOMO) and the lowest unoccupied molecular orbital (LUMO), which has a significant impact on the material's optoelectronic capabilities. The crystal structure and optoelectronic properties of a hybrid pervoskite, incorporating 5,5‴-bis(aminoethyl)-2,2′:5′,2″:5″,2‴-quaterthiophene in a layered lead halide pervoskite, were examined.[43] The single crystal structure of this pervoskite is displayed in Figure 3.3. This special organic oligothiophene is very useful, owing

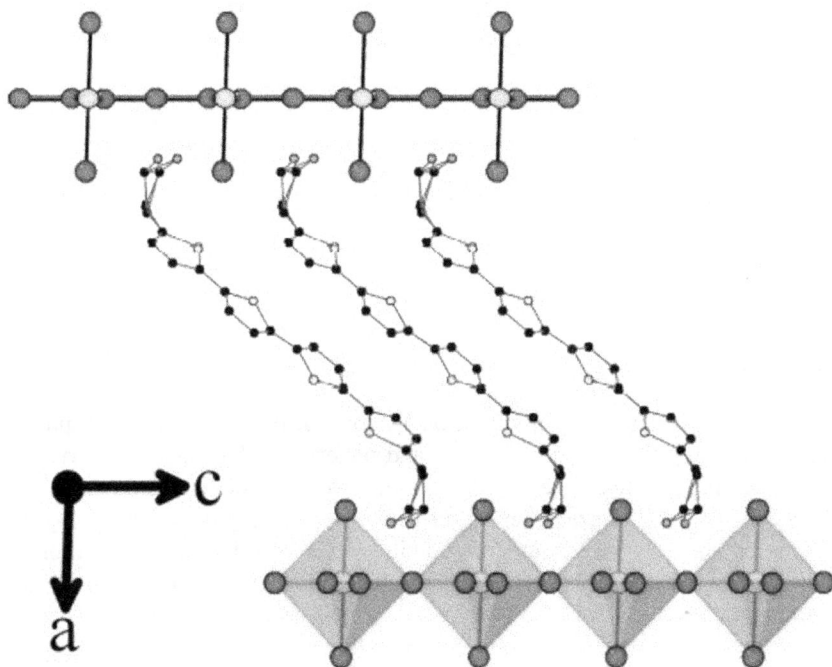

FIGURE 3.3 Combined skeletal (top) and polyhedral (bottom, shaded) view of the crystal structure of AEQTPbBr4, where AEQT = 5,5‴-bis(aminoethyl)-2,2′:5′,2‴:5″,2‴-quaterthiophene, consisting of self-assembling and ordered organic and inorganic layers. (Reproduced with permission from Mitzi et al.[43] Copyright 2016 American Chemical Society.)

TABLE 3.1

Variation of HOMO-LUMO Separation With Increasing n

n (number of α-linked thiophene rings within the chain)	HOMO-LUMO Separation	λ_{max} (absorption peak, nm)	Reference
2	4.1	302	43
6	2.9	432	43
4	3.2	380	44

to the fact that the HOMO-LUMO gap can be varied very easily by controlling the length of the thiophene chain (n). The variation in HOMO-LUMO gap for an inorganic molecule is not so easy. The variation of the HOMO-LUMO gap and absorption peak (λ_{max}) for different values of n are provided in Table 3.1.

Two types of charge transfer or energy transfer process are observed when a functional organic cation is part of the hybrid pervoskite. One is Förster resonant energy transfer (FRET) and another is Dexter energy transfer (DET).[45,46] FRET is the outcome of nonradiative dipole-dipole coupling and it is proportional to r^{-6}, where r is the distance between donor and acceptor. DET is the result of nonradiative transfer and excited electron from donor to acceptor or from acceptor to donor, and it is proportional to e^{-r}.

There are many examples of charge transfer between the organic and inorganic parts of hybrid pervoskite. Functional organic moieties such as alkylammonium cation, naphthalene, pyrene, 3–2-(aminoethyl)indole ($C_8NH_6CH_2CH_2NH_2$), azobenezene, n-(3-aminopropyl)imidazole, incorporated in metallic halides have been studied and reported.[47–58] These studies have allowed the matchless flexibility of charge transfer between inorganic and organic parts in a synergistic way.

3.3.2 EFFECT OF INTERCALATION IN PERVOSKITE

Intercalated pervoskites are important in the areas of photoluminescence, photoinduced photocatalysis ionic conductivity and battery development.[59] The very first work on an intercalated system was reported using the intercalation of 1-chloron aphthalene/o-dichlorobenzene/hexane in pervoskite already containing long chain aliphatic cations. The intercalated groups interacted with the aliphatic cations via weak van der Waal interaction. However, this intercalated composite is unstable and study of its structure was tricky using single crystal X-ray diffraction studies.[60] The very first explicit structure determination was done for hybrid perovoskite $(C_6H_5C_2H_4NH_3)_2SnI_4$ using benzene, hexafluorobenzene and their analogues. The film of $(C_6H_5C_2H_4NH_3)_2SnI_4$ deposited from methanol solution results in an interlayer distance of 16.3Å. Surprisingly, when the same film was dipped into a solution of hexaflurobenzene, the interlayer separation increased by 4.3 Å, owing to the fact that total interlayer separation is 20.3Å, as evident from x-ray diffraction studies.

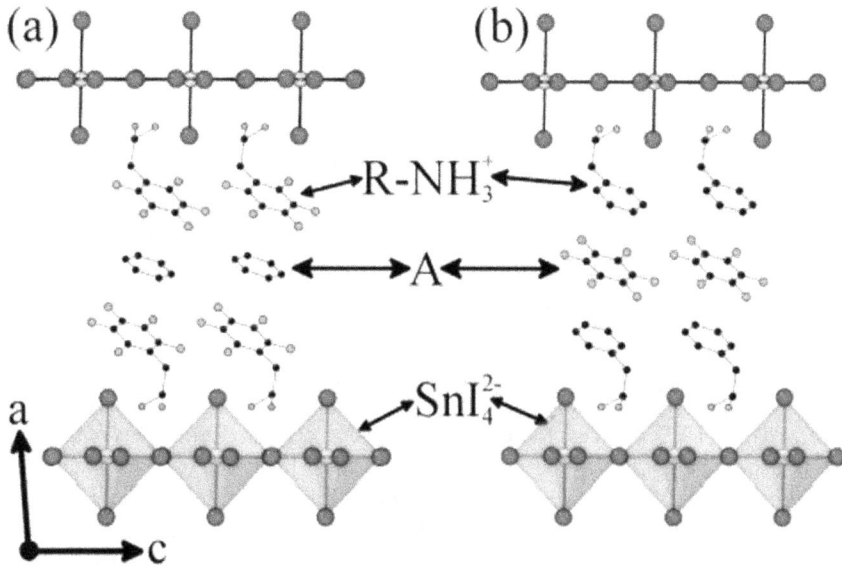

FIGURE 3.4 Combined skeletal (top) and polyhedral (bottom, shaded) views of the crystal structures of (a) $(C_6F_5C_2H_4NH_3)_2SnI_4 \cdot (C_6H_6)$ and (b) $(C_6H_5C_2H_4NH_3)_2SnI_4 \cdot (C_6F_6)$ along the b-axis, showing the details of the stacking of the intercalated molecules within the perovskite framework. For clarity, the fluorine atoms are colored green and hydrogen atoms are omitted. (Reproduced with permission from Mitzi et al.[61] Copyright 2016 American Chemical Society.)

The noncovalent interaction between fluoroaryl and aryl was playing a very important role in increasing the interlayer distance, as shown in Figure 3.4.

The intercalation was successfully achieved with tin halide but using the same process, $(C_6H_5C_2H_4NH_3)_2CuCl_4$, this perovskite cannot be intercalated owing to unfavorable orientation of the phenyl ring.[61] In the layered pervoskite there are possibilities of intercalation of other inorganic/organic groups that can result in the increase of interlayer separation and simultaneously change of utility or generation of a new application. Based on the above precedence, the intercalation of organic cation can assist the molecular transport in the pervoskite films.

3.4 DEVICE APPLICATION OF PEROVSKITE AS LEDs

There are various parameters for the evaluation of device performance of LEDs. Some of these are lifetime, power conversion efficiency, wall-plug efficiency, internal quantum efficiency and exterior quantum efficiency (EQE). In perovskite LEDs, EQE is the most widely used and accepted metric to assess device performance. EQE is nothing more than the ratio of the number of photons emitted from the device to the number of electrons passing through the device. In 2014, the first hybrid perovskite-based LEDs were announced. They functioned at room temperature and had an adequate maximum EQE (EQEmax) of 0.76% in the infrared region and 0.1%

in the green region.[62] These EQE values are far lower than those of inorganic LEDs and OLEDs, but this study has inspired the research community to pursue new avenues for enhancing device performance.[63–66] A major development has been achieved by a green $CH_3NH_3PbBr_3$ hybrid perovskite LED with EQE value of 8.53%.[9] It was done by decreasing the presence of metallic impurity and simultaneously increasing the ratio of CH_3NH_3Br. Another factor is grain size. Smaller grains are preferred owing to their property of limiting the exciton diffusion length, consequently reducing the possibility of exciton dissociation before recombination.[67] After the discovery of this hybrid pervoskite LED, different groups rekindled the value of EQE greater than 20% in the green, red and infrared light regions.[68–71]

3.5   DEVICE PATTERN OF PERVOSKITE AS LED

In perovoskite LED devices, there are three layers: a light emitting layer (LEL), a hole transport layer (HTL) and an electron transport layer (ETL). The function of pervoskite is to act as a light emitting layer and it lies in between the hole transport layer and electron transport layer. The fabrication of LED devices is done in two ways: normal and inverted structures, as shown in Figure 3.5(a) and (b). The working mechanism of perovskite LED is shown in Figure 3.5(c).

The electrons are injected via the electron transport layer, while holes are injected via the hole transport layer. The radiative recombination of holes and electrons takes place in the perovskite light emitting layer. In the perovskite layer photons are formed and emitted outside the device. The first hybrid pervoskite as OLED is designed similar to the solar cell. In this device a 15 nm $CH_3NH_3PbI_{3-x}Cl_x$ perovskite emitter is inserted between TiO_2, functioning as an electron transport layer and poly(9,9'-dioctylfluorene), working as a hole transport layer.[72] This is also the first paper about inverted structured (n-i-p) perovskite-LEDs. As organic materials are very poor carriers, it is kept very thin. As in OLED, the EML layer needs to be very thin because of the poor carrier mobility of the organic materials. However, as the charge mobility of perovskite materials is higher than that of the pure organic materials, it seems that the perovskite LEDs are not sensitive to the thickness of the light emitting layer. There are many high-performance perovoskite LEDs reported, in which the thickness of the light emitting layer is more than 100 nm.[65–67]

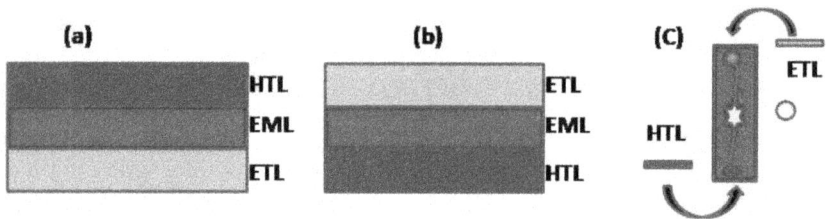

FIGURE 3.5   Cell configuration diagram of (a) the inverted LED structure of ETL/LEL/HTL (n-i-p), (b) normal LED structure of HTL/LEL/ETL (p-i-n), and (c) operation mechanism diagram of both structures. EML = emission layer; ETL = electron transport layer; HTL = hole transport layer; LEL = light emitting layer.

An efficient green bright hybrid perovskite is reported with external quantum efficiency (EQE) value of 8.53%. Its hole transport layer is a self-organized conducting polymer [poly(3,4-ethylenedioxythiophene): poly(styrenesulfonate) PEDOT: PSS and perfluorinated ionomer, PFI] and light emitting layer is a CH_3N-H_3PbBr_3 perovskite, and, 2,2′,2″-(1,3,5-benzinetriyl)-tris (1-phenyl-1-H-benzimidazole) serves as an electron transport layer.[9] Selection of the material for the hole transport and electron transport layers depends on the following: (a) The energy levels of the charge transporter layers should be comparable with the perovskite layer. (b) The preparation method for the charge transporter layers should be such that it causes no disturbance, that is, it causes no harm to the perovskite layer. (c) The charge transport mobility of the hole transport and electron transport materials need to be close to each other. One more example is a p-i-n device, in which PEDOT:PSS serves as the hole transport layer and bis-4,6-(3,5-di-3-pyridylphenyl)-2-methylpyrimidine (B3PYMM) serves as the electron transport layer; the EQE value is reported to be 20.3%.[67] Huang et al. studied and reported an n-i-p device structure with the EQE value of 20.7%, in which the role of electron transport layer was played by ZnO/PEIE and the role of hole transport layer was played by TFB polymer.[73] Many devices with three-layer fabrication have been reported; however, attaining higher EQE is possible by including more layers, such as hole injection layer, hole-blocking layer, electron injection layer, electron-blocking layer, and many more.[74]

3.6 HYBRID PEROVSKITE NANOCRYSTAL OR QUANTUM DOTS

The discussion of hybrid perovskite as LED includes perovskite nanocrystals or quantum dots. Perovskite colloidal nanocrystals or quantum dots (QDs) are a group of perovskites with fascinating properties of light emission. The very first QDs, $MAPbBr_3$, were synthesized by ligand-assisted reprecipitation with a mean size of 6 nm, and their photoluminescence quantum yield (PLQY) falls around 20%.[75] But it has the capacity of maximum luminescence (L_{max}) 1 Cd/m^2. This value is so small that it cannot be used in day-to-day life. One more group has changed the strategy of synthesis and PLQY falls around 70% for $MAPbX_3$ (X = Br, I, Cl). In addition to this, this colloidal hybrid perovskite with amorphous structure has the PLQY more than 70%.[76] The nanoparticle of amorphous $MAPbBr_3$ has produced L_{max} of 11830 cd/m^2.[77] The PLQY of these quantum dots after 1 week increased from 70% to 80% in solution and 35% to 71% in film form. This increase in PLQY was attributed to increased crystallization of perovskite QDs. Rand et al. fabricated self-assembled, nanosized perovskite crystallites by using butylamine halide as ligand in the in situ ligand-assisted reprecipitation process (LARP).[78] The butyl ammonium halide inhibited the growth of 3D pervoskite grains during film formation; consequently, 10 nm nanocrystallites were obtained. These nanosized perovskite grains with a long chain organic cation coating produces highly efficient emitters, resulting in pervoskite LEDs with EQE_{max} of 10.4% and L_{max} of 20 cd/m^2 for the $MAPbI_3$ system (EL 748 nm) and EQE_{max} of 9.3% and L_{max} of 1200 cd/m^2 for the $MAPbBr_3$ system (EL 513 nm). Zhong et al. demostrated the in situ LARP design of highly luminescent $FAPbBr_3$ nanocrystal thin films by using 3,3-diphenylpropylamine bromide as ligand.[79] The resulting films are uniform and composed of 5–20 nm $FAPbBr_3$ perovoskite nanocrystals, and the PLQY was improved to 78%. As a result, they achieved highly efficient pure green (EL 526 nm) pervoskite-LEDs with EQE_{max} of 16.3% and L_{max} of 13970 cd/m^2. Recently, they developed a general hybrid ligand strategy to passivate $CsPbX_3$ QDs for highly efficient hybrid pervoskite LEDs.[80] There are a number of metal halides such as $ZnBr_2$, $MnBr_2$, $GaBr_3$ and $InBr_3$ that can possibly be used to passivate nonradiative defects in pervoskite QDs. The introduced metal bromide inorganic ligands could enhance the luminescent features and effective carrier injection originating from increase of radiative recombination and their intrinsic high conductivity. As a consequence, the hybrid ligand strategy enhanced perovoskite-LEDs performance up to 40% improvement. The best EQE_{max} of 16.48% and L_{max} of 100080 cd/m^2 were accomplished based on $ZnBr_2$ and $MnBr_2$ passivation; this is also the highest EQE reported in QD perovskite LEDs with green emission (EL 518 nm). A bidentate ligand containing two carboxylic groups has been used to passivate the $CsPbI_3$ QD surface.[81] The study suggested that the bidentate ligand binds firmly to perovskite surfaces. It does so by two carboxylic groups. It reduces the surface traps and injects the extra electron in QDs. The passivated $CSPBI_3$ QDs exhibited a PLQY value close to unity and improved stability. Oleylammonium iodide (OAMI) and aryl-based aniline hydroiodide (AnHI) were used in $CsPbBr_3$ to fabricate red perovoskite.[69] With the inclusion of OAMI and AnHI the PLQY values are 80% and 69%, while it was only 38% for pristine $CSPBBr_3$. The EQE_{max} and L_{max} are 21.3% and 500 cd/m^2, respectively, for OAMI-based red perovoskite (EL 653 nm). ANHI-based QD LED (EL 645 nm) displayed EQE_{max} of 14.1% and L_{max} of 794 cd/m^2. Furthermore, the operational stability is higher for ANHI-based QD LED than for OAMI-based QD LEDs.

3.7   PEROVSKITE QUANTUM WELLS AND DEVICE APPLICATIONS

This type of pervoskite is truncated and restricted along only one axis of 3D pervoskite, similar to 2D perovskite. The 3D perovskite is cut along one-layer-thick slices along the $\langle 100 \rangle$ or $\langle 110 \rangle$ direction. In addition to this, the small organic molecules or ions are replaced by long-chain amines. In this process, the monolayer or multilayer pervoskite slab is sandwiched by long chain organic amines. These types of hybrid structure are known as hybrid perovskite quantum wells (QWs). Multilayer pervoskite QWs are fabricated with the help of phenylethylammonium (PEA) in $CH_3NH_3PbI_3/MAPb_3$ and results in $PEA_2(MA)_{n-1}Pb_nI_{3n+1}$.[82] For this pervoskite QWs, under low excitation the PLQY value falls around 10%. It was also observed that as the excitation intensity rises it promotes the PLQY value. It is found that due to the transport of excited carriers in the funneling mechanism, the concentration of these carriers increases and they result in effective radiative combination by overcoming the trap-mediated nonradiative recombination. This leads to the development of pervoskite QWs with 8.8% of EQE_{max} and 80 W $sr^{-1}m^{-2}$ of maximum radiance, operating at near-infrared wavelength of 760 nm. Another example of pervoskite multiple QWs was designed by the combination of 1-naphthylmethylamine iodide (NMAI) and a pervoskite $FAPbI_3$.[63] The hybrid QWs film (EL 763 nm) produces PLQY of 60%, and with 11.7% EQE_{max} 82 $Wsr^{-1}m^{-2}$.

The transference of energy takes place from the high band gap domain to small and less-populated lower band gap domains. In view of this, radiative recombination wins over nonradiative recombinations.[83] Sargent and coworkers pointed out a really important fact of domain distribution. They used a solvent method for the domain distribution in perovskite QWs $PEA_2(MA)_{n-1}Pb_nBr_{3n+1}$, which resulted in 60% PLQY, with 7.4% EQE_{max} and 84000 cd/m^2 of L_{max}. Trioctylphosphine oxide (TPPO) was used for surface passivation and it increased the PLQY value to 73%, which was only 57% before coating. It was again due to decrement in nonradiative recombination.[65] The similar treatment to green LEDs (EL 532 nm) based on the passivated perovskite (EL 532 nm) results in 14.6% of EQE_{max} and 9120 cd/m^2 of L_{max}. To reduce the crystallization and to attain an improved domain size of PEABr, a long alkyl chain of 1,4,7,10,13,16-hexaoxacyclooctadecane was added in the $PEA_2Cs_{n-1}PbnBr_{3n+1}$ perovskite QWs[84] and this surface passivation upgraded PLQY up to 70%, EQE_{max} (EL 514 nm) of 15.5% and 20,000 cd/m^2 of L_{max}. Perovskite QWs paved the pathway to accomplish blue perovskite LEDs attributed to the quantum confinement effect. It was found that, as the number of inorganic layers (n) decreases, an increasing band gap was associated with pervoskite QWs. The hybrid pervoskite QWs are also found spanning deep blue, pure blue, sky blue and pure green by changing the n value in $(OLA)_2(MA)_{n-1}Pb_nBr_{3n+1}$ (OLA, oleylamine) but, its EQE and luminance value are too low. In place of OLA, one group has chosen POEA(2-phenoxyethylamine) and studied the structural and property changes of the resulting pervoskite QWs, $(POEA)_2(MA)_{n-1}Pb_nBr_{3n+1}$.[85] The EQE_{max}/L_{max} for sky blue and blue pervoskite are 1.1%/19.25 cdm^{-2} and 0.06%/0.07 cdm^{-2}, respectively. Most of the EL spectrum that have been reported have single emission EL; a hybrid pervoskite on $(EA)_2(MA)_{n-1}Pb_nBr_{3n+1}$ (EA, ethylamine), with multiple emission has also been reported.[86] It has the EQE_{max} of 2.6% and L_{max} of 200 cd/m^2. A long-chain cation combination was used to change the crystallization of perovskites $(PEA/IPA)_2(MA/Cs)_{n-1}Pb_nBr_{3n+1}$QWs (IPA, isopropyl ammonium).[87] This pervoskite QWs resulted in magnificient PLQY value of 88% with reduced n value and a single emission peak at 477 nm. The developed QW LEDs have L_{max} of 2480 cdm^{-2} and EQE_{max} of 1.5%, and the EL wavelength can be varied by different hole transport layers.

3.8 PEROVSKITE BULK

The performance of bulk 3D pervoskite-LED is limited due to excess nonradiative defects, low PLQY, poor surface morphology, unbalanced charges injection rate, when film is generated using the spin coating method.[9,88] A new method has been used in which the different solubility of $CsPbBr_3$ perovskite and an additive, MABr, was utilized. $CSPbBr_3$ was mixed with MABr and, owing to the different solubility behaviour, the resulting compound yields cuboids of micrometre size with a quasi-core structure.

Due to this strategy high luminescence and balance charge injection were achieved.[67] The high luminescence was achieved as the MABr shell deactivates the nonradiative defects, which results in high PLQY. Simultaneously, it capped the layer to provide balance charge injection. This perovskite has EQE_{max} of 20.3

and L_{max} 14,000 cdm^{-2}, having electroluminescence at 525 nm. During the same period, there was a report of near-infrared pervoskite-LEDs (EL 803 nm). It was synthesised using an amino acid derivative in the precursor solution.[69] The fabricated perovskite films were composed of submicrometre-scale particles. These particles efficiently extract light from the device and retain wavelength and viewing-angle-independent electroluminescence efficiently. Finally, pervoskite LEDs with an EQE_{max} of 20.7% and luminescence value of up to 390 W sr^{-1} m^{-2} were accomplished.

3.9   CONCLUSION AND FUTURE OF HYBRID PEROVSKITE

There are still great scope for and interest in developing new hybrid perovskite. The useful properties of organic materials impart luminescence, optical response, electrical transport, flexible processability and mechanical softness. The inorganic components have attributes such as high electrical mobility, tunable conductivity, magnetic and dielectric responsiveness, and thermal/mechanical robustness. Both components have broad possibilities in structure. Organic molecules provide strong emission/absorbtion and long-range charge transport will be supplemented by the inorganic counterpart. Electron and holes will be segregated to different layers to minimize the recombination effect. On the other way absorption, emission, charge transport may be restricted to the inorganic layer and the organic part is exploited mechanically/chemically to protect the active layers from degradation.[89–91].

REFERENCES

1. Hosker, E. World Energy Outlook; International Energy Agency; Paris, 2021.
2. Schubert, E. F.; Kim, J. K. Solid-State Light Sources Getting Smart. Science. 2005, 308, 1274–1279.
3. Crawford, M. H. LEDs for Solid-State Lighting: Performance Challenges and Recent Advances. IEEE J. Sel. Top. Quantum Electron. 2009, 15, 1764–1769.
4. McKittrick, J.; Shea-Rohwer, L. E. Review: Down Conversion Materials for Solid-State Lighting. J. Am. Ceram. Soc. 2014, 97, 1764–1769.
5. Pimputkar, S.; Speck, J. S.; Denbaars, S. P.; Nakamura, S. Prospects for LED Lighting. Nature Photon. 2009, 3, 2–4.
6. Song, Z.; Zhao J.; Liu, Q. Luminescent Perovskites: Recent Advances in Theory and Experiments. Inorg. Chem. Front. 2019, 6, 2969–3011.
7. Erdem, T.; Demir, H. V. Color Science of Nanocrystal Quantum Dots for Lighting and Displays. Nanophotonics. 2013, 2, 1764–1769.
8. Kim, Y.-H.; Cho, H.; Heo, J. H.; Kim, T.-S.; Myoung, N.; Lee, C.-L.; Im, S. H.; Lee, T.-W. Multicolored Organic/Inorganic Hybrid Perovskite Light-Emitting Diodes. Adv. Mater. 2015, 27, 1248–1254.
9. Cho, H.; Jeong, S.-H.; Park, M.-H.; Kim, Y.-H.; Wolf, C.; Lee, C.-L.; Heo, J. H.; Sadhanala, A.; Myoung, N.; Yoo, S.; et al. Overcoming the Electroluminescence Efficiency Limitations of Perovskite Light-Emitting Diodes. Science. 2015, 350, 1222–1225.
10. Yin, W.-J.; Shi, T.; Yan, Y. Unique Properties of Halide Perovskites as Possible Origins of the Superior Solar Cell Performance. Adv. Mater. 2014, 26, 4653–4658.
11. Tan, Z.-K.; Moghaddam, R. S.; Lai, M. L.; Docampo, P.; Higler, R.; Deschler, F.; Price, M.; Sadhanala, A.; Pazos, L. M.; Credgington, D.; et al. Bright Light-Emitting Diodes based on Organometal Halide Perovskite. Nat. Nanotechnol. 2014, 9, 687–692.

12. Weber, D. $CH_3NH_3PbX_3$, a Pb(II)-System with Cubic Perovskite Structure. Z. Naturforsch. B: J. Chem. Sci. 1978, B33, 1443–1445.

13. Weber, D. The Perovskite System $CH_3NH_3[Pb_nSn_{1-n}X_3]$ (X = Cl, Br, I). Z. Naturforsch. 1979, B34, 939–941.

14. Yamada, K.; Kawaguchi, H.; Matsui, T.; Okuda, T.; Ichiba, S. Structural Phase Transition and Electrical Conductivity of the Perovskite $CH_3NH_3Sn_{1-x}Pb_xBr_3$ and CsSnBr3. Bull. Chem. Soc. Jpn. 1990, 63, 2521–2525.

15. Mitzi, D. B.; Feild, C. A.; Schlesinger, Z.; Laibowitz, R. B. Transport, Optical, and Magnetic Properties of the Conducting Halide Perovskite $CH_3NH_3SnI_3$. J. Solid State Chem. 1995, 114, 159–163.

16. Mitzi, D. B.; Liang, K. Synthesis, Resistivity, and Thermal Properties of the Cubic Perovskite $NH_2CH = NH_2SnI_3$ and Related Systems. J. Solid State Chem. 1997, 134, 376–381.

17. Mitzi, D. B. Synthesis, Structure and Properties of Organic-Inorganic Perovskites and Related Materials. Prog. Inorg. Chem. 1999, 48, 1–121.

18. Mitzi, D. B. Templating and Structural Engineering in Organic-Inorganic Perovskites. J. Chem. Soc., Dalton Trans. 2001, 1–12.

19. Goldschmidt, V. M. Die Gesetze der Krystallochemie. Naturwissenschaften. 1926, 14, 477–485.

20. Mitzi, D. B.; Liang, K. Preparation and Properties of (C4H9NH3)2EuI4: A Luminescent Organic–Inorganic Perovskite with a Divalent Rare-Earth Metal Halide Framework. Chem. Mater. 1997, 9, 2990–2995.

21. Van Aken, B. B.; Palstra, T. T. M.; Filippetti, A.; Spaldin, N. A. The Origin of Ferroelectricity in Magnetoelectric YMnO3. Nat. Mater. 2004, 3, 164–170.

22. Kieslich, G.; Sun, S.; Cheetham, A. K. Solid-State Principles Applied to Organic-Inorganic Perovskites: New Tricks for an Old Dog. Chem. Sci. 2014, 5, 4712–4715.

23. Grimm, J.; Suyver, J. F.; Beurer, E.; Carver, G.; Güdel, H. U. Light-Emission and Excited-State Dynamics in Tm^{2+} Doped $CsCaCl_3$, $CsCaBr_3$, and $CsCaI_3$. J. Phys. Chem. B. 2006, 110, 2093–2101.

24. Seifert, H. J.; Haberhauer, D. Ü ber die Systeme Alkalimetallbromid/Calciumbromid. Z. Anorg. Allg. Chem. 1982, 491, 301–307.

25. Swainson, I.; Chi, L.; Her, J.-H.; Cranswick, L.; Stephens, P.; Winkler, B.; Wilson, D. J.; Milman, V. Orientational Ordering, Tilting and Lone-Pair Activity in the Perovskite Methylammonium Tin Bromide, CH3NH3SnBr3. Acta Crystallogr., Sect. B Struct. Sci. 2010, 66, 422–429.

26. Hu, K.-L.; Kurmoo, M.; Wang, Z.; Gao, S. Metal–Organic Perovskites: Synthesis, Structures, and Magnetic Properties of $[C(NH_2)_3][M^{II}(HCOO)_3]$ (M= Mn, Fe, Co, Ni, Cu, and Zn; $C(NH_2)_3$= Guanidinium). Chem. Eur. J. 2009, 15, 12050–12064.

27. Wang, Z.; Zhang, B.; Otsuka, T.; Inoue, K.; Kobayashi, H.; Kurmoo, M. Anionic NaCl-type Frameworks of $[MnII(HCOO)_3^-]$, Templated by Alkylammonium, Exhibit Weak Ferromagnetism. Dalton Trans. 2004, 2209–2216.

28. Wang, X.-Y.; Gan, L.; Zhang, S.-W.; Gao, S. Perovskite-like Metal Formates with Weak Ferromagnetism and as Precursors to Amorphous Materials. Inorg. Chem. 2004, 43, 4615–4625.

29. Bayrammurad, S.; David, B. Mitzi Organic-Inorganic Perovskites: Structural Versatility for Functional Materials Design. Chem. Rev. 2016, 116, 4558–4596.

30. Mitzi, D. B. Templating and Structural Engineering in Organic-Inorganic Perovskites. J. Chem. Soc., Dalton Trans. 2001, 1–12.

31. Barman, S.; Venkataraman, N. V.; Vasudevan, S.; Seshadri, R. Phase Transitions in the Anchored Organic Bilayers of Long-Chain Alkylammonium Lead Iodides (CnH2n+1NH3)2PbI4; n = 12, 16, 18. J. Phys. Chem. B. 2003, 107, 1875–1883.

32. Naik, V. V.; Vasudevan, S. Melting of an Anchored Bilayer: Molecular Dynamics Simulations of the Structural Transition in $(C_nH_{2n+1}NH_3)_2PbI_4$ (n = 12, 14, 16, 18). J. Phys. Chem. C. 2010, 114, 4536–4543.

33. Tang, Z.; Guan, J.; Guloy, A. M. Synthesis and Crystal Structure of New Organic-Based Layered Perovskites with 2,2′-Biimidazolium Cations. J. Mater. Chem. 2001, 11, 479–482.

34. Mitzi, D. B.; Wang, S.; Feild, C. A.; Chess, C. A.; Guloy, A. M. Conducting Layered Organic-inorganic Halides Containing < 110>- Oriented Perovskite Sheets. Science. 1995, 267, 1473–1476.

35. Wang, S.; Mitzi, D. B.; Feild, C. A.; Guloy, A. Synthesis and Characterization of [NH$_2$C(I):NH$_2$]$_3$MI$_5$ (M = Sn, Pb): Stereochemical Activity in Divalent Tin and Lead Halides Containing Single < 110> Perovskite Sheets. J. Am. Chem. Soc. 1995, 117, 5297–5302.

36. Mousdis, G. A.; Gionis, V.; Papavassiliou, G. C.; Raptopoulou, C. P.; Terzis, A. Preparation, Structure and Optical Properties of [CH$_3$SC(NH$_2$)NH$_2$]$_3$PbI$_5$, [CH$_3$SC(NH$_2$) NH$_2$]$_4$Pb$_2$Br$_8$ and [CH$_3$SC- (NH$_2$)NH$_2$]$_3$PbCl$_5$*CH$_3$SC(NH$_2$)NH$_2$Cl. J. Mater. Chem. 1998, 8, 2259–2262.

37. Lee, B.; Stoumpos, C. C.; Zhou, N.; Hao, F.; Malliakas, C.; Yeh, C.-Y.; Marks, T. J.; Kanatzidis, M. G.; Chang, R. P. H. Air-Stable Molecular Semiconducting Iodosalts for Solar Cell Applications: Cs2SnI6 as a Hole Conductor. J. Am. Chem. Soc. 2014, 136, 15379–15385.

38. Peresh, E. Y.; Sidei, V. I.; Zubaka, O. V.; Stercho, I. P. K2(Rb2,Cs2,Tl2)TeBr6(I6) and Rb$_3$(Cs$_3$)Sb$_2$(Bi$_2$)Br$_9$(I$_9$) Perovskite Compounds. Inorg. Mater. 2011, 47, 208–212.

39. Abriel, W.; du Bois, A. The Crystal Structures of Compounds A2TeX6 (A = K, NH4, Rb, Cs; X = Cl, Br, I). Z. Naturforsch. B J. Chem. Sci. 1989, B44, 1187–1194.

40. Mitzi, D. B.; Feild, C. A.; Harrison, W. T. A.; Guloy, A. M. Conducting Tin Halides with a Layered Organic-Based Perovskite Structure. Nature. 1994, 369, 467–469.

41. Guan, J.; Tang, Z.; Guloy, A. M. Alpha-[NH$_3$(CH$_2$)$_5$NH$_3$]SnI$_4$: A New Layered Perovskite Structure. Chem. Commun. 1999, 1833– 1834.

42. Mitzi, D. B.; Chondroudis, K.; Kagan, C. R. Design, Structure, and Optical Properties of Organic-Inorganic Perovskites Containing an Oligothiophene Chromophore. Inorg. Chem. 1999, 38, 6246–6256.

43. Grebner, D.; Helbig, M.; Rentsch, S. Size-Dependent Properties of Oligothiophenes by Picosecond Time-Resolved Spectroscopy. J. Phys. Chem. 1995, 99, 16991–16998.

44. Förster, T. 10th Spiers Memorial Lecture. Transfer Mechanisms of Electronic Excitation. Discuss. Faraday Soc. 1959, 27, 7.

45. Dexter D. L. A Theory of Sensitized Luminescence in Solids. J. Chem. Phys. 1953, 21, 836.

46. Zhu, X.-H.; Mercier, N.; Frere, P.; Blanchard, P.; Roncali, J.; Allain, M.; Pasquier, C.; Riou, A. Effect of Mono-versus Diammonium Cation of 2,2′-Bithiophene Derivatives on the Structure of Organic-Inorganic Hybrid Materials Based on Iodo Metallates. Inorg. Chem. 2003, 42, 5330–5339.

47. Rentsch, S.; P. Yang, J.; Paa, W.; Birckner, E.; Schiedt, J.; Weinkauf, R. Size Dependence of Triplet and Singlet States of Alphaoligothiophenes. Phys. Chem. Chem. Phys. 1999, 1, 1707–1714.

48. Agranovich, V. M.; La Rocca, G. C.; Bassani, F. Efficient Electronic Energy Transfer from a Semiconductor Quantum Well to an Organic Material. JETP Lett. 1997, 66, 748–751.

49. Agranovich, V. M.; Basko, D. M.; Rocca, G. C. L.; Bassani, F. Excitons and Optical Nonlinearities in Hybrid Organic-Inorganic Nanostructures. J. Phys. Condens. Matter. 1998, 10, 9369.

50. Blumstengel, S.; Sadofev, S.; Xu, C.; Puls, J.; Henneberger, F. Converting Wannier into Frenkel Excitons in an Inorganic/Organic Hybrid Semiconductor Nanostructure. Phys. Rev. Lett. 2006, 97, 237401.

51. Ema, K.; Inomata, M.; Kato, Y.; Kunugita, H.; Era, M. Nearly Perfect Triplet-Triplet Energy Transfer from Wannier Excitons to Naphthalene in Organic-Inorganic Hybrid Quantum-Well Materials. Phys. Rev. Lett. 2008, 100, 257401.

52. Era, M.; Maeda, K.; Tsutsui, T. Enhanced Phosphorescence from Naphthalene-Chromophore Incorporated into Lead Bromide Based Layered Perovskite Having Organic-Inorganic Superlattice Structure. Chem. Phys. Lett. 1998, 296, 417–420.

53. Braun, M.; Tuffentsammer, W.; Wachtel, H.; Wolf, H. C. Tailoring of Energy Levels in Lead Chloride Based Layered Perovskites and Energy Transfer Between the Organic and Inorganic Planes. Chem. Phys. Lett. 1999, 303, 157–164.

54. Braun, M.; Tuffentsammer, W.; Wachtel, H.; Wolf, H. C. Pyrene as Emitting Chromophore in Organic-Inorganic Lead Halide Based Layered Perovskites with Different Halides. Chem. Phys. Lett. 1999, 307, 373–378.

55. Zheng, Y.-Y.; Wu, G.; Deng, M.; Chen, H.-Z.; Wang, M.; Tang, B.-Z. Preparation and Characterization of a Layered Perovskite-type Organic-Inorganic Hybrid Compound $(C_8NH_6CH_2CH_2NH_3)_2CuCl_4$. Thin Solid Films. 2006, 514, 127–131.

56. Era, M.; Miyake, K.; Yoshida, Y.; Yase, K. Orientation of Azobenzene Chromophore Incorporated into Metal Halide-Based Layered Perovskite Having Organic-Inorganic Superlattice Structure. Thin Solid Films. 2001, 393, 24–27.

57. Era, M.; Shimizu, A. Incorporation of Bulky Chromophore into PbBr-Based Layered Perovskite Organic/Inorganic Superlattice by Mixing of Chromophore-Linked Ammonium and Alkyl Ammonium Molecules. Mol. Cryst. Liq. Cryst. Sci. Technol., Sect. A. 2001, 371, 199–202.

58. Clearfield, A. Role of Ion Exchange in Solid-State Chemistry. Chem. Rev. 1988, 88, 125–148.

59. Gamble, F. R.; Osiecki, J. H.; Cais, M.; Pisharody, R.; DiSalvo, F. J.; Geballe, T. H. Intercalation Complexes of Lewis Bases and Layered Sulfides: A Large Class of New Superconductors. Science. 1971, 174, 493–497.

60. Mitzi, D. B.; Medeiros, D. R.; Malenfant, P. R. L. Intercalated Organic-Inorganic Perovskites Stabilized by Fluoroaryl-Aryl Interactions. Inorg. Chem. 2002, 41, 2134–2145.

61. Kamminga, M. E.; Fang, H.-H.; Loi, M. A.; Brink, G. H. T.; Blake, G. R.; Palstra, T. T. M.; Elshof J. E. T. Micropatterned 2D Hybrid Perovskite Thin Films with Enhanced Photoluminescence Lifetimes. ACS Appl. Mater. Interfaces. 2018, 10, 12878–12885.

62. Wang, N.; Cheng, L.; Ge, R.; Zhang, S.; Miao, Y.; Zou, W.; Yi, C.; Sun, Y.; Cao, Y.; Yang, R. Perovskite Light-Emitting Diodes Based on Solution-Processed Self-Organized Multiple Quantum Wells. Nat. Photon. 2016, 10, 699–704.

63. Xiao, Z. G.; Kerner, R. A.; Zhao, L. F.; Tran, N. L.; Lee, K. M.; Koh, T. W.; Scholes, G. D.; Rand, B. P. Efficient Perovskite Lightemitting Diodes Featuring Nanometre-Sized Crystallites. Nat. Photon. 2017, 11, 108–115.

64. Yang, X.; Zhang, X.; Deng, J.; Chu, Z.; Jiang, Q.; Meng, J.; Wang, P.; Zhang, L.; Yin, Z.; You, J. Efficient Green Light-Emitting Diodes Based on Quasi-Two-Dimensional Composition and Phase Engineered Perovskite with Surface Passivation. Nat. Commun. 2018, 9, 570.

65. Zhang, L.; Yang, X.; Jiang, Q.; Wang, P.; Yin, Z.; Zhang, X.; Tan, H.; Yang, Y. M.; Wei, M.; Sutherland, B. R.; et al. Ultra-Bright and Highly Efficient Inorganic Based Perovskite Light-Emitting Diodes. Nat. Commun. 2017, 8, 15640.

66. Lin, K.; Xing, J.; Quan, L. N.; de Arquer, F. P. G.; Gong, X.; Lu, J.; Xie, L.; Zhao, W.; Zhang, D.; Yan, C.; et al. Perovskite Light Emitting Diodes with External Quantum Efficiency Exceeding 20%. Nature. 2018, 562, 245–248.

67. Chiba, T.; Hayashi, Y.; Ebe, H.; Hoshi, K.; Sato, J.; Sato, S.; Pu, Y.-J.; Ohisa, S.; Kido, J. Anion-Exchange Red Perovskite Quantum Dots with Ammonium Iodine Salts for Highly Efficient Light-Emitting Devices. Nat. Photon. 2018, 12, 681–687.

68. Cao, Y.; Wang, N.; Tian, H.; Guo, J.; Wei, Y.; Chen, H.; Miao, Y.; Zou, W.; Pan, K.; He, Y.; et al. Perovskite Light-Emitting Diodes Based on Spontaneously Formed Submicrometre-Scale Structures. Nature. 2018, 562, 249–253.

69. Zhao, B.; Bai, S.; Kim, V.; Lamboll, R.; Shivanna, R.; Auras, F.; Richter, J. M.; Yang, L.; Dai, L.; Alsari, M.; High-Efficiency Perovskite-Polymer Bulk Heterostructure Light-Emitting Diodes. Nat. Photon. 2018, 12, 783–789.
70. Smith, I. C.; Hoke, E. T.; Solis-Ibarra, D.; McGehee, M. D.; Karunadasa, H. I. A Layered Hybrid Perovskite Solar-Cell Absorber with Enhanced Moisture Stability. Angew. Chem. 2014, 126, 11414–11417.
71. Tan, Z.-K.; Moghaddam, R. S.; Lai, M. L.; Docampo, P.; Higler, R.; Deschler, F.; Price, M.; Sadhanala, A.; Pazos, L. M.; Credgington, D.; Hanusch, F.; Bein, T.; Snaith, H. J.; Friend, R. H. Bright Light-Emitting Diodes Based on Organometal Halide Perovskite. Nat. Nanotechnol. 2014, 9, 687–692.
72. Zhang, F.; Zhong, H.; Chen, C.; Wu, X.-G.; Hu, X.; Huang, H.; Han, J.; Zou, B.; Dong, Y. Brightly Luminescent and Color-Tunable Colloidal $CH_3NH_3PbX_3$ (X = Br, I, Cl) Quantum Dots: Potential Alternatives for Display Technology. ACS Nano. 2015, 9, 4533–4542.
73. Wei, Z.; Xing, J. The Rise of Perovskite Light-Emitting Diodes. J. Phys. Chem. Lett. 2019, 10, 3035–3042.
74. Schmidt, L. C.; Pertegás, A.; González-Carrero, S.; Malinkiewicz, O.; Agouram, S.; Mínguez Espallargas, G.; Bolink, H. J.; Galian, R. E.; Perez-Prieto, J. Nontemplate Synthesis of' $CH_3NH_3PbBr_3$ Perovskite Nanoparticles. J. Am. Chem. Soc. 2014, 136, 850–853.
75. Xing, J.; Yan, F.; Zhao, Y.; Chen, S.; Yu, H.; Zhang, Q.; Zeng, R.; Demir, H. V.; Sun, X.; Huan, A.; et al. High-Efficiency Light Emitting Diodes of Organometal Halide Perovskite Amorphous Nanoparticles. ACS Nano. 2016, 10, 6623–6630.
76. Yan, F.; Xing, J.; Xing, G.; Quan, L.; Tan, S. T.; Zhao, J.; Su, R.; Zhang, L.; Chen, S.; Zhao, Y.; et al. Highly Efficient Visible Colloidal Lead-Halide Perovskite Nanocrystal Light-Emitting Diodes. Nano Lett. 2018, 18, 3157–3164.
77. Xiao, Z. G.; Kerner, R. A.; Zhao, L. F.; Tran, N. L.; Lee, K. M.; Koh, T. W.; Scholes, G. D.; Rand, B. P. Efficient Perovskite Light Emitting Diodes Featuring Nanometre-Sized Crystallites. Nat. Photon. 2017, 11, 108–115.
78. Han, D.; Imran, M.; Zhang, M.; Chang, S.; Wu, X.-G.; Zhang, X.; Tang, J.; Wang, M.; Ali, S.; Li, X.; et al. Efficient Light-Emitting Diodes Based on in Situ Fabricated $FAPbBr_3$ Nanocrystals: The Enhancing Role of the Ligand-Assisted Reprecipitation Process. ACS Nano. 2018, 12, 8808–8816.
79. Song, J.; Fang, T.; Li, J.; Xu, L.; Zhang, F.; Han, B.; Shan, Q.; Zeng, H. Organic–Inorganic Hybrid Passivation Enables Perovskite QLEDs with an EQE of 16.48%. Adv. Mater. 2018, 30, 1805409.
80. Pan, J.; Shang, Y.; Yin, J.; De Bastiani, M.; Peng, W.; Dursun, I.; Sinatra, L.; El-Zohry, A. M.; Hedhili, M. N.; Emwas, A.-H. Bidentate Ligand-passivated $CsPbI_3$ Perovskite Nanocrystals for Stable Near-unity Photoluminescence Quantum Yield and Efficient Red Light-Emitting Diodes. J. Am. Chem. Soc. 2018, 140, 562–565.
81. Yuan, M.; Quan, L. N.; Comin, R.; Walters, G.; Sabatini, R.; Voznyy, O.; Hoogland S.; Yongbiao, Z.; Beauregard, E. M.; Kanjanaboos, P.; Lu, Z.; Kim, D. H.; Sargent, E. H.; Perovskite Energy Funnels for Efficient Light-Emitting Diodes. Nat Nanotechnol. 2016, 11, 872–877.
82. Quan, L. N.; Zhao, Y.; García de Arquer, F. P.; Sabatini, R.; Walters, G.; Voznyy, O.; Comin, R.; Li, Y.; Fan, J. Z.; Tan, H.; et al. Tailoring the Energy Landscape in Quasi-2D Halide Perovskites Enables Efficient Green-Light Emission. Nano Lett. 2017, 17, 3701–3709.
83. Ban, M.; Zou, Y.; Rivett, J. P. H.; Yang, Y.; Thomas, T. H.; Tan, Y.; Song, T.; Gao, X.; Credgington, D.; Deschler, F.; et al. Solution Processed Perovskite Light Emitting Diodes with Efficiency Exceeding 15% Through Additive-Controlled Nanostructure Tailoring. Nat. Commun. 2018, 9, 3892.

84. Chen, Z.; Zhang, C.; Jiang, X.-F.; Liu, M.; Xia, R.; Shi, T.; Chen, D.; Xue, Q.; Zhao, Y.-J.; Su, S.; Yip, H.-L.; Cao, Y. High-Performance Color-Tunable Perovskite Light Emitting Devices Through Structural Modulation from Bulk to Layered Film. Adv. Mater. 2017, 29, 1603157.

85. Wang, Q.; Ren, J.; Peng, X.-F.; Ji, X.-X.; Yang, X.-H. Efficient Sky-Blue Perovskite Light-Emitting Devices Based on Ethylammonium Bromide Induced Layered Perovskites. ACS Appl. Mater. Interfaces. 2017, 9, 29901–29906.

86. Xing, J.; Zhao, Y.; Askerka, M.; Quan, L. N.; Gong, X.; Zhao, W.; Zhao, J.; Tan, H.; Long, G.; Gao, L. Color-Stable Highly Luminescent Sky-Blue Perovskite Light-Emitting Diodes. Nat. Commun. 2018, 9, 3541.

87. Wei, Z.; Perumal, A.; Su, R.; Sushant, S.; Xing, J.; Zhang, Q.; Tan, S. T.; Demir, H. V.; Xiong, Q. Solution-Processed Highly Bright and Durable Cesium Lead Halide Perovskite Light-Emitting Diodes. Nanoscale. 2016, 8, 18021–18026.

88. Li, X.; Ibrahim Dar, M.; Yi, C.; Luo, J.; Tschumi, M.; Zakeeruddin, S. M.; Nazeeruddin, M. K.; Han, H.; Gratzel, M. Improved Performance and Stability of Perovskite Solar Cells by Crystal Crosslinking with Alkylphosphonic Acid ω-ammonium Chlorides. Nat. Chem. 2015, 7, 703–711.

89. Cao, D. H.; Stoumpos, C. C.; Farha, O. K.; Hupp, J. T.; Kanatzidis, M. G. 2D Homologous Perovskites as Light-Absorbing Materials for Solar Cell Applications. J. Am. Chem. Soc. 2015, 137, 7843–7850.

90. Quan, L. N.; Yuan, M.; Comin, R.; Voznyy, O.; Beauregard, E. M.; Hoogland, S.; Buin, A.; Kirmani, A. R.; Zhao, K.; Amassian, A.; et al. Ligand-Stabilized Reduced-Dimensionality Perovskites. J. Am. Chem. Soc. 2016, 138, 2649–2655.

4 Stretchable and Flexible Materials for OLEDs

Barnali Jana and Sushovan Paladhi

4.1 INTRODUCTION: BACKGROUND OF OLEDs

When organic materials are placed between two electrodes of a conventional diode or light emitting diode (OLED)—instead of *p*-type and *n*-type semiconductors—a semiconductor diode is formed, named an organic light emitting diode (OLED). An emissive electroluminescent organic layer can emit light in response to an electrical current by creating electrons and holes in the organic molecules, and became known in display technology as indoor lighting and displays in various consumer electronics, such as cell phones, digital cameras, and ultra-high-definition televisions. The developments of electroluminescence in organic materials during 1950–1990 were periodically done by high voltages in air to acridine orange, ohmic dark-injecting electrode by Martin Pope (to explain the needs for hole and electron electrode contacts), direct current electroluminescence under vacuum on one crystal of anthracene by Pope's group, double injection recombination electroluminescence in an anthracene crystal through hole and electron electrodes by W. Helfrich and W. G. Schneider, electroluminescence of polymer films (poly n-vinyl carbazole) by Roger Partridge and green light-emitting polymer device by J.H. Burroughs. However, in 2000, the Nobel Prize in Chemistry was awarded for the contribution in the finding of organic semiconductors by Alan Heeger, Alan MacDiarmid, and Hideki Shirakawa in the mid-1970s.

Following the report by Tang and Van Slyke in 1987,[1] OLEDs have been considered a very interesting and attractive target in both academia and industry for applications in full-color display panels and eco-friendly lighting sources. The delocalization of pinegatron originated by conjugation over half or the whole molecule make the organic molecules square measure (panel build of organic carbon-based materials) electrically semiconductive. The main important variable features of OLEDs are excellent image quality, sensible colours, infinite distinction, quick response rate and wide viewing angles. Further the preparation of OLEDs is economical as well as easier due to the absence of backlight and filters (unlike digital display), which directly affects the brightness (brighter) as well as battery life (consumes less battery energy).

OLEDs can be classified mainly into six different categories: passive matrix OLEDs (PMOLEDs), active matrix OLEDs (AMOLEDs), top-emitting OLEDs, foldable OLEDs and white OLEDs. In general, six different thin layers exist in a simple OLED, where the top and bottom layers are abbreviated as seal and substrate, respectively, with a negative terminal and a positive terminal between the layers

DOI: 10.1201/9781003260417-4

63

OLED STRUCTURE

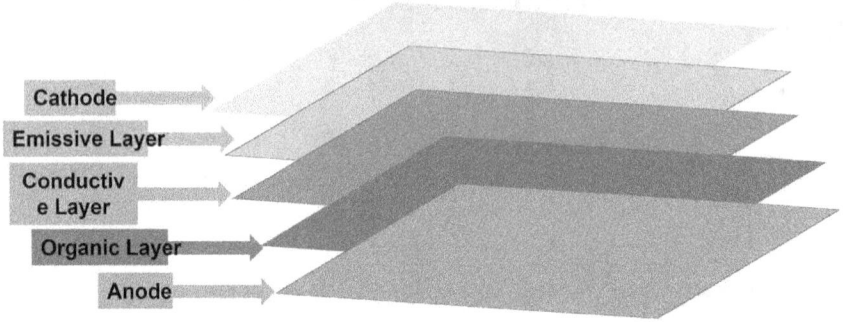

FIGURE 4.1 Basic OLED structure.

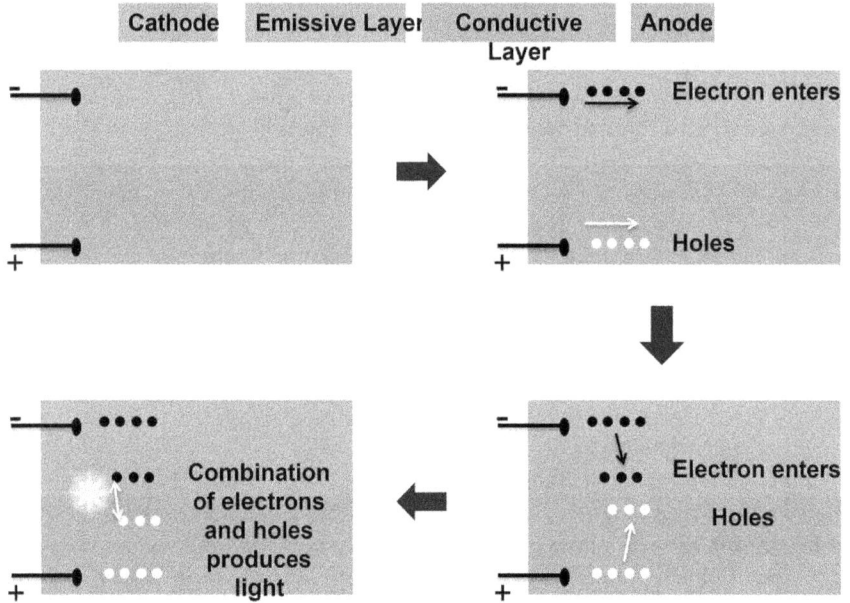

FIGURE 4.2 Working principle of OLED devices.

(Figure 4.1).[2] Currently modified multilayer OLEDs are being developed, where to improve the efficiency of the devices, more layers are incorporated utilizing several organic polymer materials.

The working principal of OLEDs can be explained with a few simple steps (Figure 4.2).[3]

1. Applying voltage on OLED creates an anode with a positive charge and a cathode with a negative charge.

2. Electrons enter into the lowest unoccupied molecular orbital (LUMO) of the organic layer at the cathode and are removed from the highest occupied molecular orbital (HOMO) at the anode; consequently, a current of electron flow is generated in the device from cathode to anode.
3. The holes created in the anode move towards the emissive layer from the conductive layer.
4. Next, the light of the photon is released after the merging of the holes with the negative electrons spontaneously.

Depending upon the design of the devices, the OLEDs follow either singlet emission, doublet emission or triplet emission.[4] OLEDs are classified into seven major categories, depending upon their manufacture and the nature of applications: passive matrix OLED (PMOLED), active matrix OLED (AMOLED), transparent OLED, top-emitting OLED, foldable OLED, white OLED and phosphorescent organic light emitting diodes (PHOLEDs).

In this chapter the main focus will be on the development of stretchable and flexible materials in OLEDs.

4.2   STRETCHABLE AND FLEXIBLE MATERIALS IN OLEDs

The advantages of lower power consumption, improved mechanical flexibility, lighter and thinner, and better image quality make flexible OLEDs promising devices over liquid crystal displays (LCDs) and other LEDs. The first flexible OLED with small-molecule organic nonpolymetric thin-film materials from tris-(8-hydroxyquinoline) aluminum (Alq3) and N, N'diphenyl-N, N'-bis(3-methylphenyl)1–1'biphenyl-4,4'diamine (NPB) was reported in 1997 by Forrest et al.[5,6] The new reported flexible OLEDs were found to have comparable properties after repeated bending also with that of conventional OLEDs. However, the brittle properties of indium tin oxide (ITO) electrodes make them incompatible for flexible devices, although they are mostly implemented in transparent conductive electrodes (TCEs).[7,8] So the conducting polymers, graphene and carbon nanotubes (CNT) based materials, thin metal films, metal nanowires, and mostly their composites became the alternatives to ITO.[9] The key features for the developments of a good flexible OLED (FOLED) are based on either fundamental properties, such as substrate flexibility, electrodes, fabrication technologies, encapsulation, or efficiency improvement by manipulation of the surface morphology electrodes, energy level modification of the electrode/organic interface, or light extraction and encapsulation.[10]

In this chapter, systematic developments from the year 2000 of flexible OLEDs by modifying substrates, electrodes, fabrication technologies, etc., will be discussed thoroughly.

In 2000 Hea and Kanicki reported a thin, flexible plastic substrate for organic polymer light-emitting heterostructure devices where an aluminium cathode was fabricated on plastic material for application in flat panel displays.[11] Benzothiadiazole-fluorene and amine-fluorene copolymer were used as organic polymer substrate for the experiment. The reported flexible device has properties for green light emission with luminance of more than 2000 cd m^{-2}, with up to 56.2 cd A^{-1} emission, ~9.0 lm W^{-1} luminous and ~15% external quantum efficiency.

In 2002, Weaver et al. reported polyethylene terephthalate (PET) plastic substrate coated organic–inorganic multilayered barrier film as long-lasting organic light-emitting materials for flexible displays, with a comparable rate of degradation to ITO-coated glass-based light-emitting devices.[12]

In 2003, Anna B. Chwang and coworkers developed a highly efficient electrophosphorescent OLED for FOLED with a passive matrix of 80 dpi resolution and grown on 178-mm-thick barrier coated PET substrates.[13] The device was hermetically sealed with an optically transmissive multilayer barrier coating, which was prepared to achieve a half initial luminance (L$_0$ ~ 100 cd m^{-2}) of order 200 h and minimal damage was observed after the device was flexed 1000 times around a 1 in. diameter cylinder.

Z. Xie et al. in the same year reported a thin spin-on-glass (SOG)-coated steel foil based top-emitting FOLED with organic stack of NPB and Alq$_3$. Silver (Ag) and Sm were used as electrodes in the anode and cathode, respectively, with an ultrathin plasma-polymerized hydrocarbon film (CF$_x$).[14] The reported device (Figure 4.3) was found to be very efficient i (4.4 cd A^{-1}) and flexible (a curvature around 2 cm^{-1}).

The same type of steel-based FOLEDs were also reported in 2004 by Z.Y. Xie et al., where the reported device was found to be suitable with 3 V turn-on voltage and 0.56 cd A^{-1} light-emitting efficiency.[15]

In the same year, Yongtaek Hong and coworkers compared several properties of FOLEDs on plastic substrates, such as optical transmission, surface roughness, gas-blocking properties, patterning of the electrode and polymer adhesion to the transparent conducting electrode.[16] After the comparative studies, they came to the conclusion that the substrates may be very useful for different flexible OLED applications.

In 2005, Yanqing Li et al. developed a top-emitting FOLED based on aluminium-laminated PET by increasing surface morphology and the substrate anode adhesion.[17] Several organic polymeric substances were used for the hole transporting layer, light-emitting polymer such as polysstyrene sulfonated-doped polys3,4-ethylene

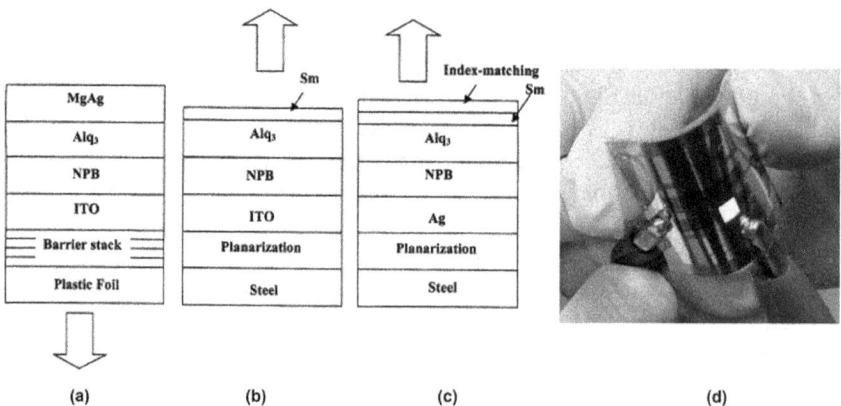

FIGURE 4.3 FOLED display developed by Z. Xie et al. (Reprinted with permission from Xie et al.[14] Copyright © 2003 *Elsevier B.V.*)

dioxythiophened and phenyl-substituted polysp-phenylenevinylened. The electrodes for this flexible device were made with Ag/CFX and Ag/indium-tin oxide bilayer, as well as an anode and semitransparent top cathode, which was found to be suitable with 7.5 V operating voltage to obtain 4.56 cd A^{-1} luminous efficiency (Figure 4.4).

Ching-Ming Hsu and coworkers incorporated Al/Alq$_3$-NPB/nickel-embedded ITO/PET in a film structure to prepare a FOLED by reducing the turn-on voltage by 2.3 V.[18] This reported FOLED device was found to be suitable in respect of optoelectrical characteristics, as well as in lifetime, with a reduced surface roughness of ITO from 2.60 to 0.36 nm after surface polishing (Figure 4.5).

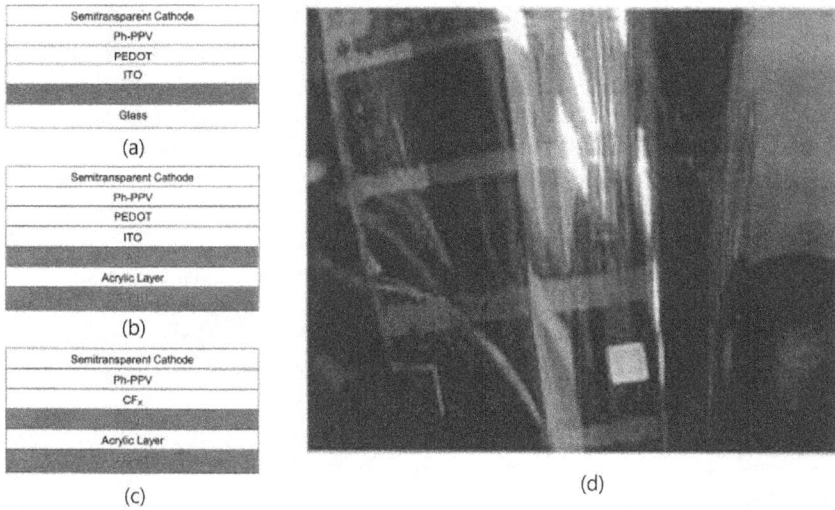

FIGURE 4.4 FOLED display developed by Yanqing Li et al. (Reprinted with permission from Li et al.[17] Copyright © 2005 American Institute of Physics.)

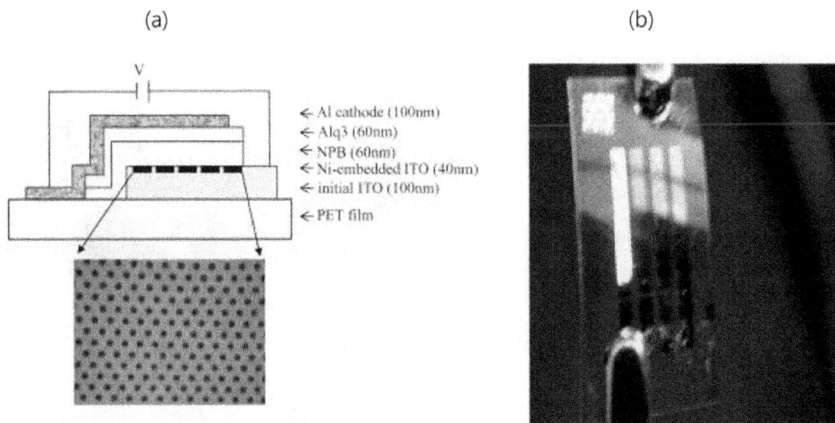

FIGURE 4.5 FOLED display developed by Ching-Ming Hsu and coworkers. (Reprinted with permission from Hsu et al.[18] Copyright © 2006 American Institute of Physics.)

A similar Ni-ITO cosputter approach was prepared by Ching-Ming Hsu and cow-orkers for the preparation of FOLED by combination of Al/Alq$_3$, NPB/ITO/Arton film.[19] The device was found to be very effective in terms of lowering of threshold voltage and turn-on voltage (by 3.4 V and 2.6 V, respectively).

In the same year, Dong-Sing Wuu et al. reported on a FOLED prepared by the deposit of silicon oxide (SiO$_x$) and silicon nitride (SiN$_x$) based thin films via parylene/SiO$_x$/SiN$_x$ . . . parylene/SiO$_x$/SiN$_x$ multilayers, deposited onto flexible polycarbonate (PC) substrates.[20] The main feature of the device was the increased transparency in the visible light region by combining SiOx and SiNx films to protect the SiO$_x$ film from moisture. Further addition of parylene layers protects the inorganic layers from being scratched.

Simultaneously, Hsin Her Yu and coworkers studied the properties of the ITO film on thickness by applying an inverse target sputtering system to deposit ITO sheets on cyclic olefin copolymer substrate.[21] It was observed that the thickness plays an important role in sheet resistance and resistivity, optical properties, turn-on voltage, luminance and current density. The reported data indicates that with the increasing of thickness low turn-on voltage is required.

In 2007, Myung-Gyu Kang et al. developed Cu-based electrodes instead of ITO electrodes for the design of FOLED by utilizing polydimethylsiloxane stamp and nanoimprint lithography for the incorporation of a Cu-electrode into a PET substrate at 30 psi pressure and 100°C temperature.[22] The main feature that was observed was the high transmittance (up to 75% average transmittance) as well as good electrical conductivity (Figure 4.6).

In 2008, the utilization of stamp printing and ion-assisted deposition method for the synthesis of FOLEDs was reported by Jianfeng Li and coworkers using ITO-modified single-walled carbon nanotube (SW-CNT) transparent electrodes through fabrication on PET.[23] The device showed high performance in terms of flexibility and high brightness (8900 cd m^{-2} with a current efficiency of 4.5 cd A^{-1}).

Kwang-Hyuk Choi et al. also prepared a PET-based FOLED using very thin mul-tilayers combining indium zinc tin oxide/Ag/indium zinc tin oxide.[24] The thin device developed (~30 nm) was very effective for its low resistance (4.99 Ω sq.$^{-1}$), high transparency (86%) and very high flexibility.

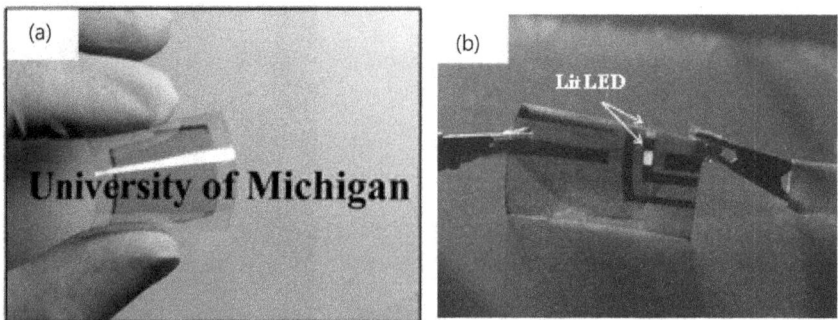

FIGURE 4.6 FOLED display developed by Myung-Gyu Kang et al. (Reprinted with permission from Kang et al.[22] Copyright © 2007 American Vacuum Society.)

The use of SW-CNT as electrodes in polymer-based blue emission foldable OLEDs was reported by Zhibin Yu and coworkers in 2009.[25] The main characteristics of the developed device was its low turn-on voltage (3.8 V), high transparency (transmittance above 70%), high working efficiency (2.2 cd A^{-1} at 480 cd m^{-2}), high brightness (1400 cd m^{-2} at 10 V) and foldable up to 2.5 mm radius without any damage of the device (Figure 4.7).

Yoko Okahisa et al. developed a FOLED of wood-cellulose nanocomposites that was flexible, with a low coefficient of thermal expansion and optically transparent.[26] The use of wood-cellulose as the most abundant biomass resources with high flexibility and ductile properties opens a broad scope to researchers to explore for commercialization in flexible, transparent displays (Figure 4.8).

Fulvia Villani and coworkers utilized an inkjet-printed polymer (poly(9,9-dihexyl-9H-fluorene-2,7-diyl) (PF$_6$)) drop film morphology for ITO treatment to prepare FOLED on PET.[27] The electrical and optical properties were studied after the fabrication of the sheets of PET/ITO/PF$_6$/Alq$_3$/Al. Modifying the ITO surface wettability

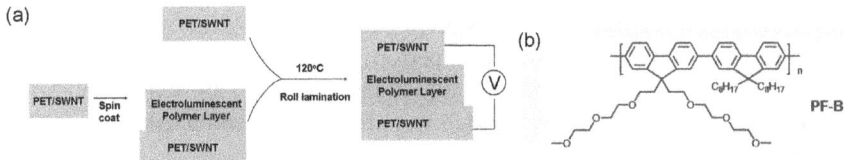

FIGURE 4.7 Polymer-based FOLED device developed by Zhibin Yu and coworkers. (reprinted/reproduced from Ref.,[25] copyright © 2009 *American Institute of Physics*).

FIGURE 4.8 Wood-cellulose nanocomposite FOLED device developed by Yoko Okahisa et al. (Reprinted with permission from Okahisa et al.[26] Copyright © 2009 Elsevier Ltd.)

increased surface energy, and resulted in good electrooptical performance, with better charge carrier injection through the interface.

Mitsunori Suzuki et al. also reported FOLEDs based on inkjet printing where phosphorescent polymers were used in combination of mixed solvents.[28] This system performance makes it suitable for moving color images at a field frequency of 60 Hz.

Guang-Feng Wang and coworkers prepared poly(3,4-ethylenedioxythiophene): poly(styrene sulfonate) (PEDOT:PSS) particles as anodes for FOLEDs to provide a performance similar to ITO anodes.[29] The PEDOT:PSS particles were prepared by the thermal treatment of PEDOT:PSS in dimethyl sulfoxide (DMSO) and investigated the characteristics using field emission scanning electron microscopy (FE-SEM) and X-ray photoelectron spectroscopy (XPS). This new system was found to be very effective in terms of lower turn voltage and higher luminous intensity.

Hyunsu Cho et al. reported preparing a FOLED by replacing the conventional ITO electrode with $ZnS/Ag/WO_3$ (ZAW) multilayer transparent electrodes to obtain high conductivity and ductility with enhanced optical transmission and/or carrier-injection compatibility.[30]

V. L. Calil and coworkers developed poly(etherimide) (PEI)-based ITO thin films through annealing at 423–523 K for the preparation of a FOLED.[31] This newly developed device showed high performance, with low resistivity (3.04 × 10^{-4} Ω cm for a thickness of 150 nm), 80% transmittance in the visible range, with working efficiency of 2.2 cd A^{-1}, which were found to be comparable with that of glass OLEDs.

In 2010, Liangbing Hu et al. compared the properties of a single-walled nanotube (SWNT) networks-based electrode with an ITO-based electrode in FOLEDs, including sheet resistance effects, morphological effects, network wetting, lifetime of sheets and mechanical properties.[32] They concluded that, compared to conventional ITO electrodes, the performance of the SWNT-based electrode is very similar in terms of lifetime performances, but the performance of the SWNT-based electrode under bending was better.

Gi-Seok Heo prepared Zn-In-Sn-O (ZITO) films from ZnO and ITO to fabricate with a flexible polyethersulfone (PES) substrate for the preparation of a FOLED.[33] The developed device was found to be more efficient in terms of resistivity (1.22 × 10^{-3} Ω cm) and optical transparency compared to Sn-doped ITO films. Further, it was found that the mechanical durability was also better than for conventional ITO films.

Knowing the higher color rendering index (CRI) of MoO_3 (40 nm)/Ag (17 nm)/ MoO_3 (40 nm) cathode (MAM cathode) than ITO-based OLEDs, in 2011 Wenyu Ji and coworkers developed a new type of flexible top light emitting OLED using an MAM cathode.[34] The device was prepared by fabrication of MAM with PET substrate to produce high transmittance in the visible range (>84%) and low resistivity of 11 Ω h^{-1}. The key factors such as high CRI and low correlated color temperature (CCT) make the device a prominent lighting source.

Mansu Kim et al. reported the polymerization of PEDOT:FTS in the presence of N, N-dimethylacetamide (DMAc-PEDOT:FTS) for the preparation of FOLEDs (Figure 4.9).[35] The new device was shown to be more efficient than normal PEDOT:FTS film in terms of 0.3 eV greater work function (reported work function 4.67 eV). Further, luminous efficiency and max luminance were found to be higher in the

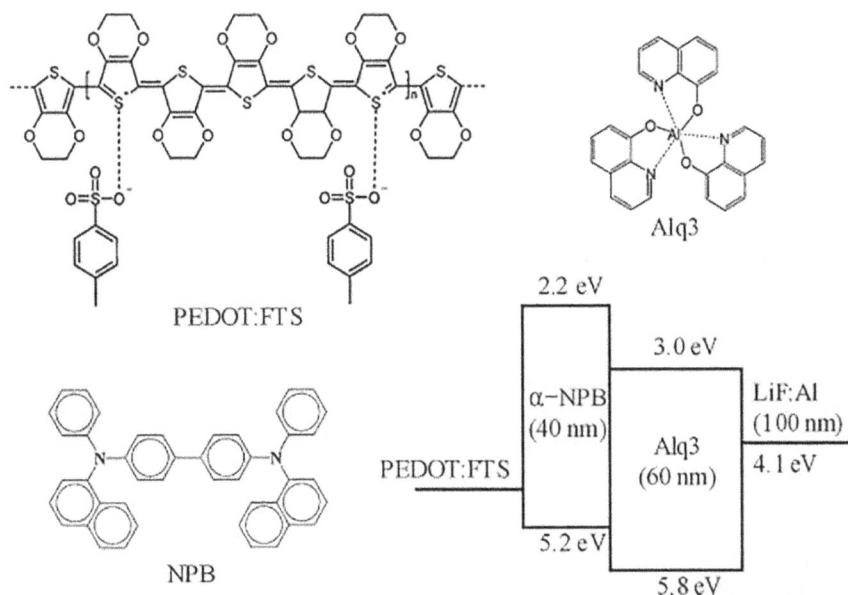

FIGURE 4.9 FOLED device with polymerized DMAc-PEDOT:FTS films, developed by Mansu Kim et al. (Reprinted with permission from Kim et al.[35] Copyright © 2011 Elsevier B.V.)

DMAc-PEDOT:FTS film anode in contrast to the normal PEDOT:FTS film anode in the Alq3-based FOLED.

In 2012, Junwei Xu and coworkers developed a FOLED based on nickel–chromium (Ni-Cr) alloy dropped on the Al anode and Al-ITO cathode through fabrication of PES substrate for better performance.[36] Organic substances 4,4′,4″-tris[3-methylphenyl(phenyl)amino]triphenylamine (m-MTDATA):2,3,5,6-tetrafluoro-7,7,8,8-tetracyanop-quinodimethane (F4-TCNQ), NPB and 10-(2-benzothiazolyl)-1,1,7,7-tetramethyl-2,3,6,7-tetrahydro-1H,5H,11H-[1] benzopyrano [6,7,8-ij] quinolizin-11-one (C545T) dye doped Alq$_3$ were used as organic hole injection layer, hole transport layer and emitting/electron-transport layer, respectively, for green light emission. The flexible system developed was found to be 24 times higher than the device without Ni-Cr alloy, with 15,000 cd m^{-2} luminance at 9.5 V.

E. Najafabadi et al. developed glass and PES substrates coated with PEDOT:PSS green-emitting electrophosphorescent top-emitting OLEDs with aluminum/lithium fluoride (Al/LiF) bottom cathodes and an organic electron transport layer 1,3,5-tri(m-pyrid-3-yl-phenyl)benzene (TpPyPB) with current efficacy of 60.6 cd A^{-1} at a luminance of 1073 cd/m^2.[37] However, addition of N,N′-Di-[(1-naphthyl)-N,N′-diphenyl]-(1,10-biphenyl)-4,40-diamine (a-NPD) into the anode made the device more efficient to achieve the current efficiency 96.3 cd A^{-1} at a luminance of 1387 cd m^{-2}.

S. Ummartyotin and coworkers reported FLOEDs based on nanocomposite film prepared with bacterial cellulose (10–50 wt.%) and polyurethane (PU)-based resin, which was incorporated through fabrication (Figure 4.10).[38] The device was typically set up with a Cu (200 nm)/molybdenum trioxide (MoO_3) (1.5 nm)/4,4'-*N*,*N*'-dicarbazole—biphenyl (CBP) (50 nm)/Alq_3 (50 nm)/LiF) (1 nm/Al (100 nm) layer to show high light transmittance of up to 80%, thermal stability up to 150°C and dimensional stability as low as 18 ppm/K in terms of coefficient of thermal expansion (CTE).

Gun Woo Hyung et al. developed an ITO-free nickel/silver/nickel (Ni/Ag/Ni) anode and aluminum/silver/aluminum (Al/Ag/Al) cathode in a device for top-emitting FOLED for improved current density, maximum luminance efficiency, and maximum external quantum efficiency (EQE).[39]

Sandström and coworkers developed a FOLED using a slot-die roll-coating apparatus and flexible PET substrate, precoated with ZnO-on-ITO cathodic stripes coated with diluted PEDOT:PSS (Figure 4.11) to show the highest recorded current efficacy 0.6 cd A^{-1} at a brightness of B = 50 cd m^{-2}.[40]

Tae-Hee Han et al. prepared a modified graphene anode by transferring graphene film to a flexible PET substrate in support of a poly(methyl methacrylate) (PMMA) polymer or thermal release tape for the synthesis of a FOLED. The synthesized

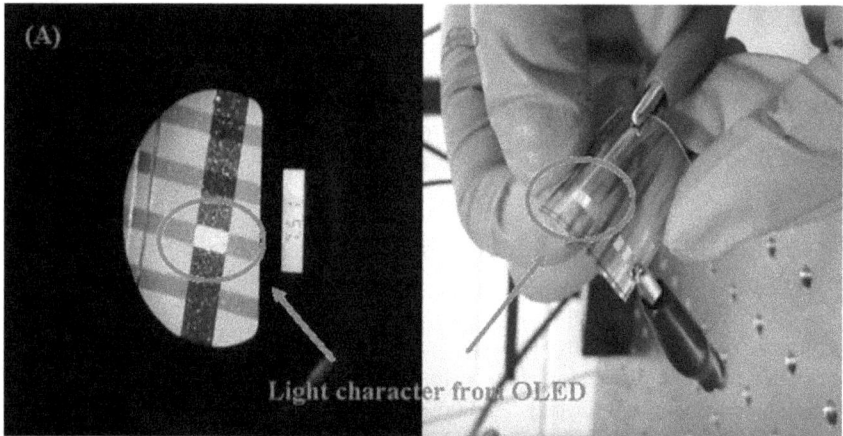

FIGURE 4.10   FOLED on bacterial cellulose nanocomposite developed by Yue-Feng Liu and coworkers. (Reprinted with permission from Ummartyotinet al.[38] Copyright © 2012 Elsevier B.V.)

FIGURE 4.11   FOLED prepared using a slot-die roll-coating apparatus by Sandström and coworkers. (Reprinted with permission from Sandström et al.[40] Copyright © 2012 Macmillan Publishers Limited.)

FOLED showed extremely high luminous efficiencies compared to ITO-based devices (obtained 37.2 and 102.7 lm W^{-1} in fluorescent and phosphorescent OLEDs, respectively).[41]

Y.-F. Liu and coworkers synthesized an ultrasmooth Ag-anode-based top-emitting FOLED by using a template-stripping process to obtain high mechanical robustness of the flexibility with 60% more efficiency (Figure 4.12).[42]

The device was designed with the structure of Ag (80 nm)/MoO_3 (4 nm)/m-MTDATA (30 nm)/NPB (20 nm)/Alq_3 (50 nm)/LiF (1 nm)/Al (1 nm)/Ag (20 nm). The methodology was further extended for the inverted top-emitting FOLEDs to obtain maximum current efficiency of 9.72 cd A^{-1}.[43]

Jeong-Hwan Lee et al. reported a p-doped copper phthalocyanine (CuPc)/n-doped 4,7-diphenyl-1,10-phenanthroline (Bphen) layer as the reverse bias in the p–n junction to generate electrons and holes as an efficient inverted FOLED.[44] The typical setup was made with an ITO (cathode)/5 mol% rhenium oxide (ReO_3) doped CuPc (15 nm)/15 wt% rubidium carbonate (Rb_2CO_3) doped Bphen (15 nm)/undoped Bphen (20 nm)/1 wt% C545T doped CBP (20 nm)/undoped 1,1-bis-(4-bis(4-methyl-phenyl)-amino-phenyl)-cyclohexane (TAPC) (30 nm)/8 wt% ReO_3 doped TAPC (20 nm)/Al to generate a very efficient charge of 100 mA cm^{-2} at 0.3 V.

Sung-Min Lee and coworkers replaced ITO with a very low resistive stacked Ag/ZnO/Ag multilayer transparent electrode (1.6 Ω sq^{-1}) to show high transmittance (75%) for FOLEDs.[45]

In 2013 Chen Shu-Fen et al. constructed top emitting FOLEDs using MoO_x film, deposited on a PET substrate.[46] The device was constructed with a Sm/Ag semitransparent cathode in combination with the MoO_x film, with a setup of PET/MoO_x (50/100 nm)/Ag (70 nm)/MoOx (3 nm)/mMTDATA:1.8 wt% F_4-TCNQ (25 nm)/NPB (10 nm)/CBP: 0.5 wt% bis(2-methyldibenzo[f,h]quinoxaline)(acetylacetonate) iridium(III) (Ir(MDQ)$_2$(acac), 5 nm)/CBP (3 nm)/CBP: 10 wt% bis (3,5-difluoro-2-(2-pyridyl) phenyl-(2-carboxypyridyl) iridium(III) (FIrpic, 25 nm)/Bphen (40 nm)/Sm (8 nm)/Ag (16 nm) to obtain maximum current efficiency of 4.64 cd A^{-1} and a power efficiency of 1.9 lm W^{-1}.

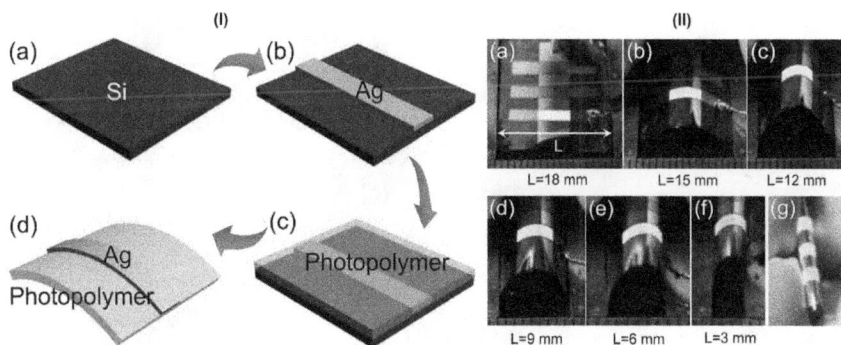

FIGURE 4.12 Ultrasmooth Ag-anode-based top-emitting FOLED developed by Y.-F. Liu and coworkers. (Reprinted with permission from Liu et al.[42] Copyright © 2012 Optical Society of America.)

Ja-Ryong Koo and coworkers reported a Ni/Ag/Ni anode bottom-emitting white FOLED with a combination of ET/Ni/Ag/Ni (3/6/3 nm)/NPB (50 nm)/mCP (10 nm)/7% FIrpic:mCP (10 nm)/3% Ir(pq)$_2$acac:TPBi (5 nm)/7% FIrpic:TPBi (5 nm)/ TPBi (10 nm)/8-hydroxyquinolinolato-lithium (Liq, 2 nm)/Al (100 nm) to show a maximum value for luminous efficiency of 13.13 cd A^{-1}, with a quantum efficiency of 5.85% (Figure 4.13).[47]

Fushan Li et al. developed FOLEDs based on transparent conductive graphene/Ag/ AZO films with a set up of graphene/Ag/aluminum-doped zinc oxide (AZO)/NPB/ Alq$_3$/LiF/Al to show a current efficiency of 1.46 cd A^{-1} with excellent light-emitting stability during the bending (Figure 4.14).[48]

Keith A. Knauer and coworkers constructed stacked inverted top-emitting FOLEDs based on glass and flexible glass substrates with either single or double units to show the maximum current efficacy of 46.8 cd A^{-1} at a luminance of 1215 cd m^{-2} and 97.8 cd A^{-1} at a luminance of 1119 cd m^{-2}, respectively (Figure 4.15).[49]

Woohyun Kim et al. reported soft fabric-based high-performance FOLEDs using PU and poly(vinyl alcohol) (PVA) layers, to show a high current efficiency of around

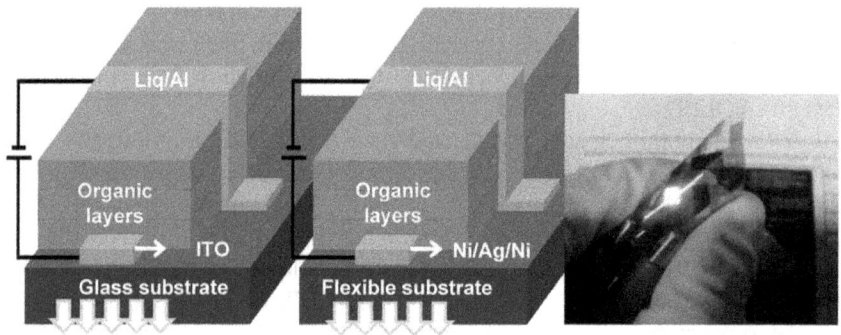

FIGURE 4.13 Ni/Ag/Ni anode bottom-emitting white FOLED developed by Ja-Ryong Koo and coworkers. (Reprinted with permission from Koo et al.[47] Copyright © 2013 Optical Society of America.)

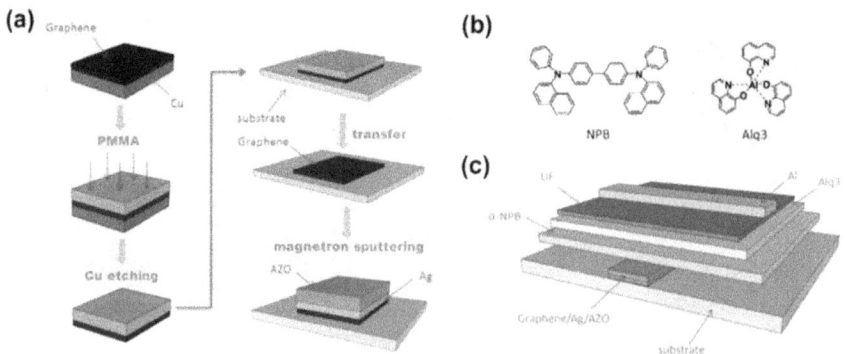

FIGURE 4.14 FOLED based on graphene/Ag/AZO films developed by Fushan Li et al. (Reprinted with permission from Li et al.[48] Copyright © 2013 Elsevier B.V.)

(I)

(a)

| Au 20 nm |
| MoO₃ 15 nm |
| TAPC 35 nm |
| CBP:Ir(ppy)₃ 20 nm |
| TpPyPB 40 nm |
| LiF 2.5 nm |
| Al 50 nm |
| PEDOT:PSS 40 nm |
| Glass |

(b)

| Au 20 nm |
| MoO₃ 15 nm |
| TAPC 35 nm |
| CBP:Ir(ppy)₃ 20 nm |
| TpPyPB 40 nm |
| LiF 2.5 nm |
| Al 1.0 nm |
| HAT-CN 10 nm |
| TAPC 35 nm |
| CBP:Ir(ppy)₃ 20 nm |
| TpPyPB 40 nm |
| LiF 2.5 nm |
| Al 50 nm |
| PEDOT:PSS 40 nm |
| Glass |

(II)

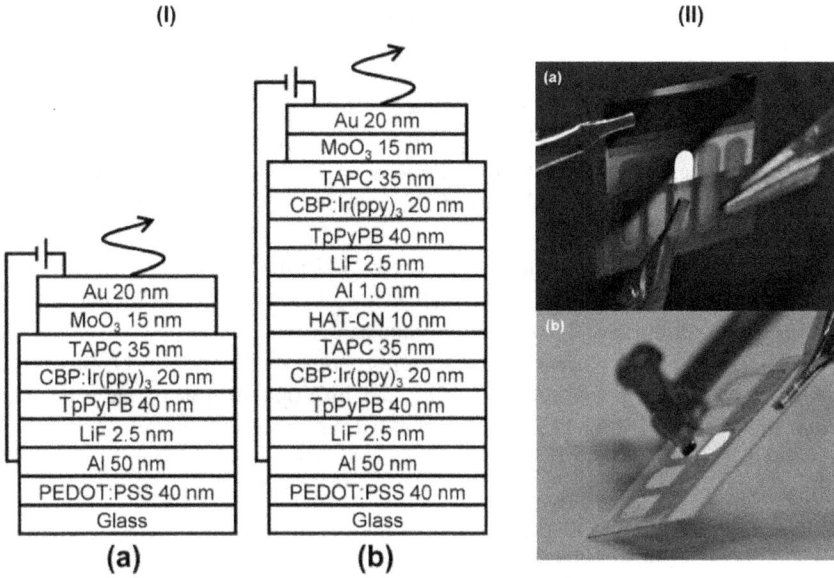

FIGURE 4.15 FOLEDs based on glass and flexible glass substrates in either single or double units developed by Keith A. Knauer and coworkers. (Reprinted with permission from Knauer et al.[49] Copyright © 2013 Elsevier B.V.)

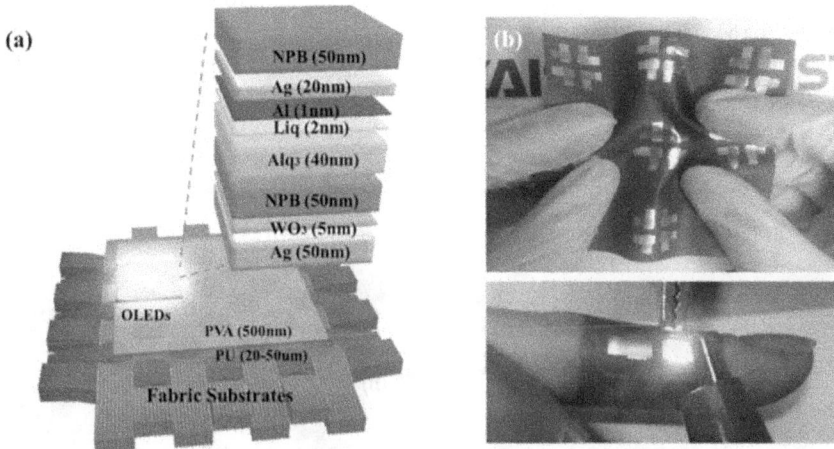

FIGURE 4.16 Soft fabric-based high-performance FOLEDs developed by Woohyun Kim et al. (Reprinted with permission from Kim et al.[50] Copyright © 2013 Elsevier B.V.)

8 cd m^{-2} which is compatible for 1000 cyclic bending testing with a bending radius of 5 mm (Figure 4.16).[50]

Shufen Chen and coworkers developed thin white top-emitting FOLEDs based on tris(phenypyrazole)iridium in a PET substrate to show an efficiency of 9.9 cd A^{-1}.

The effectiveness of the device was confirmed with electroluminescent spectra, the time-resolved transient photoluminescence decay lifetimes of the phosphors and the tunneling phenomenon.[51]

Kim and Park increased the surface roughness of FOLEDs through the fabrication of amorphous and crystalline ITO on a Teflon-based polymer substrate. This fabrication makes the surface unidirectionally wavy, with a 500% increase in the surface area, which was tolerable to 10,000 cycles at a bending radius of 10 mm (Figure 4.17).[52]

Yun Cheol Han et al. introduced ITO-free ZnS (25 nm)/Ag (7 nm)/MoO$_3$ (5 nm) multilayer films for FOLEDs to show satisfactory performance in efficiency, along with high transmittance and low sheet resistance.[53]

Kiyeol Kwak and coworkers reported amorphous ZnO-doped In$_2$O$_3$ (a-IZO) in a FOLED to obtain low sheet resistance and high optical transparency by combining the sheet as a-IZO/PEDOT:PSS/poly[(9,9-di-n-octylfluorenyl-2,7-diyl)-alt-(benzo[2,1,3]thiadiazol-4,8-diyl)] (F8BT)/LiF/Al. The sheet was found to be highly flexible, with good retention of the efficient luminescent characteristics after bending 1000 times.[54]

In 2014, Hahn-Gil Cheong et al. prepared transparent conductive electrodes using flexible Ag nanowire (AgNW) networks welded with transparent conductive oxide (TCO) based on ITO for FOLEDs. The excellent welding of the wire junctions of the electrode lowered the junction resistance and made the device highly flexible.[55]

Ran Ding and coworkers developed a FOLED based on a Au anode and Ca/Ag cathode fabricated with 2,5-bis(4-biphenyl)bithiophene (BP2T) and 1,4-bis(4-methylstyryl)benzene (BSB-Me) on a Si/SiO$_2$ substrate through template stripping (Figure 4.18).[56] The device was found to be very bright and showed polarized EL emission with high flexibility (no obvious deterioration was found after 100 bending cycles).

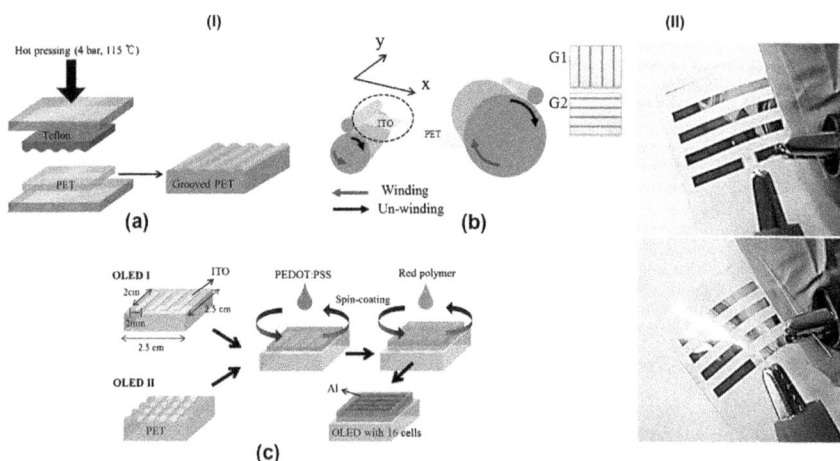

FIGURE 4.17 ITO-based FOLEDs developed by Jung-Hoon Kim and J.-W. Park. (Reprinted with permission from Kim and Park.[52] Copyright © 2013 Elsevier B.V.)

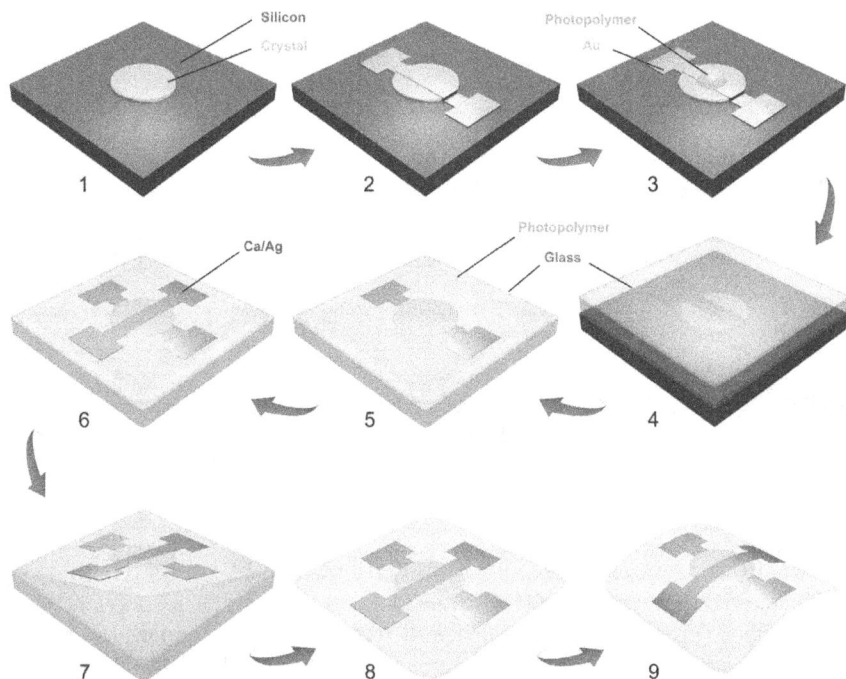

FIGURE 4.18 FOLEDs developed by Ran Ding and coworkers through template stripping. (Reprinted with permission from Ding et al.[56] Copyright © 2014 WILEY-VCH Verlag GmbH & Co. KGaA, Weinheim.)

Cheng Zhang et al. reported a spray-coating technique to synthesize a double-layered graphene/PEDOT:PSS conductive film for the preparation of a FOLED by fabricating the film in combination with materials such as tris(2-phenylpyridine) iridium $(Ir(ppy)_3)$ as phosphorescent material, 5,6,11,12-tetraphenylnapthacene (Rubrene) as fluorescent dye doping with CBP host, NPB as hole-transporting layer, Bphen as electron-transporting layer and 4,4′-bis(2,2′-diphenylvinyl)-1,1′-biphenyl (DPVBi) as blue light-emitting layer (Figure 4.19).[57] The synthesized device emitted white light at 8 V and showed excellent light-emitting stability during bending at a radius of curvature of 10 mm.

Chaoxing Wu and coworkers reported single layer graphene (SLG)-based electrodes in fabrication with PEDOT:PSS on a SiO_2/PET substrate via chemical vapour deposition (CVD) for FOLEDs, which showed high conductivity and transmittance with excellent mechanical robustness and bendable performances.[58]

A top-emitting, bottom-emitting and transparent FOLED was prepared by Shihao Liu et al. based on an ultrathin silver/germanium/silver (AGA) electrode (Figure 4.20).[59] The device was prepared by a combination of PET substrate/anode/ MoO_3/TAPC)/4,4′,4″-tris (carbazol-9-yl)-triphenylamine (TCTA)/emission layer (EML)/Bphen/Liq/cathode, where the AGA electrode was used as either the anode or cathode, 10 wt% $Ir(ppy)_3$ doped 2,7-bis (diphenylphosphoryl)- 9-[4-(N,N-diphenylamino) phenyl]-9-phenylfluorene (POAPF) for green OLED or FIrpic or iridium(III)

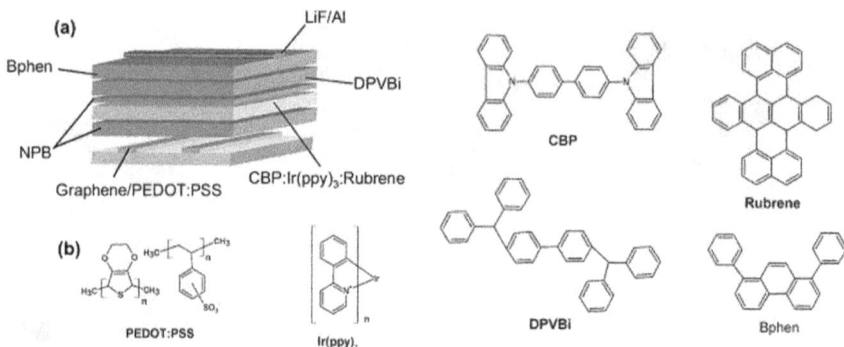

FIGURE 4.19 FOLEDs developed by Cheng Zhang et al. (Reprinted with permission from Zhang et al.[57] Copyright © 2014 Elsevier B.V.)

bis(4-phenylthieno[3,2-c]pyridinato-*N*,C2′) acetylacetonate (PO-01) doped POAPF for white OLED.

Yuxin Li and coworkers reported an ITO-free FOLED based on poly(3,4-ethylenedioxythiophene) (PEDOT) flexible films via the vapor phase polymerization (VPP) method (Figure 4.21).[60] The device showed excellent performance in terms of conductivity when dipped in H_2SO_4 solution and performed well in bending.

A laser lift-off method was used by Kisoo Kim et al. for the preparation of a FOLED from amorphous gallium oxide (α-GaO$_x$) which can be used further for fabrication of flexible electronic devices at high temperature (Figure 4.22).[61]

Wenya Xu and coworkers reported a FOLED based on flexible print top-gated thin-film transistors (TFTs) from semiconducting SW-CNT inks derived by arc-discharge SW-CNTs with poly[2,7-(9,9-dioctylfluorene)-alt-4,7-bis(thiophen-2-yl) benzo-2,1,3-thiadiazole] (PFO-DBT) aided by sonication and centrifugation in tetrahydrofuran.[62]

Sumit Purandare et al. prepared FOLEDs based on phosphorescent emitters fabricated on flexible and transparent reconstituted cellulose obtained from wood pulp to obtain current efficiencies up to 47 cd A^{-1}, emission efficiencies up to 20 lm W^{-1} and a maximum brightness of 10,000 cd m^{-2}.[63]

Hongying Shi et al. reported a top-emitting white FOLED to show very high current as well as power efficiency with a blue/red/blue sandwiched triemission layer for color alteration.[64]

A template stripping process with dropping H_2SO_4 treatment was applied for the synthesis of an ITO-free FOLED by Yue-Feng Liu and coworkers, using a PEDOT:PSS film on a glass substrate which is free of cracks and dark spots under small bending radius.[65]

Dae Young Yan et al. developed a FOLED based on ZnS/1st Ag/ZnO/2nd Ag/ WO_3 (ZAZAW) multilayer electrodes on PET substrate to show very high luminous transmittance (>80% in the visible range) and very low sheet resistance with good mechanical flexibility.[66]

Xiaoxiao Wu and coworkers reported a graphene/PEDOT:PSS based green FOLED using a spray-coating technique with a PET/graphene/PEDOT:PSS/NPB(64

Green Devices

White Devices

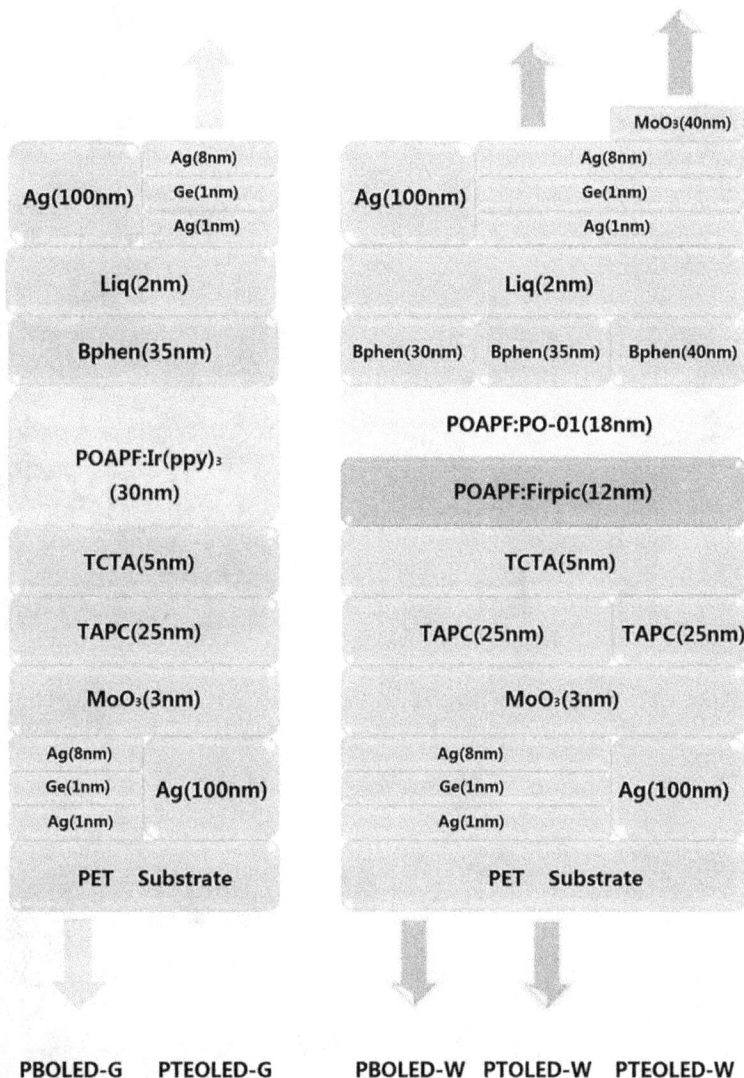

| Devices | PBOLED-G | PTEOLED-G | PBOLED-W | PTOLED-W | PTEOLED-W |

FIGURE 4.20 FOLEDs developed by Shihao Liu et al. (Reprinted with permission from Liu et al.[59] Copyright © 2014 The Royal Society of Chemistry.)

nm)/Alq$_3$(105 nm)/LiF (0.4 nm)/Al (100 nm) combination which upon bending produced green emission.[67]

In 2015, Junwei Xu et al. developed a highly efficient FOLED based on Al/multiwall carbon nanotube (MWCNT)/Al multilayered electrode, bendable up to 120° without cracking, with power efficiencies up to 22 lm W^{-1} at luminances up to 4000 cd m^{-2} (Figure 4.23).[68]

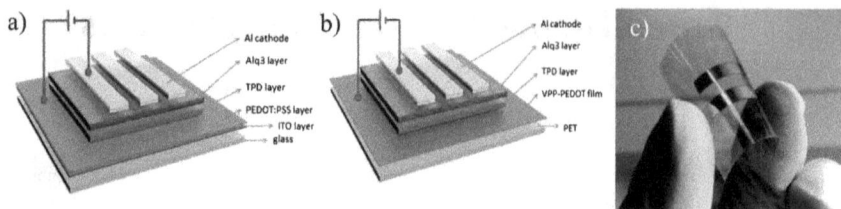

FIGURE 4.21 FOLEDs developed by Yuxin Li and coworkers via the vapor phase polymerization method. (Reprinted with permission from Li et al.[60] Copyright © 2014 The Royal Society of Chemistry.)

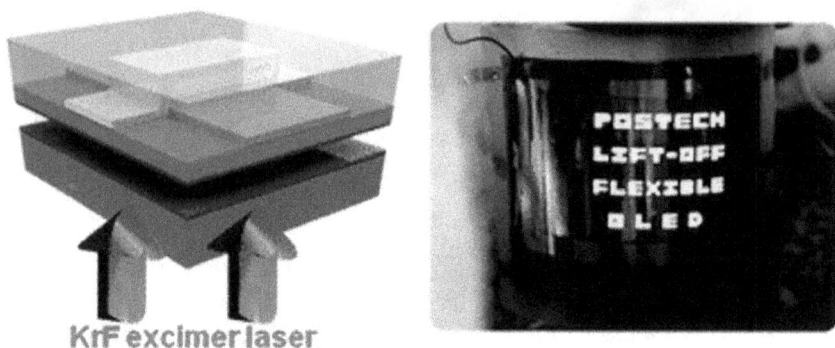

FIGURE 4.22 FOLEDs developed by Kisoo Kim and coworkers via a laser lift-off method. (Reprinted with permission from K. Kim et al.[61] copyright © 2014 The Royal Society of Chemistry.)

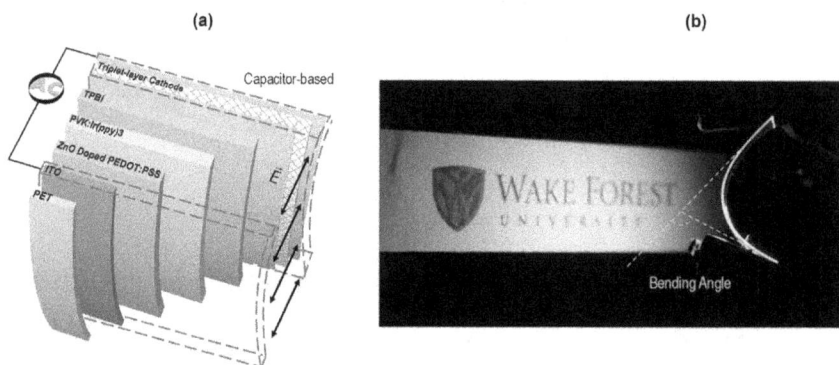

FIGURE 4.23 FOLEDs developed by Junwei Xu et al. based on Al/MWCNT/Al. (Reprinted with permission from Xu et al.[68] Copyright © 2015 WILEY-VCH Verlag GmbH & Co. KGaA, Weinheim.)

Dong-Young Kim and coworkers developed multilayer electrodes based on ZnS (24 nm)/Ag (7 nm)/MoO_3 (5 nm) as anode and ZnS (3 nm)/Cs_2CO_3 (1 nm)/Ag (8 nm)/ZnS (22 nm) as cathode, which were used for the design of a transparent FOLED by fabricating on a PET substrate to obtain 74.22% peak transmittance at 550 nm.[69]

Ya-Hui Duan et al. reported an alternative to ITO for FOLEDs by using silver nanowire (AgNW)-photopolymer (NOA63) film to show very high current efficiency (13 cd A^{-1}) with high stability during bending.[70]

E. R. P. Pinto and coworkers used bacterial cellulose (BC) and castor oilbased PU composites as excellent and flexible thin films for FOLEDs, with good thermal stability (>250°C).[71]

Chih-Hao Chang et al. reported blue, green and red FOLEDs based on molybdenum-doped gallium zinc oxide (MGZO) on a glass substrate to provide high transmittance, work function with low sheet resistance and an adequate mobility.[72]

Po-Wen Sze and coworkers prepared moisture and oxygen protected FOLEDs by applying a liquid phase deposition method to the plastic-coated silicon dioxide (SiO$_2$) films.[73]

Al$_2$O$_3$/ZnO (AZO) nanolaminated film were used for the synthesis of a FOLED by Yuan-Yu Lin et al. with high efficiency. The device showed an air-storage lifetime of >10,000 h.[74]

FOLEDs reported by Michael P. Gaj and co-workers, based on biocompatible shape memory polymer (SMP) substrates prepared from 1,3,5-triallyl-1,3,5-triazine-2, 4,6(1H,3H,5H)-tri-one (TATATO), trimethylolpropane tris(3-mercaptopropionate) (TMTMP) and tricyclo[5.2.1.02,6]de-canedimethanol diacrylate (TCMDA), showed excellent current and power efficacy.[75]

Lu Li et al. prepared green FOLEDs based on a plastic substrate composite with SW-CNT and AgNW dispersed with barium strontium titanate nanoparticles to obtain very high current efficiency (118 cd A^{-1} at 10,000 cd m^{-2}). The device retained its electroluminescent performance during bending up to 3 mm.[76]

Ki-Hun Ok and coworkers developed a blue FOLED based on transparent AgNW embedded in a colorless polyimide (cPI) electrode to show high efficacy during bending of 30 mm bending radius, with a slightly reduced performance (<3%) (Figure 4.24).[77]

Yuqiang Liu et al. reported a FOLED based on biocompatible and biodegradable natural silk fibroin (SF) films embedded AgNWs mesh to show sheet resistance similar to ITO-based polymeric substrates with a lower surface roughness and high current efficiency (19 cd A^{-1}).[78]

In 2016, Monica Morales-Masis and coworkers replaced an ITO-based FOLED with zinc oxide (ZnO) and tin oxide (SnO$_2$) electrodes to show better performance in terms of optical activity in low or high temperature.[79]

Xicheng Liu et al. improved the brightness of the FOLED by introducing two hole-transporting materials based on triphenylamine in combination with poly[2-(4–30,70-dimethyloctyloxy)-phenyl]-p-phenylenevinylene) as emitting and electron transporting layers to obtain very high luminance (up to 103690 cd m^{-2}) with good bending stability.[80]

Yan-Gang Bi and coworkers used Au film as an alternate to ITO for FOLEDs to show high current efficiency, transparency with comparable sheet resistance and good bending ability.[81]

An electrode based on a graphene oxide/graphene vertical heterostructure was used by S. Jia et al. in a FOLED to show excellent optical transmittance, work function and stability with compatibility. The other materials used in this

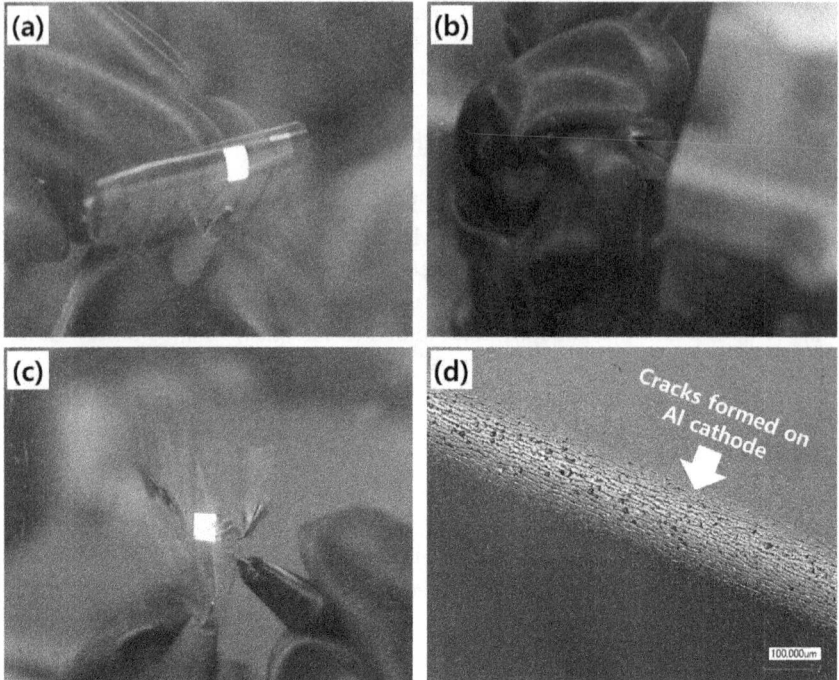

FIGURE 4.24 FOLEDs developed by Ki-Hun Ok and coworkers based on AgNW embedded in a colorless polyimide (cPI) electrode. (Reprinted with permission from Ok et al.[77] Copyright © 2015 The Author(s).)

device were MoO_3 as hole injection layers, TAPC as hole transportation layer, two bis(2-phenylpyridine)(acetylacetonate)iridium(III) [$Ir(ppy)_2(acac)$] doped with TCTA and Bphen as a light emission layer for green emission, $(Ir(MDQ)_2$ (acac) doped with NPB for red emission, two 10 nm FIrpic doped with TCTA and 2,6-bis(3-(9H-carbazol-9-yl)- phenyl)pyridine (26DCZ) for blue emission and bis(2-phenylbenzothiazolato)(acetylacetonate)iridium(III) doped with 4P-NPB for white emission.[82]

Yu-shan Liu and coworkers reported a FOLED based on a PEDOT:PSS/AgNW composite electrode through a template stripping method to show increased optical activity by improving the surface roughness, sheet resistance and transmittance properties compared to the conventional PEDOT:PSS-based electrodes.[83]

An Al_2O_3/ZnO nanostratified structure with S-H nanocomposite organic layer was fixed via atomic layer deposition by Eun Gyo Jeong et al. for encapsulating FOLEDs to show very good optical property with high bending stability.[84] The device was composed with ZnS (25 nm)/Ag (7 nm)/MoO_3 (5 nm)/NPB (50 nm)/Alq_3(50 nm)/Liq (1 nm)/Al (100 nm) layers.

Xue Li and coworkers reported FOLED using an anodic aluminium oxide (AAO) nanoimprint lithography technique to improve device lifetime. The device was made up of layers of Ag (80 nm)/MoO_x(2 nm)/*N,N,N′,N′*-tetrakis (4-Methoxy-phenyl) benzidine (MeOTPD): 3 wt% F4-TCNQ (30 nm)/MeO-TPD (10 nm)/TCTA: FIrpic (8 wt%)/TCTA:Ir(MDQ)$_2$ (acac) [3 wt%]/TCTA:FIrpic (8 wt%)/Bphen (10 nm)/ Bphen: 3 wt% Li (20 nm)/Sm (4 nm)/Ag (14 nm).[85]

Yun Cheol Han and coworkers reported a FOLED with high bending properties made with silica nanoparticle-embedded sol-gel organic-inorganic hybrid nanocomposite and Al$_2$O$_3$.[86]

In 2017, Eonseok Oh et al. found a FOLED based on a combination of NPB/hexaazatriphenylene hexacarbonitrile (HATCN)/grapheme compared well with a MoO$_3$-based FOLED.[87]

Yue-Feng Liu and coworkers developed a FOLED based on silk substrate by lanarizing the silk substrate with photopolymer NOA63 to show very high luminance and current efficiency with perfect bending properties.[88]

A dual-scale AgNW electrode was introduced by Jinhwan Lee et al. for improved flexibility as well as optical properties.[89]

Eun Gyo Jeong and coworkers used the Griffith crack model to investigate the defect suppression mechanism in a FOLED with layers of Ag (30 nm)/MoO$_3$ (5 nm)/NPB (75 nm)/bis(10-hydroxybenzo[h] quinolinato)beryllium (Bebq$_2$): tris(1-phenylisoquinoline) iridium (Ir(piq)$_3$) (30 nm)/Bebq$_2$ (40 nm)/Liq (1 nm)/Al (100 nm).[90]

Single-layer graphene (SLG) with thin silver layers were stacked together for the preparation of TCE in a FOLED developed by Kun Li et al. to show high transmittance and bending stability with low sheet resistance.[91]

Kyung Min Lee and coworkers reported FOLEDs based on a polyimide/AgNW composite to show high thermal and mechanical stability with very high optical activities.[92]

Jun Li et al. used the argon (Ar) plasma treatment method in a AgNWs-based FOLED, to eliminate the PVP layer and increase the contact between AgNWs for better transmittance and low resistance.[93]

Zheng Chen and coworkers applied plasma enhanced atomic layer deposition (PEALD) and molecular layer deposition to fabricate hybrid zirconium inorganic/ organic nanolaminates for a FOLED to increase the lifetime of the device.[94]

In 2018, again a PEDOT:PSS anode-based FOLED was reported by Lu Zhou et al. to show high optical transparency with low sheet resistance and excellent bending properties.[95]

In 2019, Yi-Ning Lai and coworkers reported AZO and ITO composite on mica for blue-, green-, and red-emitting FOLEDs to show excellent electroluminescence efficiencies.[96]

Finally, in 2020, a plasmonic ultrathin Au grid electrode was used by Fang-Shun Yi et al. for a FOLED as a substitute for ITO for better current efficiency.[97]

4.3 CONCLUSION

In summary, this chapter presented a concise overview of the preparation of FOLEDs using different types of electrodes, conducting materials, several composite, organic polymers. The progress of ITO-based electrodes to ITO-free electrodes with good mechanical stability were also elaborated. The chapter will be helpful to the researcher for further exploration in the area of FOLEDs.

REFERENCES

1. Tang, C. W.; VanSlyke, S. A. Organic electroluminescent diodes. *Appl. Phys. Lett.*, 1987, *51*, 913.
2. www.explainthatstuff.com/how-oleds-and-leps-work.html
3. www.theengineeringknowledge.com/full-form-of-oled/
4. Zou, S.-J.; Shen, Y.; Xie, F.-M.; Chen, J.-D.; Li, Y.-Q.; Tang, J.-X. Recent advances in organic light-emitting diodes: toward smart lighting and displays. *Mater. Chem. Front.*, 2020, *4*, 788–820.
5. Gu, G.; Burrows, P. E.; Venkatesh, S.; Forrest, S. R. Vacuum-deposited, nonpolymeric flexible organic light-emitting devices. *Opt. Lett.*, 1997, *22*, 172–174.
6. Han, T.-H.; Jeong, S.-H.; Lee, Y.; Seo, H.-K.; Kwon, S.-J.; Park, M.-H.; Lee, T.-W. Flexible transparent electrodes for organic light-emitting diodes. *J. Inf. Disp.*, 2015, *16*, 71–84.
7. Chen, Z.; Cotterell, B.; Wang, W.; Guenther, E.; Chua, S. A. A mechanical assessment of flexible optoelectronic devices. *Thin Solid Films*, 2001, *394*, 201–205.
8. Kim, S.; Kwon, H.-J.; Lee, S.; Shim, H.; Chun, Y.; Choi, W.; Kwack, J.; Han, D.; Song, M.; Kim, S. et al. Low-power flexible organic light-emitting diode display device. *Adv. Mater.*, 2011, *23*, 3511–3516.
9. Kumar, A.; Zhou, C. The race to replace tin-doped indium oxide: which material will win? *ACS Nano*, 2010, *4*, 11–14.
10. Liu, Y.-F.; Feng, J.; Bi, Y.-G.; Yin, D.; Sun, H.-B. Recent developments in flexible organic light-emitting devices. *Adv. Mater. Technol.*, 2018, *4*, 1800371.
11. Hea, Y.; Kanicki, J. High-efficiency organic polymer light-emitting heterostructure devices on flexible plastic substrates. *Appl. Phys. Lett.*, 2000, *76*, 661–663.
12. Weaver, M. S.; Michalski, L. A.; Rajan, K.; Rothman, M. A.; Silvernail, J. A.; Brown, J. J.; Burrows, P. E.; Graff, G. L.; Gross, M. E.; Martin, P. M.; Hall, M.; Mast, E.; Bonham, C.; Bennett, W.; Zumhoff, M. Organic light-emitting devices with extended operating lifetimes on plastic substrates. *Appl. Phys. Lett.*, 2002, *81*, 2929–2931.
13. Chwang, A. B.; Rothman, M. A.; Mao, S. Y.; Hewitt, R. H.; Weaver, M. S.; Silvernail, J. A.; Rajan, K.; Hack, M.; Brown, J. J.; Chu, X.; Moro, L.; Krajewski, T.; Rutherford, N. Thin film encapsulated flexible organic electroluminescent displays. *Appl. Phys. Lett.* 2003, *83*, 413–415.
14. Xie, Z.; Hung, L.-S.; Zhu, F. A flexible top-emitting organic light-emitting diode on steel foil. *Chem. Phys. Lett.*, 2003, *381*, 691–696.
15. Xie, Z. Y.; Li, Y. Q.; Wong, F. L.; Hung, L. S. Fabrication of flexible organic top-emitting devices on steel foil substrates. *Mater. Sci. Eng. B*, 2004, *106*, 219–223.
16. Hong, Y.; He, Z.; Lennhoff, N. S.; Banach, D. A.; Kanicki, J. Transparent flexible plastic substrates for organic light-emitting devices. *J. Electron. Mater.*, 2004, *33*, 312–320.
17. Li, Y.; Tan, L.-W.; Hao, X.-T.; Ong, K. S.; Zhu, F.; Hung, L.-S. Flexible top-emitting electroluminescent devices on polyethylene terephthalate substrates. *Appl. Phys. Lett.* 2005, *86*, 153508.
18. Hsu, C.-M.; Tsai, C.-L.; Wu, W.-T. Selective light emission from flexible organic light-emitting devices using a dot-nickel embedded indium tin oxide anode. *Appl. Phys. Lett.*, 2006, *88*, 083515.
19. Hsu, C.-M.; Liu, C.-F.; Cheng, H.-E.; Wu, W.-T. Low-temperature nickel-doped indium tin oxide anode for flexible organic light-emitting devices. *J. Electron. Mater.* 2006, *35*, 383–387.
20. Wuu, D.-S.; Chen, T.-N.; Wu, C.-C.; Chiang, C.-C.; Chen, Y.-P.; Horng, R.-H.; Juang, F.-S. Transparent barrier coatings for flexible organic light-emitting diode applications. *Chem. Vap. Deposition*, 2006, *12*, 220–224.
21. Yu, H. H.; Hwang, S.-J.; Tseng, M.-C.; Tseng, C.-C. The effect of ITO films thickness on the properties of flexible organic light emitting diode. *Opt. Commun.*, 2006, *259*, 187–193.

22. Kang, M.-G.; Guo, L. J. Semitransparent Cu electrode on a flexible substrate and its application in organic light emitting diodes. *J. Vac. Sci. Technol. B*, **2007**, *25*, 2637–2641.

23. Li, J.; Hu, L.; Liu, J.; Wang, L.; Marks, T. J.; Grüner, G. Indium tin oxide modified transparent nanotube thin films as effective anodes for flexible organic light-emitting diodes. *Appl. Phys. Lett.*, **2008**, *93*, 083306.

24. Choi, K.-H.; Nam, H.-J.; Jeong, J.-A.; Cho, S.-W.; Kim, H.-K.; Kang, J.-W.; Kim, D.-G.; Cho, W.-J. Highly flexible and transparent InZnSnOx/Ag/InZnSnOx multilayer electrode for flexible organic light emitting diodes. *Appl. Phys. Lett.*, **2008**, *92*, 223302.

25. Yu, Z.; Hu, L.; Liu, Z.; Sun, M.; Wang, M.; Grüner, G.; Pei, Q. Fully bendable polymer light emitting devices with carbon nanotubes as cathode and Anode. *Appl. Phys. Lett.*, **2009**, *95*, 203304.

26. Okahisa, Y.; Yoshida, A.; Miyaguchi, S.; Yano, H. Optically transparent wood–cellulose nanocomposite as a base substrate for flexible organic light-emitting diode displays. *Compos. Sci. Technol.*, **2009**, *69*, 1958–1961.

27. Villani, F.; Vacca, P.; Nenna, G.; Valentino, O.; Burrasca, G.; Fasolino, T.; Minarini, C.; della Sala, D. Inkjet printed polymer layer on flexible substrate for OLED applications. *J. Phys. Chem. C*, **2009**, *113*, 13398–13402.

28. Mitsunori, S.; Hirohiko, F.; Yoshiki, N.; Toshimitsu, T.; Tatsuya, T.; Toshihiro, Y.; Shizuo, T. A 5.8-in. phosphorescent color AMOLED display fabricated by ink-jet printing on plastic substrate. *J. Soc. Inf. Disp.* **2009**, *17*, 1037–1042.

29. Wang, G.-F.; Tao, X.-M.; Xin, J. H.; Fei, B. Modification of conductive polymer for polymeric anodes of flexible organic light-emitting diodes. *Nanoscale Res. Lett.*, **2009**, *4*, 613–617.

30. Cho, H.; Yun, C.; Park, J.-W.; Yoo, S. Highly flexible organic light-emitting diodes based on ZnS/Ag/WO$_3$ multilayer transparent electrodes. *Org. Electron.*, **2009**, *10*, 1163–1169.

31. Calil, V. L.; Legnani, C.; Moreira, G. F.; Vilani, C.; Teixeira, K. C.; Quirino, W. G.; Machado, R.; Achete, C. A.; Cremona, M. Transparent thermally stable poly(etherimide) film as flexible substrate for OLEDs. *Thin Solid Films*, **2009**, *518*, 1419–1423.

32. Hu, L.; Li, J.; Liu, J.; Gruner, G.; Marks, T. Flexible organic light-emitting diodes with transparent carbon nanotube electrodes: problems and solutions. *Nanotechnology*, **2010**, *21*, 155202.

33. Heo, G.-S.; Matsumoto, Y.; Gim, I.-G.; Lee, H.-K.; Park, J.-W.; Kim, T.-W. Transparent conducting amorphous Zn-In-Sn-O anode for flexible organic light-emitting diodes. *Solid State Commun.*, **2010**, *150*, 223–226.

34. Ji, W.; Zhao, J.; Sun, Z.; Xie, W. High-color-rendering flexible top-emitting warm-white organic light emitting diode with a transparent multilayer cathode. *Org. Electron.*, **2011**, *12*, 1137–1141.

35. Kim, M.; Lee, Y. S.; Kim, Y. C.; Choi, M. S.; Lee, J. Y. Flexible organic light-emitting diode with a conductive polymer electrode. *Synth. Met.*, **2011**, *161*, 2318–2322.

36. Xu, J.; Zhang, L.; Zhong, J.; Lin, H. High-luminance flexible top emission organic light emitting diode with nickel–chromium alloy thin film on aluminum anode. *Opt. Rev.*, **2012**, *19*, 358–360.

37. Najafabadi, E.; Knauer, K. A.; Haske, W.; Fuentes-Hernandez, C.; Kippelen, B. Highly efficient inverted top-emitting green phosphorescent organic light-emitting diodes on glass and flexible substrates. *Appl. Phys. Lett.*, **2012**, *101*, 023304.

38. Ummartyotin, S.; Juntaro, J.; Sain, M.; Manuspiya, H. Development of transparent bacterial cellulose nanocomposite film as substrate for flexible organic light emitting diode (OLED) display. *Ind. Crops Prod.*, **2012**, *35*, 92–97.

39. Hyung, G. W.; Park, J.; Koo, J. R.; Lee, S. J.; Lee, H. W.; Kim, Y. H.; Kim, W. Y.; Kim, Y. K. Flexible top-emitting organic light-emitting diodes using semi-transparent multi-metal layers. *J. Nanosci. Nanotechnol.*, **2012**, *12*, 5444–5448.

40. Sandström, A.; Dam, H. F.; Krebs, F. C.; Edman, L. Ambient fabrication of flexible and large-area organic light-emitting devices using slot-die coating. *Nat. Commun.*, **2012**, *3*, 1002.

41. Han, T.-H.; Lee, Y.; Choi, M.-R.; Woo, S.-H.; Bae, S.-H.; Hong, B. H.; Ahn, J.-H.; Lee, T.-W. Extremely efficient flexible organic light-emitting diodes with modified graphene anode. *Nat. Photon*, **2012**, *6*, 105–110.

42. Liu, Y. F.; Feng, J.; Yin, D.; Bi, Y. G.; Song, J. F.; Chen, Q. D.; Sun, H. B. Highly flexible and efficient top-emitting organic light-emitting devices with ultrasmooth Ag anode. *Opt. Lett.*, **2012**, *37*, 1796–1798.

43. Liu, Y.-F.; Feng, J.; Cui, H.-F.; Zhang, Y.-F.; Yin, D.; Bi, Y.-G.; Song, J.-F.; Chen, Q.-D.; Sun, H.-B. Fabrication and characterization of Ag film with subnanometer surface roughness as a flexible cathode for inverted top-emitting organic light-emitting devices. *Nanoscale*, **2013**, *5*, 10811–10815.

44. Lee, J.-H.; Kim, J. W.; Kim, S.-Y.; Yoo, S.-J.; Lee, J.-H.; Kim, J.-J. An organic p–n junction as an efficient and cathode independent electron injection layer for flexible inverted organic light emitting diodes. *Org. Electron.*, **2012**, *13*, 545–549.

45. Lee, S.-M.; Choi, C. S.; Choi, K. C.; Lee, H.-C. Low resistive transparent and flexible ZnO/Ag/ZnO/Ag/WO$_3$ electrode for organic light-emitting diodes. *Org. Electron.*, **2012**, *13*, 1654–1659.

46. Shu-Fen, C.; Xu, G.; Qiang, W.; Xiao-Fei, Z.; Ming, S.; Wei, H. Flexible white top-emitting organic light-emitting diode with a MoO$_x$ roughness improvement layer. *Chin. Phys. B*, **2013**, *22*, 128506.

47. Koo, J. R.; Lee, S. J.; Lee, H. W.; Lee, D. H.; Yang, H. J.; Kim, W. Y.; Kim, Y. K. Flexible bottom-emitting white organic light emitting diodes with semitransparent Ni/Ag/Ni anode *Opt. Express*, **2013**, *21*, 11086–11094.

48. Li, F.; Lin, Z.; Zhang, B.; Zhang, Y.; Wu, C.; Guo, T. Fabrication of flexible conductive graphene/Ag/Al-doped zinc oxide multilayer films for application in flexible organic light-emitting diodes. *Org. Electron.*, **2013**, *14*, 2139–2143.

49. Knauer, K. A.; Najafabadi, E.; Haske, W.; Gaj, M. P.; Davis, K. C.; Fuentes-Hernandez, C.; Carrasco, U.; Kippelen, B. Stacked inverted top-emitting green electrophosphorescent organic light-emitting diodes on glass and flexible glass substrates. *Org. Electron*, **2013**, *14*, 2418–2423.

50. Kim, W.; Kwon, S.; Lee, S.-M.; Kim, J. Y.; Han, Y.; Kim, E.; Choi, K. C.; Park, S.; Park, B.-C. Soft fabric-based flexible organic light-emitting diodes. *Org. Electron.*, **2013**, *14*, 3007–3013.

51. Chen, S.; Zhao, X.; Wu, Q.; Shi, H.; Mei, Y.; Zhang, R.; Wang, L.; Huang, W. Efficient, color-stable flexible white top-emitting organic light-emitting diodes. *Org. Electron.*, **2013**, *14*, 3037–3045.

52. Kim, J.-H.; Park, J.-W. Improving the flexibility of large-area transparent conductive oxide electrodes on polymer substrates for flexible organic light emitting diodes by introducing surface roughness. *Org. Electron.*, **2013**, *14*, 3444–3452.

53. Han, Y. C.; Lim, M. S.; Park, J. H.; Choi, K. C. ITO-free flexible organic light-emitting diode using ZnS/Ag/MoO$_3$ anode incorporating a quasi-perfect Ag thin film. *Org. Electron.*, **2013**, *14*, 3437–3443.

54. Kwak, K.; Cho, K.; Kim, S. Stable bending performance of flexible organic light-emitting diodes using IZO anodes. *Sci. Rep.*, **2013**, *3*, 2787.

55. Cheong, H.-G.; Triambulo, R. E.; Lee, G.-H.; Yi, I.-S.; Park, J.-W. Silver nanowire network transparent electrodes with highly enhanced flexibility by welding for application in flexible organic light-emitting diodes. *ACS Appl. Mater. Interfaces*, **2014**, *6*, 7846–7855.

56. Ding, R.; Feng, J.; Zhang, X.-L.; Zhou, W.; Fang, H.-H.; Liu, Y.-F.; Chen, Q.-D.; Wang, H.-Y.; Sun, H.-B. Fabrication and Characterization of organic single crystal-based light-emitting devices with improved contact between the metallic electrodes and crystal. *Adv. Funct. Mater.*, **2014**, *24*, 7085–7092.

57. Wu, X.; Li, F.; Wu, W.; Guo, T. Flexible white phosphorescent organic light emitting diodes based on multilayered graphene/PEDOT:PSS transparent conducting film. *Appl. Surf. Sci.*, **2014**, *295*, 214–218.
58. Wu, C.; Li, F.; Wu, W.; Chen, W.; Guo, T. Liquid-phase exfoliation of chemical vapor deposition-grown single layer graphene and its application in solution-processed transparent electrodes for flexible organic light emitting devices. *Appl. Phys. Lett.*, **2014**, *105*, 243509.
59. Liu, S.; Liu, W.; Yu, J.; Zhang, W.; Zhang, L.; Wen, X.; Yin, Y.; Xie, W. Silver/germanium/silver: an effective transparent electrode for flexible organic light-emitting devices. *J. Mater. Chem. C*, **2014**, *2*, 835–840.
60. Li, Y.; Hu, X.; Zhou, S.; Yang, L.; Yan, J.; Sun, C.; Chen, P. A facile process to produce highly conductive poly(3,4-ethylenedioxythiophene) films for ITO-free flexible OLED devices. *J. Mater. Chem. C*, **2014**, *2*, 916–924.
61. Kim, K.; Kim, S. Y.; Lee, J.-L. Flexible organic light-emitting diodes using laser lift-off method, *J. Mater. Chem. C*, **2014**, *2*, 2144–2149.
62. Xua, W.; Zhaoa, J.; Qiana, L.; Hanb, X.; Wub, L.; Wua, W.; Songa, M.; Zhoua, L.; Sua, W.; Wanga, C.; Niea, S.; Cui, Z. Sorting of large-diameter semiconducting carbon nanotube and printed flexible driving circuit for organic light emitting diode (OLED). *Nanoscale*, **2014**, *6*, 1589–1595.
63. Purandare, S.; Gomez, E. F.; Steckl, A. J. High brightness phosphorescent organic light emitting diodes on transparent and flexible cellulose films. *Nanotechnology*, **2014**, *25*, 094012.
64. Shi, H.; Deng, L.; Chen, S.; Xu, Y.; Zhou, H.; Cheng, F.; Li, X.; Wang, L.; Huang, W. Flexible top-emitting warm-white organic light-emitting diodes with highly luminous performances and extremely stable chromaticity. *Org. Electron.*, **2014**, *15*, 1465–1475.
65. Liu, Y.-F.; Feng, J.; Zhang, Y.-F.; Cui, H.-F.; Yin, D.; Bi, Y.-G.; Song, J.-F.; Chen, Q.-D.; Sun, H.-B. Improved efficiency of indium-tin-oxide-free flexible organic light-emitting devices. *Org. Electron.*, **2014**, *15*, 478–483.
66. Yang, D. Y.; Lee, S.-M.; Jang, W. J.; Choi, K. C. Flexible organic light-emitting diodes with ZnS/Ag/ZnO/Ag/WO₃ multilayer electrode as a transparent anode. *Org. Electron.*, **2014**, *15*, 2468–2475.
67. Wu, X.; Li, F.; Wu, W.; Guo, T. Flexible organic light emitting diodes based on double-layered graphene/PEDOT:PSS conductive film formed by spray-coating. *Vacuum*, **2014**, *101*, 53–56.
68. Xu, J.; Smith, G. M.; Dun, C.; Cui, Y.; Liu, J.; Huang, H.; Huang, W.; Carroll, D. L. Layered, nanonetwork composite cathodes for flexible, high-efficiency, organic light emitting devices. *Adv. Funct. Mater.*, **2015**, *25*, 4397–4404.
69. Kim, D.-Y.; Han, Y. C.; Kim, H. C.; Jeong, E. G.; Choi, K. C.; Highly Transparent and flexible organic light-emitting diodes with structure optimized for anode/cathode multilayer electrodes. *Adv. Funct. Mater.*, **2015**, *25*, 7145–7153.
70. Duan, Y.-H.; Duan, Y.; Wang, X.; Yang, D.; Yang, Y.-Q.; Chen, P.; Sun, F.-B.; Xue, K.-W.; Zhao, Y. Highly flexible peeled-off silver nanowire transparent anode using in organic light-emitting devices. *Appl. Surf. Sci.*, **2015**, *351*, 445–450.
71. Pinto, E. R. P.; Barud, H. S.; Silva, R. R.; Palmieri, M.; Polito, W. L.; Calil, V. L.; Cremona, M.; Ribeiro, S. J. L.; Messaddeq, Y. Transparent composites prepared by bacterial cellulose and castor oil based polyurethane as substrate for flexible OLEDs. *J. Mater. Chem. C*, **2015**, *3*, 11581–11588.
72. Chang, C.-H.; Huang, J.-L.; Wu, S.-W. Mo-doped GZO films used as anodes or cathodes for highly efficient flexible blue, green and red phosphorescent organic light-emitting diodes. *J. Mater. Chem. C*, **2015**, *3*, 12048–12055.
73. Sze, P.-W.; Chen, K.-L.; Huang, C.-J.; Kang, C.-C.; Meen, T.-H. Effect of Passivation layers by liquid phase deposition (lpd) on moisture and oxygen protection for flexible organic light-emitting diode (FOLED). *Microelectron. Eng.*, **2015**, *148*, 17–20.

74. Yuan-Yu, L.; Yi-Neng, C.; Ming-Hung, T.; Ching-Chiun, W.; Feng-Yu, T. Air-stable flexible organic light-emitting diodes enabled by atomic layer deposition. *Nanotechnology*, **2015**, *26*, 024005.
75. Gaj, M. P.; Wei, A.; Fuentes-Hernandez, C.; Zhang, Y. D.; Reit, R.; Voit, W.; Marder, S. R.; Kippelen, B. Organic light-emitting diodes on shape memory polymer substrates for wearable electronics. *Org. Electron*, **2015**, *25*, 151–155.
76. Li, L.; Liang, J.; Chou, S.-Y.; Zhu, X.; Niu, X.; Yu, Z.; Pei, Q. A solution processed flexible nanocomposite electrode with efficient light extraction for organic light emitting diodes. *Sci. Rep.*, **2015**, *4*, 4307.
77. Ok, K.-H.; Kim, J.; Park, S.-R.; Kim, Y.; Lee, C.-J.; Hong, S.-J.; Kwak, M.-G.; Kim, N. Han, C. J.; Kim, J.-W. Ultra-thin and smooth transparent electrode for flexible and leakage-free organic light-emitting diodes. *Sci. Rep.*, **2015**, *5*, 9464.
78. Yuqiang, L.; Yuemin, X.; Yuan, L.; Tao, S.; Ke-Qin, Z.; Liangsheng, L.; Baoquan, S. Flexible organic light emitting diodes fabricated on biocompatible silk fibroin substrate. *Semicond. Sci. Technol.*, **2015**, *30*, 104004.
79. Morales-Masis, M.; Dauzou, F.; Jeangros, Q.; Dabirian, A.; Lifka, H.; Gierth, R.; Ruske, M.; Moet, D.; Hessler-Wyser, A.; Ballif, C. An indium-free anode for large-area flexible OLEDs: Defect-free transparent conductive zinc tin oxide. *Adv. Funct. Mater.*, **2016**, *26*, 384–392.
80. Liu, X. C.; You, J.; Xiao, Y.; Wang, S. R.; Gao, W. Z.; Peng. J. B.; Li, X. G. Film-forming hole transporting materials for high brightness flexible organic light-emitting diodes. *Dyes Pigm.*, **2016**, *125*, 36–43.
81. Bi, Y.-G.; Feng, J.; Ji, J.-H.; Chen, Y.; Liu, Y.-S.; Li, Y.-F.; Liu, Y.-F.; Zhang, X.-L.; Sun, H.-B. Ultrathin and ultrasmooth Au films as transparent electrodes in ITO-free organic light-emitting devices. *Nanoscale*, **2016**, *8*, 10010–10015.
82. Jia, S.; Sun, H. D.; Du, J. H.; Zhang, Z. K.; Zhang, D. D.; Ma, L. P.; Chen, J. S.; Ma, D. G.; Cheng, H. M.; Ren, W. C. Graphene oxide/graphene vertical heterostructure electrodes for highly efficient and flexible organic light emitting diodes. *Nanoscale*, **2016**, *8*, 10714–10723.
83. Liu, Y.-S.; Feng, J.; Ou, X.-L.; Cui, H.-F.; Xu, M.; Sun, H.-B. Ultrasmooth, highly conductive and transparent PEDOT:PSS/silver nanowire composite electrode for flexible organic light-emitting devices. *Org. Electron.*, **2016**, *31*, 247–252.
84. Jeong, E. G.; Han, Y. C.; Im, H.-G.; Bae, B.-S.; Choi, K. C. Highly reliable hybrid nano-stratified moisture barrier for encapsulating flexible OLEDs. *Org. Electron.*, **2016**, *33*, 150–155.
85. Li, X.; Yuan, X.; Shang, W.; Guan, Y.; Deng, L.; Chen, S. Lifetime improvement of organic light-emitting diodes with a butterfly wing's scale-like nanostructure as a flexible encapsulation layer. *Org. Electron.*, **2016**, *37*, 453–457.
86. Han, Y. C.; Jeong, E. G.; Kim, H.; Kwon, S.; Im, H.-G.; Bae, B.-S.; Choi, K. C. Reliable thin-film encapsulation of flexible OLEDs and enhancing their bending characteristics through mechanical analysis. *RSC Adv.*, **2016**, *6*, 40835–40843.
87. Oh, E.; Park, S.; Jeong, J.; Kang, S. J.; Lee, H.; Yi, Y. Energy level alignment at the interface of NPB/HAT-CN/graphene for flexible organic light-emitting diodes. *Chem. Phys. Lett.*, **2017**, *668*, 64–68.
88. Wang, Y. J.; Lu, J. G.; Shieh, H. P. D. Flexible efficient top-emitting organic light-emitting devices on a silk substrate. *IEEE Photonics J.*, **2016**, *8*, 1.
89. Lee, J.; An, K.; Won, P.; Ka, Y.; Hwang, H.; Moon, H.; Kwon, Y.; Hong, S.; Kim, C.; Lee, C.; Ko, S. H. A dual-scale metal nanowire network transparent conductor for highly efficient and flexible organic light emitting diodes. *Nanoscale*, **2017**, *9*, 1978–1985.
90. Jeong, E. G.; Kwon, S.; Han, J. H.; Im, H.-G.; Bae, B.-S.; Choi, K. C. Mechanically enhanced hybrid nano-stratified barrier with defect suppression mechanism for highly reliable flexible OLEDs. *Nanoscale*, **2017**, *9*, 6370–6379.

91. Kun, L.; Hu, W.; Huiying, L.; Ye, L.; Guangyong, J.; Lanlan, G.; Mazzeo, M.; Yu, D. Highly-flexible, ultra-thin, and transparent single-layer graphene/silver composite electrodes for organic light emitting diodes. *Nanotechnology*, **2017**, *28*, 315201.

92. Lee, K. M.; Fardel, R.; Zhao, L.; Arnold, C. B.; Rand, B. P. Enhanced outcoupling in flexible organic light-emitting diodes on scattering polyimide substrates. *Org. Electron.*, **2017**, *51*, 471–476.

93. Li, J.; Tao, Y.; Chen, S.; Li, H.; Chen, P.; Wei, M.-Z.; Wang, H.; Li, K.; Mazzeo, M.; Duan, Y. A flexible plasma-treated silver nanowire electrode for organic light-emitting devices. *Sci. Rep.*, **2017**, *7*, 16468.

94. Chen, Z.; Wang, H.; Wang, X.; Chen, P.; Liu, Y.; Zhao, H.; Zhao, Y.; Duan, Y. Low-temperature remote plasma enhanced atomic layer deposition of ZrO_2/zircone nanolaminate film for efficient encapsulation of flexible organic light-emitting diodes. *Sci. Rep.*, **2017**, *7*, 40061.

95. Zhou, L.; Yu, M.; Chen, X.; Nie, S.; Lai, W. Y.; Su, W.; Cui, Z.; Huang, W. Screen-printed poly(3,4-ethylenedioxythiophene): poly(styrenesulfonate) grids as ITO-free anodes for flexible organic light-emitting diodes. *Adv. Funct. Mater.*, **2018**, *28*, 1705955.

96. Lai, Y.-N.; Chang, C.-H.; Wang, P.-C.; Chu, Y.-H. Highly efficient flexible organic light-emitting diodes based on a high temperature durable mica substrate. *Org. Electron.*, **2019**, *75*, 105442.

97. Yi, F.-S.; Bi, Y.-G.; Gao, X.-M.; Wen, X.-M.; Zhang, X.-L.; Liu, Y.-F.; Yin, D.; Feng, J.; Sun, H.-B. Plasmonic ultrathin metal grid electrode induced optical outcoupling enhancement in flexible organic light-emitting device. *Org. Electron.*, **2020**, *87*, 105960.

5 Metal–Dielectric Composites for OLEDs Applications

Rajeev Kumar

5.1 INTRODUCTION: BACKGROUND AND CURRENT SCENARIO

Organic light emitting diodes (OLEDs) have attracted substantial active research in the decades since their invention in 1987.[1] OLEDs find usage in flat screens and solid-state lightning owing to their self-luminescence, energy efficiency, high mechanical flexibility and broad colour gamut.[2,3] Lately, transparent and flexible organic light emitting devices have emerged as a notable technology for foldable, curved and portable electronics. It is also required that they have superlative white light emission to deliver a colour rendering index (CRI) value higher than 80, along with correlated colour temperature (CCT) range between 6500 and 2500 K, that usually utilizes three or four colours (red, green, blue and/or yellow) to generate the full emission spectrum rather than just two complementary colours.[3] One major challenge for their development is obtaining transparent conductive electrodes (TCE) with reasonably high optical, mechanical and electrical attributes.[2] In older devices glass substrates were used primarily, but with new-generation light-weight devices and the requirement of flexibility, they are being replaced by plastic substrates.[4] It is now well known that conductive metals such as Ag, Cu, Au or Al can be made transparent by keeping the thickness between 5 and 15 nm. Concerning avoidance of high reflectance in the visible-light regime, semiconductors can be used, although they generally have lower charge carrier density than metals. In-doped SnO_2 (ITO) is the conventional transparent conductive oxide (band gap 3.7 eV), owing to its low resistivity (sheet resistance <10 Ω/square) with high light transmission (>80%) on glass substrates, but is limited in its usage on flexible substrates (e.g. PET) due to its brittle nature and restricted light out coupling efficiency from acute trapping loss of interiorly emitted photons.[2,5] ITO rather has high sheet resistance on PET (>40 Ω sq^{-1}), possibly due to lower substrate temperature during deposition. Upon bending, ITO develops cracks, which rapidly increase the resistance. Furthermore, the demand for ITO-free devices has escalated in recent years because ITO contains an expensive and rare element (indium).[4,5] ITO contains 74% indium, which is sporadic and generally extracted in very low concentrations as a by-product of Zn mining.[4]

In the OLEDs, the valence band pertains to the highest occupied molecular orbital (HOMO) level, whereas the conduction band pertains to the lowest unoccupied molecular orbital (LUMO) level. Conventionally, the OLED consists of an organic

DOI: 10.1201/9781003260417-5

semiconductor (OSC) which is squeezed between two electrodes (Figure 5.1). The OSC, anode and cathode electrodes lie within the flat band regime. In this context, the flat band regime represents the flattened region of LUMO and HOMO under a fairly small applied forward bias. The bias leads to an increase in potential at the cathode side, followed by the positioning of the energy levels within the device (including LUMO and HOMO), forming a triangular shape. It should be noted that the work functions of the components don't change. The electrons and holes can tunnel into the OSC through this triangle shaped energy barrier. The injected electrons and holes move closer inside the active region within the electric field gradient and the coulombic interactions generate an exciton.[6] From this exciton state the electrons and holes recombine and emit light. The photon energy of the emitted light is proportional to the band gap.

The features of a current-voltage-luminance (J-V-L) plot of an OLED provide details of the rectification of the device. It is desired to have an exponential increase in current from turn-on voltage under a forward bias. Conversely, it should be as low as possible under reverse bias and also prior to turn-on voltage. Any deviations indicate shunts in the light emitting layer. A small turn-on voltage is indicative of favourable energy band-matching of the various component layers in the device emanating in competent charge carrier inoculation. The efficiency of a device is evaluated by dividing values of luminance by current density. More precisely, device efficiency is elucidated by power efficiency, which also considers the operating bias of the device.

Highly efficient OLEDs show internal quantum efficiency (IQE) close to 100% (due to radiative recombination of the singlet and triplet types of excitons that are formed during electrical injection) but the external quantum efficiency (EQE) remains restricted to ~20%–30% in standard device architecture.[2,7] The connections between the organic layers and electrodes are being optimized either with various doping strategies or by introducing buffer layers at the interfaces to realize barrier-free carrier paths, which helps to minimize the ohmic losses and keeps a low operating voltage.[3] The main reason for the low EQE originates in the ill-matched refractive indices of the organic section (n ≈ 1.6–1.8), transparent conductive oxide (e.g., In-doped SnO_2, n ≈ 1.9–2.1) and the substrate (e.g., plastic, n ≈ 1.5).[2] The EQE can be boosted considerably by using various out coupling strategies, such as using diffraction grates, arrays of micro lens and coupling layers with either high or low refractive indices.[3] Therefore, it is pertinent to develop alternative flexible TCEs using

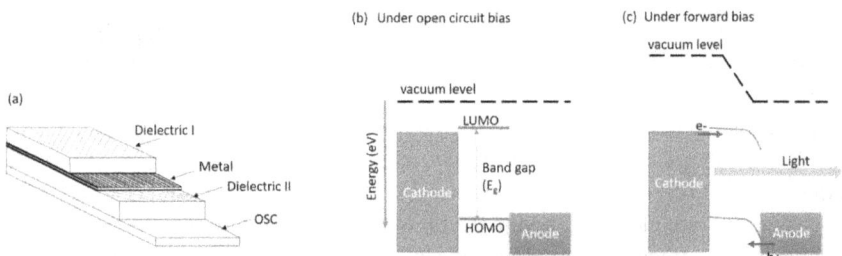

FIGURE 5.1 (a) Typical dielectric/metal/dielectric device architecture. Energy levels (HOMO: highest occupied molecular orbital, LUMO: lowest unoccupied molecular orbital) of regular OLED architecture under (b) open circuit bias and (c) forward bias.

materials such as metal-based nanostructures (e.g., conducting oxides, nanowires, mesh, topological insulators, metal-dielectric multilayers), carbonaceous materials (e.g., nanotubes, graphene, conductive polymers) and metal-dielectric composite electrodes (MDCEs).[2] The Haacke figure of merit (FOM), φ_{TC} considers both transmittance and sheet resistance and is evaluated as $\varphi_{TC} = (T^{10}_{av}/R_s)$, where T^{10}_{av} is the averaged transmittance and R_s denotes the sheet resistance.

MDCEs have been demonstrated as excellent alternatives to ITO owing to their relatively higher performance with respect to energy efficiency and nominal angular dependence on the emission spectrum.[2] Metal-dielectric nanostructures offer high mechanical durability, optical transparency, low electrical resistivity, and scalable film uniformity, thereby appearing to be ideal candidates to replace ITO.[2] Furthermore, such structures can be fabricated by simple joule effect thermal evaporation and thus do not deteriorate the embodied organic layer.[4,8] The characteristics of the dielectric/metal/dielectric architecture (Figure 5.1a) are dependent on the width of the respective layers. Generally, the threshold width/thickness of the metallic film is ~10 nm, a value where the composite material changes its insulating behaviour to a highly conductive one.[4] But, practically, it's extremely difficult to form an ultrathin sandwiched metallic film of thickness ≤10 nm in an MDCE, arising from dewetting problem with a secluded grainy morphology that results in acute deterioration of electrical conductivity and ocular transmittance.[2] Moreover, a microcavity effect is associated with a flat configuration in MDCE, that leads to emissive features that are spectral and angular dependent.[2] Thus, to obtain a thin metal film having homogeneous morphology, several strategies are employed, namely, the use of nucleation inducers such as metal seeds, metal oxides, organic moieties, etc., or metallic alloys, or thicker film. Silver (Ag), which has the smallest value of refractive index along with the shallowest optical absorption among metal films in the visible-light region, is preferred.[2] Although simply making a thicker film can seemingly overcome the issue of homogeneity, it compromises optical transparency.[8] To mitigate these issues, one possible solution suggested is the direct integration of photonic structures into the MDCE layer, thereby reducing optical loss while still retaining the colour quality.[2] Among the dielectrics, TiO_2, WO_3, MoO_3, F-doped SnO_2(FTO), Al/Ga-doped ZnO (AZO), V_2O_5, ZnS, etc., are commonly used.

5.2 DEVICE ARCHITECTURE AND CONFIGURATION

A schematic depicting the device architecture in OLEDs is shown in Figure 5.2. The literature identifies two types of OLED architecture: regular and inverted. The configuration in which the basal electrode forms the anode and upper electrode represents the cathode is known as "regular architecture" (Figure 5.3a). Conversely, when the basal electrode becomes the cathode and the upper electrode is transformed into the anode, it is known as an "inverted architecture" (Figure 5.3b). In conventional bottom-emitting OLEDs with regular architecture, the polymer PEDOT:PSS (a hole injector) is usually spin-coated on the TCO-coated glass substrate, which forms the anode, whereas the top cathode comprises Al/Ca (electron injector). Within the organic semiconductor (OSC) region, the electrons and holes recombine to emit light. The details of the mechanism and energy levels can be understood from Figure 5.3.

(a) Regular architecture   (b) Inverted architecture

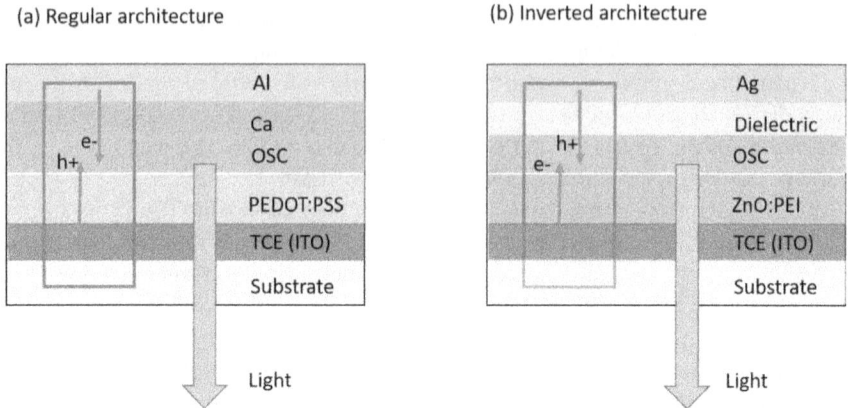

FIGURE 5.2 Schematic representing the bottom-emitting OLED formed by (a) regular architecture and (b) inverted architecture.

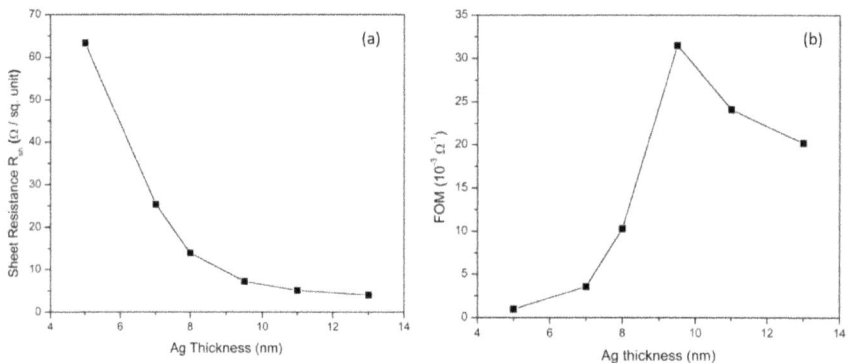

FIGURE 5.3 (a) Variation of the sheet resistance in niobium oxide/silver/niobium oxide multilayered films with respect to the silver film thickness. (b) Variation in the figure of merit of the multilayered films versus the silver film thickness. (Adapted from Dhar et al.[11] with permission.)

The regular architecture suffers from certain drawbacks. A commercial polymeric solution of PEDOT:PSS has a pH ranging from 1.5–2.5. Therefore, the underlying coating on the electrode should endure the acidic character. Furthermore, the aqueous PEDOT:PSS solution deteriorates over time, losing its conductivity. The use of Ca on the electron injector side poses degradation issues since it is typically a low-work-function reactive metal.

In contrast, for a bottom-emitting inverted architecture, the hole and electron layers are interchanged. The top electrode serves as the hole injector, while the bottom electrode performs the role of electron injector. Therefore, in such cases materials with low work functions (e.g., ZnO nanoparticles) are used instead of PEDOT:PSS. Furthermore, the solution of ZnO nanoparticles can also be variegated with polyethylenimine (PEI) prior to spin coating to boost the electron injection process in the organic semiconductor layer.[9,10] For the hole injection, a material with high work function such as MoO_x is smeared on top of the OSC.

The inverted architecture is advantageous compared to the regular architecture. Since the OSC is sandwiched between two oxide layers, it is impervious to oxygen, which enhances device stability. Along with an easier solution processability of ZnONP:PEI in IPA compared to PEDOT:PSS, the inverted structure also is analogous to the state-of-the-art active matrix technology.[11] There several recent reports on D/M/D-based OLED devices. A few of these are discussed in the following section.

5.3 SOME EXAMPLES OF METAL-DIELECTRIC ELECTRODE SYSTEMS FOR OLEDs

Dhar et al.[11] deposited $Nb_2O5/Ag/Nb_2O_5$ multilayer heterostructures aboard flexible polyethylene naphthalate substrates (125 μm thickness) by radio frequency sputtering of an Nb_2O_5 target and direct current sputtering of a silver metal target at ambient temperature. Nb_2O_5 dielectric film is highly transparent in the visible range. It is also highly stable in both acidic and alkaline media. In this work, different multilayer structures were fabricated on flexible substrates via sputtering at ambient temperature. They varied the Ag layer thickness and obtained a critical (optimal) thickness of 9.5 nm to form a ceaseless conducting layer. The optimized films had an effective Hall resistivity of 6.44×10^{-5} Ω cm^{-1}. The obtained values of carrier (electron/hole) concentration and carrier mobility were 7.4×10^{21} cm^{-3} and 13.1 cm^2 V^{-1} s^{-1}, respectively.[11] Further, the optimized multilayer stack showed a sheet resistance of only 7.2 Ω/square (Figure 5.3a) and provided a transmittance of 86% corresponding to the wavelength of 550 nm with no substrate warming or annealing procedure. The figure of merit (Haacke) for the device is remarkably high at 31.5×10^{-3} Ω^{-1} (Figure 5.3b).

Zhao et al.[12] prepared a transparent composite electrode (TCE) consisting of titanium dioxide/silver/titanium dioxide on glass at ambient temperature using sputtering at a constant pressure of 8×10^{-7} torr. TiO_2 layers were cast using RF sputtering whereas the sandwiched layer of silver metal was laid out via DC sputtering. They methodically evaluated the roles of deposition rate and thickness of the silver and titania layers on the optoelectronic properties of the transparent conductive electrode. A sufficiently high deposition rate led to formation of fewer voids, resulting in better transmittance and improved conductivity. The fabricated organic solar cell using the optimized multilayer (~42 nm TiO_2 and ~10 nm Ag layers) and P_3HT:$C_{61}BM$ showed a Haacke figure of merit two times higher than an ITO based device (Figure 5.4). The power efficiency of this device architecture was also ~100% higher than that fabricated on ITO.

Xu et al.[2] fabricated an efficient nanostructured metal dielectric composite electrode (NMDCE) for efficacious flexible OLEDs on plastic substrates with low refractive index (Figure 5.5). They utilized an ultrathin film of Ca:Ag alloy (1:1 in weight ratio) of ~8 nm thickness, to improve film wettability and homogenous growth, resulting in low resistance and low optical loss. For this purpose, a sophisticated strategy of nucleation-inducing seed layer, that is, 1 nm thick aluminum (Al) was utilized and finally covering the alloy layer by metal co-deposition (pure Ag layer of 1 nm thickness). Quasi-random nanostructures were incorporated in the device

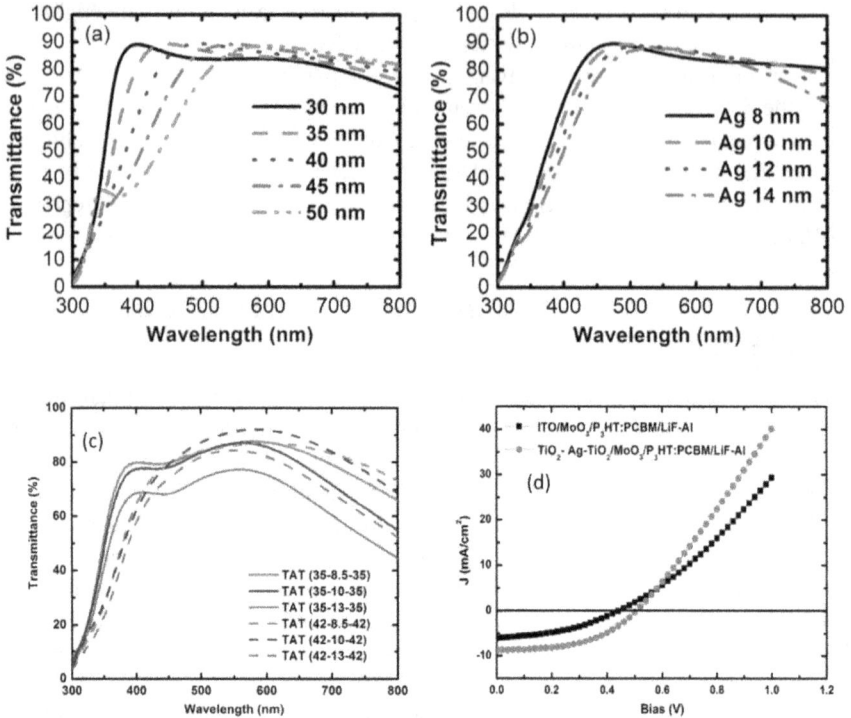

FIGURE 5.4 (a) The simulated plot of transmittance from titanium dioxide/silver (10 nm)/ titanium dioxide with variation of TiO$_2$ thicknesses. (b) The simulated plot of transmittance of a device with TiO$_2$ (thickness of 40 nm) with varying silver layer thickness. (c) The optical transmittance spectra for the device on glass with respect to thickness of titanium dioxide and silver layers. Here, glass is used as reference. (d) J-V evaluation of organic photovoltaics on ITO and titanium dioxide/silver/titanium dioxide electrodes. (Adapted from Zhao et al.[12] with permission.)

directly by engraving a layer of index matched UV-resin on the PET that successfully removed the optical microcavity effect. This enables out coupling amplification of the waveguide light which is broadband and angle-independent, with greatly reduced loss from surface plasmons.[2] This configuration provides a uniform optical transmittance of ~ 86% (on average) over the entire visible region. The fabricated white OLED generated a light out coupling efficiency greater than 2.4 times compared to a standard indium-doped SnO$_2$-based device, resulting in superlative EQE and power efficiencies of ~47% and ~112 lm W^{-1}, respectively.[2]

Cattin et al.[13] prepared and characterized MoO$_3$/Ag/MoO$_3$ structures, with Ag thickness between 5 and 15 nm, and then explored them as anodes for indium-free organic solar cells. Unlike commercial ITO, ZnO and SnO$_2$-based transparent conducting oxides, where the deposition process involves sputtering with or without annealing, which adversely affects the performance of the organic devices due to damage from these processes, thermally evaporable, highly conductive, semitransparent electrodes, deposited on substrates at ambient temperature, were developed

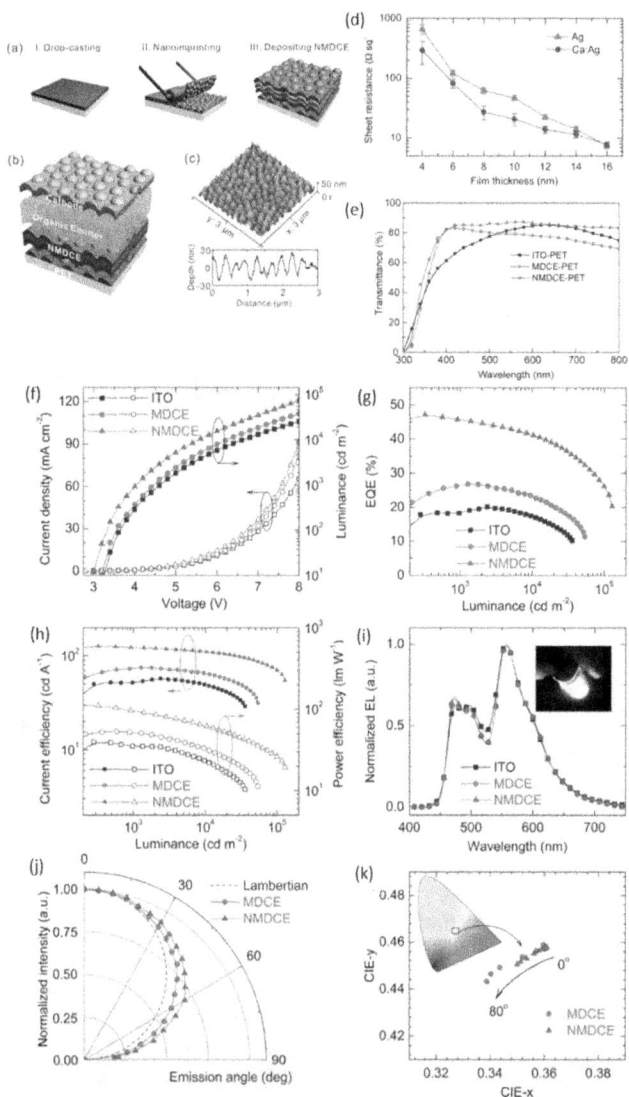

FIGURE 5.5 (a) Schematic illustrating the fabrication procedure for a nanostructured metal dielectric composite electrode (NMDCE) on polyethylene terephthalate (PET). (b) Flexible OLED architecture on PET. (c) Atomic force microscope imaging and surface profiling of the NMDCE layer. (d) Sheet resistance with respect to film thickness of only silver and Ca/Ag alloy cast on the molybdenum oxide/PET. (e) Optical transmittance of PET using various electrodes. Performance attributes of white-light-emitting flexible OLEDs. (f) Variation of current density and luminance with respect to driving voltage. (g) Exterior quantum efficiency (EQE) with respect to luminance. (h) Current efficiency and power efficiency with respect to luminance. (i) Normalized electroluminescence spectra at an intensity of 1000 cd m^{-2}. Inset shows a picture of a white emission flexible OLED. (j) Reliance of the emission angle intensity on normalized angle. Lambertian emission plot is shown as a dashed line. (k) CIE coordinates of the OLEDs at viewing angles between 0 and 80°. (Adapted from Xu et al.[2] with permission.)

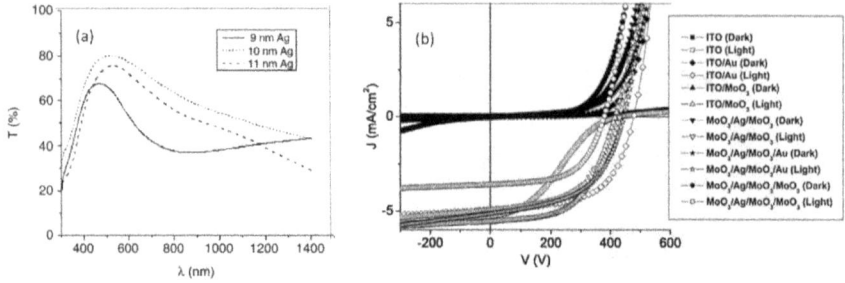

FIGURE 5.6 (a) Optical transmittance of the MoO_3/silver/MoO_3 heterostructures. (b) Typical J–V plots of the device, in dark (solid symbols) and at irradiance of AM 1.5 solar simulation (100 mW cm^{-2}) (hollow symbols). (Adapted from Cattin et al.[13] with permission.)

to overcome these limitations. They demonstrated that the characteristics of these structures are determined by the thickness of the silver interlayer, with a threshold of 10 nm. At this thickness the interlayer transforms from an insulating to a conductive one. The structure had a resistivity value of 8×10^{-5} Ω cm with an optical transmittance of ~80%, corresponding to the wavelength of 600 nm.[13] They attributed this transportation to the percolation pathway from silver. Furthermore, the transmittance increased when the Ag layer became thicker, from 8 to 10 nm, but additional increment resulted in lowering of transmittance. This phenomenon was ascribed to surface plasmon resonance. Therefore, OLED shaving highest fill factors were obtained at an optimal Ag layer thickness of 10 nm (Figure 5.6).

The performances of these optimized heterostructures allowed exploring their utility as anodes in organic photovoltaics. Incorporation of a thin molybdenum oxide layer of 3 nm thickness at the interface of the anode (In-doped SnO_2) and electron donor (copper phthalocyanine) allowed reduction of the energy barrier originating from the disparity between their work functions, allowing a superior hole-transfer efficiency between the copper phthalocyanine molecules and ITO. They compared the MoO_3/Ag/MoO_3 and ITO/MoO_3 solar cells that were fabricated in a similar manner. The devices exhibited similar efficiencies. The D/M/D hetero structures were fabricated on glass (at ambient temperature) using the thermal evaporation system. The multilayered films were consecutively cast onto the glass without breaking the vacuum. For this purpose, separate tungsten crucibles loaded with molybdenum oxide powder and silver wire were used. During the optimization of the performance parameters, the rate of deposition of the molybdenum oxide layers was set as 0.05 nm s^{-1} and that of Ag was set as 0.15 nm s^{-1}. The width of the molybdenum oxide layers was set at 45 nm and 37.5 nm for the bottom and the top layers, respectively, while the silver layer was changed from 5 nm to 12 nm. Copper phthalocyanine, fullerene and tris(8-hydroxyquinolinato)aluminium were deposited in a vacuum. The thicknesses were 35 nm in the case of copper phthalocyanine, 40 nm in the case of fullerene and 9 nm for Alq$_3$. After deposition of the organic layer, the Al upper electrodes were evaporated thermally, without breaking the vacuum, by using a mask having an active area of 2 mm × 8 mm. Then, a moisture- and oxygen-protective sheath of amorphous Se (~100 nm) was deposited via thermal evaporation without a vacuum break.

Makha et al.[14] fabricated MAM (MoO_3/silver/MoO_3) multilayered heterostructures and used them as anodes in copper phthalocyanine-based organic photovoltaic cells. The average transmittance (in the visible region) of these multilayered heterostructures was found to be 70% ± 2% with quite low sheet resistance of 3.5 ± 1.0 Ω/square. With the use of these multilayered heterostructures as the anode layer, the power conversion efficiency of the organic photovoltaics was ~1%. By incorporation of a 3 nm thin copper iodide film at the interface of the D/M/D anode and the organic layer this value increased to 50% (Figure 5.7). This improvement was ascribed to the template effect of copper iodide on the phthalocyanine molecules. The bottom and top MoO_3 layers were 20 and 35 nm, respectively while the silver layer was 10 nm. The thickness of the copper iodide layer was 3 nm, while the phthalocyanine layer was 35 nm, fullerene was 40 nm and bathocuproine (BCP) was 9 nm in thickness. The individual cell had an effective area of 0.16 cm^2. Thus, the highest value of open circuit voltage was achieved when the copper iodide buffer layer recovers the anode, which was understood by the orientation of copper phthalocyanine molecules induced by copper iodide. In this case, it was shown that the HOMO of the organic layer was adjusted by the orientation of the molecules. For copper phthalocyanine, the HOMO varies from 4.75 to 5.15 eV when the direction of the molecules is modified from being planarly perpendicular to the substrate to being parallel to it. Therefore, a higher value of open circuit voltage was obtained with CuI than with MoO_3 as the anode buffer layer could be assigned to the bigger difference of LUMO (electron acceptor)-HOMO (electron donor), when the phthalocyanine molecules are parallel to the copper iodide and substrate and perpendicular to the molybdenum oxide layer. Similarly, for the D/M/D anode, the increment in current density at zero voltage measured when the anode is masked with copper iodide can be ascribed to the higher absorption of the copper phthalocyanine films deposited on copper iodide, this increment being ascribed to the moderation by the direction of the organic molecules. Regarding the anode masked by copper iodide, an ITO anode shows higher current density due to its higher transmittance, while a D/M/D anode with high conductivity leads to a higher fill factor (FF) value. The

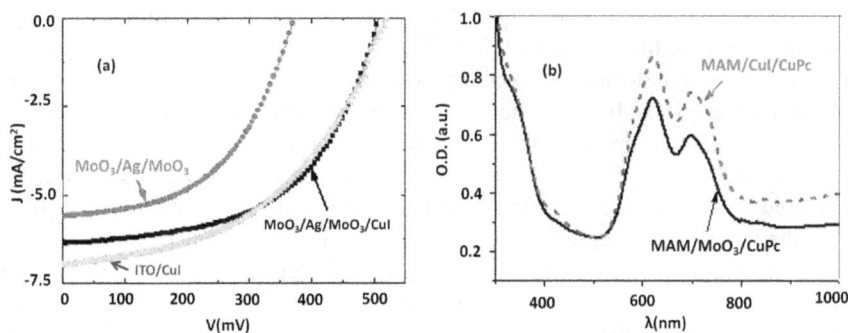

FIGURE 5.7 (a) Typical J-V plots, under AM 1.5, of the cells deposited concurrently, but with various anodes. (b) Optical density plots of copper phthalocyanine thin films deposited on ITO/molybdenum oxide (—) and ITO/copper iodide (---). (Adapted from Makha et al.[14] with permission.)

variable performances observed with cells using copper iodide modified anodes with varying silver film thickness can be traced to the difference in conductivity of the MAM structures.

Yu et al.[15] fabricated ITO-free organic photovoltaics using $WO_3/Ag/WO_3$. Compared to the ITO-based device, the D/M/D based device having an active layer 70 nm thick showed a 27.4% enhancement in efficiency but when its thickness increased to 150 nm, the efficiency was reduced by 19.9% (Figure 5.8). From the IPCE spectral analysis and transfer matrix simulations it was divulged that the optical resonance effect derived from the metal-mirror-microcavity established between the partially transparent silver layer and the p-aluminium electrode helped in improving the efficiency of the device with a 70 nm thick active layer.

For the device having a thinner active layer, the microcavity effect played a major role owing to the low absorption coefficient, but for a thicker active layer this enhancement at short wavelengths was significantly lower and couldn't compensate for the lowering of current density (J_{sc}) within the spectral range from 470 to 650 nm, that is, the lowered efficiency of the device with thicker active layer originated from the off-resonance smothering of the metal-mirror-microcavity.[15] Thus, it is an important component to consider when the D/M/D multilayer is utilized to replace the In-doped SnO_2 electrode as a transparent conductive electrode. In particular, a microcavity effect is formed due to high reflection between two metallic electrodes, which could prompt an augmentation in absorption when the active layer is weakly absorbing.

Cao et al.[16] prepared a MoO_3/thin metal/MoO_3 trilayered heterostructure to replace the In-doped SnO_2 electrode in organic solar cells. The optoelectronic characteristics of the trilayered structure were a function of the materials' nature and thickness of the sandwiched metal layer. Devices based on the polymers P3HT and PCBM delivered a power conversion efficiency up to 2.5%, whereas In-doped SnO_2 devices showed a power conversion efficiency of 3.1% under simulated AM 1.5 solar irradiation. The trilayered electrode was cast via simple thermal evaporation under vacuum, and thus was completely congruent with organic active layers and plastic substrates. The trilayered electrode not only exhibited optoelectronic properties similar to In-doped SnO_2 but also showed excellent flexibility, making it feasible to prepare the device with D/M/D electrodes on flexible substrates. These In-doped SnO_2-free devices showed a power efficiency (η_p) of 2.5% and 2.0% for gold and silver as sandwiched metal layer, respectively, under simulated AM 1.5 solar radiation. In comparison the ITO-based devices showed η_p of 3.1%. Furthermore, this device on a PET substrate showed nearly similar efficiency ($\eta_p = 2.4\%$) compared to a rigid glass substrate-based device. Furthermore, due to the flexibility of the trilayered structure, these devices demonstrated fair mechanical properties. Even with a bending radius of 1.3 cm, the efficiency of this device was only undermined by 6% from its initial value after 500 bending cycles. Therefore, the performance of these flexible devices suggests that this D/M/D electrode is an apt replacement for ITO.

A multilayered film comprising a tin oxide/gold/tin oxide transparent conductive oxide (TCO) was developed by Sharma et al.[17] using both e-beam and thermal evaporation techniques. The SnO_x layer had oxygen deficiencies with the generation of

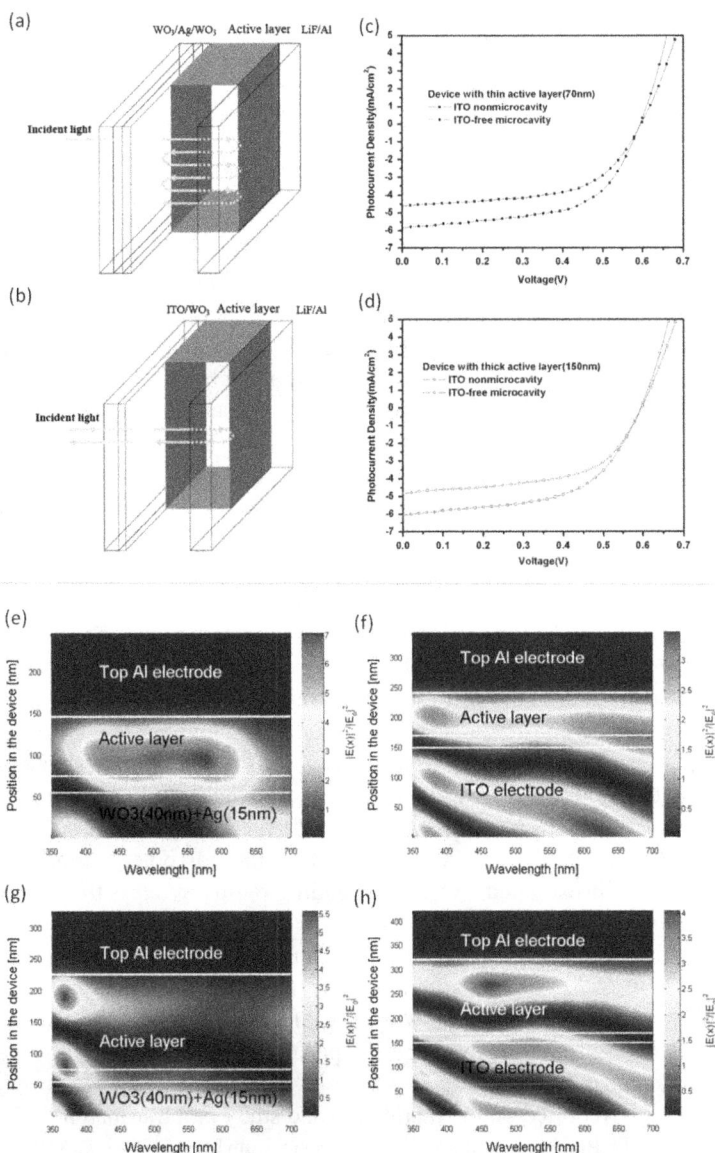

FIGURE 5.8 (a) Schematic of the In-doped SnO₂-free devices based on tungsten oxide/silver/ tungsten oxide. (b) Schematic of the standard structure with In-doped SnO₂ and tungsten oxide. J-V plots of In-doped SnO₂-free microcavity devices and In-doped tin oxidenon-microcavity devices under illumination of 100 mW cm⁻² simulated at AM 1.5 G in air with (c) thinner active layer and (d) thicker active layer. Allocation of normalized modulus square of the optical electric field: (e) for In-doped tin oxide-free microcavity device with a thin active layer; (f) for In-doped tin oxide nonmicrocavity device with a relatively thin active layer; (g) for In-doped tin oxide-free microcavity device with a relatively thick active layer and (h) for In-doped tin oxide nonmicrocavity device with a relatively thick active layer. $|E_0|^2$ is the modulus of the square of the electric field for the incoming light. (Adapted from Yu et al.[15] with permission.)

a SnO-rich phase. This specimen showed a sheet resistance and resistivity of 52 Ω/square and 3.9×10^{-4} Ω cm, respectively, with a transparency of 83% above 475 nm. The outstanding performance manifested by this multilayered stack was correlated with the low oxygen concentration in the SnO_x part, which can serve as a necessary constituent to increase the total electrical conductivity in multilayered thin films.

Li et al.[18] found that a thin layer of molybdenum oxide has a profound effect on the characteristics of the device by instigating a synergistic effect from suppression of the nonradiative surface plasmon polariton modes and modulation of the cavity resonance. This mode is suppressed from 400 nm wavelength to the 520 nm wavelength range in the vicinity of the molybdenum oxide layer. Furthermore, incorporation of a thin molybdenum oxide layer reduced the reflectance of the metal electrode, thereby leading to a lower destructive resonance in the device. Amalgamation of these effects resulted in highly efficient OLEDs, where EQE and luminous efficiency of two-component stacked white light emitting OLEDs reached 38.8% and 53.9 lm W^{-1} at an illumination of 1000 cd m^{-2}, respectively.[18] The efficiency was about 1.9 times greater than for the device excluding the molybdenum oxide layer.

Lu et al.[19] studied semitransparent electrodes with the oxide/metal-oxide/oxide configuration as follows: substrate/MoO_3/Al/MoO_3, for use as the anode in OLEDs, which was fabricated via the thermal evaporation technique. The transmittance from the metal layer was improved by casting molybdenum oxide and sandwiched aluminum layer. The thickness of the metal thin films was optimized to 15 nm for good transmittance and low resistivity. The films showed a low sheet resistance of 7 Ω/square and a high optical transmittance of 70% at 550 nm wavelength. The fabricated In-doped SnO_2-free OLEDs on a glass delivered high luminance and current efficiency of 21,750 cd/m^2 and 3.18 cd/A, respectively.[19] Furthermore, the effects of bending of the devices fabricated on the polyethersulfone (PES) substrate were also investigated. When the bending radius was less than 1 cm, cracks developed on the surface of the materials. The metal oxide masking the aluminum layer modified the electrode surface and enhanced the overall durability of the device. The surface roughness of the trilayered films was also lower than that of the bilayered films. Thus, this work demonstrates that OLEDs with an oxide/metal-oxide/oxide anode show superior performance than those with a bilayer film-based anode.

Kinner et al.[20] reported on the fabrication and study of a transparent D/M/D electrode on flexible PET substrate with polymeric interlayers. The D/M/D electrode developed using TiO_x, silver and Al-doped ZnO outperformed the cutting-edge transparent electrodes fabricated similarly on PET. It delivered an average transmittance >85% in the visible light region with a low sheet resistance less than 6 Ω/square. Having a broad spectral range in the visible region, the optical transmittance of the D/M/D electrode was much higher than the transmittance from bare PET. The superior performance was warranted by the use of an additional polymer interlayer between the substrate and the D/M/D electrode. This interlayer modified the formation of the titanium dioxide nanoparticles on the PET, analogous to its development on glass. It dramatically enhanced the optical transmittance of the D/M/D electrode and concurrently lowered the sheet resistance.

Aiming to capitalize on advantages from organic materials, Han et al.[21] examined transparent electrodes fabricated using a ZnS/Ag/WO$_3$ (ZAW) multilayered structure that led to a highly flexible and In-doped SnO$_2$-free organic photovoltaic. Exemplary J-V characteristics with fill factors up to 0.70 were acquired by using WO$_3$ as a typical dielectric layer that also functioned as an anode buffer layer. A trial on the flexible nature of these devices was also carried out, and clearly demonstrated that these multilayered transparent electrodes were effective alternatives to In-doped SnO$_2$.

Although most of the works discussed above are based on Ag and Au interlayers, which are expensive, fabrication of Cu-based transparent D/M/D electrodes is more likely to grant truly affordable photovoltaics. Copper has a larger extinction coefficient than silver in the visible light region and, thus, it is difficult to yield a power conversion efficiency equivalent to conventional devices. Nevertheless, a wide tunable transmission and reflection range of the D/M/D electrodes can be used to intensify the microcavity effect and maximize the photon absorption within the active layers, thereby neutralizing the relatively high absorption from the copper layer. A case study is presented here.

Cu-based multilayered transparent electrodes comprising a D/M/D architecture were investigated by Lim et al.[22] as a possible replacement for In-doped SnO$_2$ electrodes. The researchers were aiming for truly low-cost organic photovoltaics (OPVs) (Figure 5.9). The exterior dielectric layer was demonstrated not only to significantly modulate the optical characteristics of the electrodes but also to enhance the electrical properties of the copper layer, thus generating both high optical transmission and high conductivity. They show that optimized cells with the copper-containing electrodes could yield power conversion efficiency up to 87% of that of In-doped SnO$_2$-based standard devices, in spite of the high optical absorption by copper in the visible light region. A multilayered transparent electrode based on zinc sulfide/copper/tungsten oxide was developed as an economical, In-doped SnO$_2$-free OPV. This study disclosed that zinc sulfide served as a capping layer that enhanced the optical absorption. It also acted as a seed layer that rendered a uniform growth of thin copper layer and thereby simultaneously enhanced their optical transmittance and lowered sheet resistance. After careful optimization, they realized In-doped SnO$_2$-free OPVs with power conversion efficiency of 3.4%, which corresponds to 87% of In-doped SnO$_2$-based devices. This study showed that the D/M/D electrode efficiently suppressed the parasitic absorption of the copper layer, thus emancipating the loss of incident photons to a satisfactory level. As the cost of copper is meagre compared to indium or silver, this work provides an efficient route for the accomplishment of OPVs and OLEDs with a practically healthy cost-performance balance.

5.4   CONCLUSION

Metal-dielectric composite electrodes offer distinct advantages over ITO electrodes with respect to higher efficiency, figure of merit and cost. Several reports have demonstrated higher transmittance and mechanical stability than ITO, which is critical for future OLED applications.

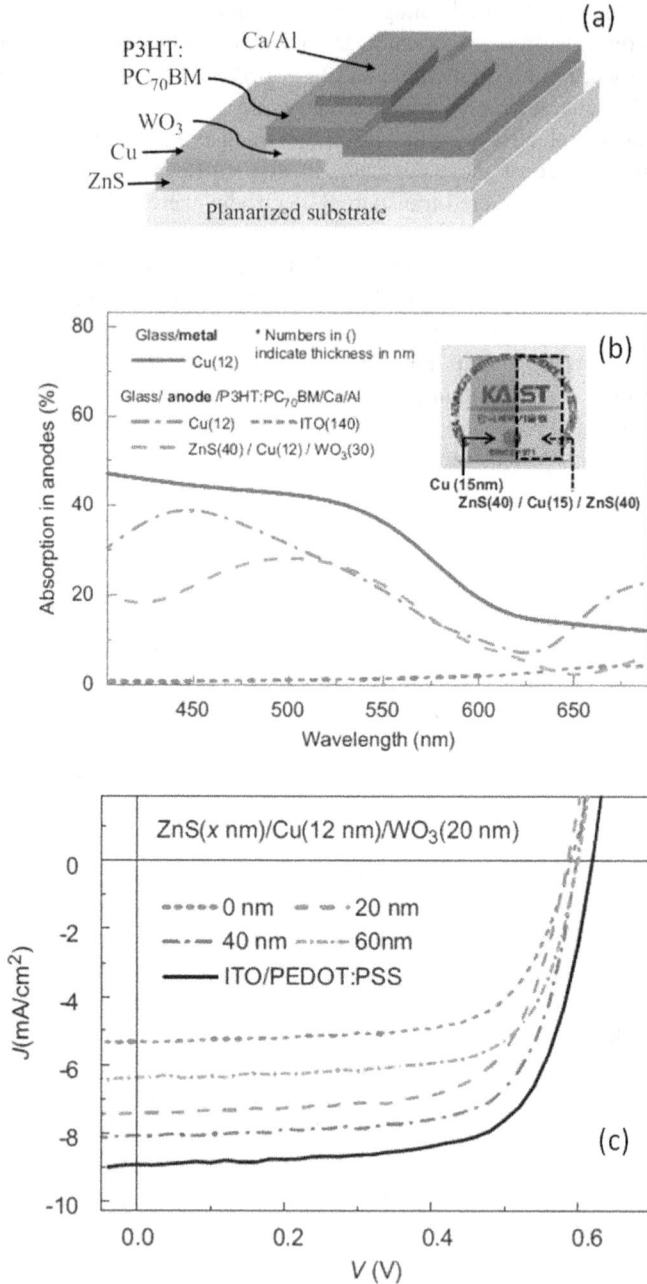

FIGURE 5.9 (a) Schematic of the In-doped SnO$_2$-free device based on zinc sulfide/copper/tungsten oxide multilayered transparent electrodes. (b) Plots of parasitic absorption in the anodes. Inset: An image of a device with a layer of copper only in the left half and a multilayer in the right half. (c) J-V characteristics of the organic photovoltaic. (Adapted from Lim et al.[22] with permission.)

REFERENCES

(1) Tang, C. W.; Vanslyke, S. A. Organic Electroluminescent Diodes. *Appl. Phys. Lett.* **1987**, *51* (12), 913–915. https://doi.org/10.1063/1.98799.

(2) Xu, L. H.; Ou, Q. D.; Li, Y. Q.; Zhang, Y. B.; Zhao, X. D.; Xiang, H. Y.; Chen, J. De; Zhou, L.; Lee, S. T.; Tang, J. X. Microcavity-Free Broadband Light Outcoupling Enhancement in Flexible Organic Light-Emitting Diodes with Nanostructured Transparent Metal-Dielectric Composite Electrodes. *ACS Nano.* **2016**, *10* (1), 1625–1632. https://doi.org/10.1021/acsnano.5b07302.

(3) Xiang, H. Y.; Li, Y. Q.; Meng, S. S.; Lee, C. S.; Chen, L. Sen; Tang, J. X. Extremely Efficient Transparent Flexible Organic Light-Emitting Diodes with Nanostructured Composite Electrodes. *Adv. Opt. Mater.* **2018**, *6* (21), 1800831. https://doi.org/10.1002/adom.201800831.

(4) Cattin, L.; Bernède, J. C.; Morsli, M. Toward Indium-Free Optoelectronic Devices: Dielectric/Metal/Dielectric Alternative Transparent Conductive Electrode in Organic Photovoltaic Cells. *Phys. Status Solidi Appl. Mater. Sci.* **2013**, *210* (6), 1047–1061. https://doi.org/10.1002/pssa.201228089.

(5) Najafi-Ashtiani, H.; Akhavan, B.; Jing, F.; Bilek, M. M. Transparent Conductive Dielectric-Metal-Dielectric Structures for Electrochromic Applications Fabricated by High-Power Impulse Magnetron Sputtering. *ACS Appl. Mater. Interfaces.* **2019**, *11* (16), 14871–14881. https://doi.org/10.1021/acsami.9b00191.

(6) Pope, M.; Swenberg, C. E. Electronic Processes in Organic Solids. *Annu. Rev. Phys. Chem.* **1984**, *35* (1), 613–655.

(7) Bae, B. H.; Jun, S.; Kwon, M. S.; Ju, B. K. Highly Efficient Flexible OLEDs Based on Double-Sided Nano-Dimpled Substrate (PVB) with Embedded AgNWs and TiO_2 Nanoparticle for Internal and External Light Extraction. *Opt. Mater.* **2019**, *92* (April), 87–94. https://doi.org/10.1016/j.optmat.2019.04.007.

(8) Huang, Q.; Liu, X.; Yan, H.; Jiao, Z.; Yang, J.; Yuan, G. P-191: Dielectric-Metal-Dielectric Structure and Its Application as Top Cathode in Highly Efficient Top-Emitting WOLEDs. *SID Symp. Dig. Tech. Pap.* **2019**, *50* (1), 1946–1949. https://doi.org/https://doi.org/10.1002/sdtp.13347.

(9) Kaçar, R.; Mucur, S. P.; Yıldız, F.; Dabak, S.; Tekin, E. Highly Efficient Inverted Organic Light Emitting Diodes by Inserting a Zinc Oxide/Polyethyleneimine (ZnO:PEI) Nano-Composite Interfacial Layer. *Nanotechnology.* **2017**, *28*, 245204. https://doi.org/10.1088/1361-6528/aa6f55.

(10) Zhou, Y.; Fuentes-Hernandez, C.; Shim, J.; Meyer, J.; Giordano, A. J.; Li, H.; Winget, P.; Papadopoulos, T.; Cheun, H.; Kim, J.; Fenoll, M.; Dindar, A.; Haske, W.; Najafabadi, E.; Khan, T. M.; Sojoudi, H.; Barlow, S.; Graham, S.; Brédas, J.; Marder, S. R.; Kahn, A.; Kippelen, B. A Universal Method to Produce Low–Work Function Electrodes for Organic Electronics. *Science.* **2012**, *336*, 327–332. https://doi.org/10.1126/science.1218829 A.

(11) Dhar, A.; Alford, T. L. Optimization of Nb_2O_5/Ag/Nb_2O_5 Multilayers as Transparent Composite Electrode on Flexible Substrate with High Figure of Merit. *J. Appl. Phys.* **2012**, *112* (10), 103113. https://doi.org/10.1063/1.4767662.

(12) Zhao, Z.; Alford, T. L. The Optimal TiO_2/Ag/TiO_2 Electrode for Organic Solar Cell Application with High Device-Specific Haacke Figure of Merit. *Sol. Energy Mater. Sol. Cells.* **2016**, *157*, 599–603. https://doi.org/10.1016/j.solmat.2016.07.044.

(13) Cattin, L.; Morsli, M.; Dahou, F.; Abe, S. Y.; Khelil, A.; Bernède, J. C. Investigation of Low Resistance Transparent MoO_3/Ag/MoO_3 Multilayer and Application as Anode in Organic Solar Cells. *Thin Solid Films.* **2010**, *518* (16), 4560–4563. https://doi.org/10.1016/j.tsf.2009.12.031.

(14) Makha, M.; Cattin, L.; Lare, Y.; Barkat, L.; Morsli, M.; Addou, M.; Khelil, A.; Bernède, J. C. $MoO_3/Ag/MoO_3$ Anode in Organic Photovoltaic Cells: Influence of the Presence of a CuI Buffer Layer between the Anode and the Electron Donor. *Appl. Phys. Lett.* **2012**, *101* (23), 233307. https://doi.org/10.1063/1.4769808.

(15) Yu, W.; Shen, L.; Meng, F.; Long, Y.; Ruan, S.; Chen, W. Effects of the Optical Microcavity on the Performance of ITO-Free Polymer Solar Cells with $WO_3/Ag/WO_3$ Transparent Electrode. *Sol. Energy Mater. Sol. Cells.* **2012**, *100*, 226–230. https://doi.org/10.1016/j.solmat.2012.01.021.

(16) Cao, W.; Zheng, Y.; Li, Z.; Wrzesniewski, E.; Hammond, W. T.; Xue, J. Flexible Organic Solar Cells Using an Oxide/Metal/Oxide Trilayer as Transparent Electrode. *Org. Electron.* **2012**, *13* (11), 2221–2228. https://doi.org/10.1016/j.orgel.2012.05.047.

(17) Sharma, V.; Vyas, R.; Bazylewski, P.; Chang, G. S.; Asokan, K.; Sachdev, K. Probing the Highly Transparent and Conducting $SnO_x/Au/SnO_x$ Structure for Futuristic TCO Applications. *RSC Adv.* **2016**, *6* (35), 29135–29141. https://doi.org/10.1039/c5ra24422f.

(18) Li, Y.; Tang, Z.; Hänisch, C.; Will, P. A.; Kovačič, M.; Hou, J. L.; Scholz, R.; Leo, K.; Lenk, S.; Reineke, S. Ultrathin MoO_3 Layers in Composite Metal Electrodes: Improved Optics Allow Highly Efficient Organic Light-Emitting Diodes. *Adv. Opt. Mater.* **2019**, *7* (3), 1–8. https://doi.org/10.1002/adom.201801262.

(19) Lu, H. W.; Huang, C. W.; Kao, P. C.; Chu, S. Y. ITO-Free Organic Light-Emitting Diodes with $MoO_3/Al/MoO_3$ as Semitransparent Anode Fabricated Using Thermal Deposition Method. *Appl. Surf. Sci.* **2015**, *347*, 116–121. https://doi.org/10.1016/j.apsusc.2015.03.188.

(20) Kinner, L.; Bauch, M.; Wibowo, R. A.; Ligorio, G.; List-Kratochvil, E. J. W.; Dimopoulos, T. Polymer Interlayers on Flexible PET Substrates Enabling Ultra-High Performance, ITO-Free Dielectric/Metal/Dielectric Transparent Electrode. *Mater. Des.* **2019**, *168*, 107663. https://doi.org/10.1016/j.matdes.2019.107663.

(21) Han, S.; Lim, S.; Kim, H.; Cho, H.; Yoo, S. Versatile Multilayer Transparent Electrodes for ITO-Free and Flexible Organic Solar Cells. *IEEE J. Sel. Top. Quantum Electron.* **2010**, *16* (6), 1656–1664. https://doi.org/10.1109/JSTQE.2010.2041637.

(22) Lim, S.; Han, D.; Kim, H.; Lee, S.; Yoo, S. Cu-Based Multilayer Transparent Electrodes: A Low-Cost Alternative to ITO Electrodes in Organic Solar Cells. *Sol. Energy Mater. Sol. Cells.* **2012**, *101*, 170–175. https://doi.org/10.1016/j.solmat.2012.01.016.

6 Organic Small Molecule Materials and Display Technologies for OLEDs

Bhawna, Shikha Jyoti Borah, Sanjeev Kumar, Ritika Sharma, Juhi Kumari, Neelu Dheer, Akanksha Gupta, Vinod Kumar, and Dhananjay Kumar

6.1 INTRODUCTION

The advent of the light bulb as a source of artificial light was a significant advancement for humanity. The transition from fluorescent tubes to the present solid-state lighting has improved the standard of living. While focusing on energy consumption, lighting consumes more than 20% of the total energy supply. This highlights the importance of energy-efficient light sources. Light sources should also require a minimum of harmful materials during production. Conventional light emitting diodes (LEDs) comprise heavy metal semiconductor materials. However, OLEDs rely on thin light-emitting films made of hydrocarbon chains. OLEDs have garnered significant interest from scientific and industrial peers owing to their lightweight and flat panel display technology. The intercalation of organic layers among electrodes is more efficient and convenient than the fluorescent tubes and LEDs. They produce brighter light without harsh shadows and consume less energy. Altering the operating current can easily manage the brightness level in OLEDs. Thus, they relax the psychological emotions of the audience by adjusting the brightness. Low-cost fabrication and energy-saving features have paved the way for OLEDs to become solid-state lighting sources for the next generation.

A typical OLED is constructed by sandwiching several organic layers between the cathode and anode. On applying voltage, recombination of excitons (electrons and holes) occurs in the emissive layer (EML). These excitons relax and produce electroluminescence. Recently, OLEDs have gained much more interest due to their advantages such as light weight, flat panel display, tenability of emission spectra, color saturation, wide view angle, etc., in the digital arena, that is, house lighting, smart TV displays, computer monitors, laptops, etc. OLEDs are not only restricted to display applications; their efficiency could be increased by incorporating a multifunctional layer in the EML region. OLEDs, in which the organic layers are composed of polymers, have been extensively used owing to their flexibility, minimum operating voltage, thinnest panel display, and many others. Similar to polymers, the use of inorganic quantum dot materials in OLEDs is due to their high conductivity, color purity, tunable spectra, and other properties.

DOI: 10.1201/9781003260417-6

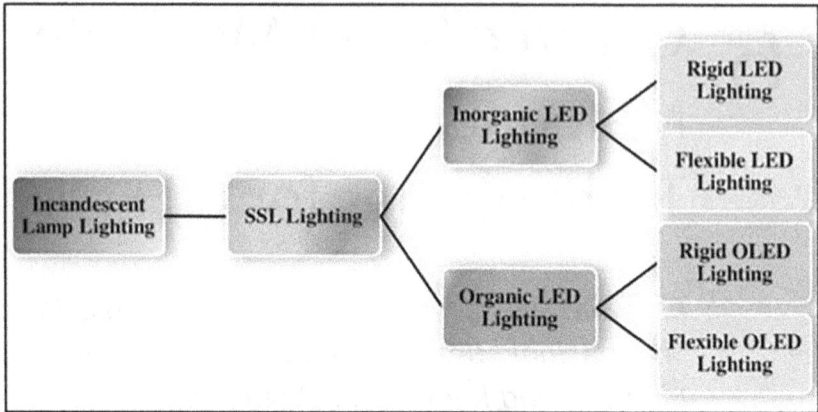

FIGURE 6.1 Block diagram showing development of lighting technologies. LED = light emitting diode; OLED = organic light emitting diode; SSL = solid-state lighting.

To ensure excellent performance, an emphasis on OLED device systematic design is necessary. High-performance devices widen utilization in various applications ranging from lighting to high-quality display screens. It accompanies the desire to evaluate and study the unique properties and technical features. Despite the outstanding progress in the last two decades, comprehensive studies to completely understand organic small-molecule materials for OLEDs are necessary for high-performance OLEDs. Yet, a few technical features confirm the generic notion of an efficient OLED. These characteristics are energy, current and quantum efficiency, operating voltage and view angle, brightness and contrast, life span and operational temperature. Among the stated features, the most challenging hurdle is the development of LEDs based on organic materials having excellent efficiency, lifetime and device stability. Figure 6.1 demonstrates OLED-based lighting technology. With progression, the device efficiency for multifunctional OLEDs was higher than for single-layer OLEDs. Furthermore, doping the different layers with appropriate material can significantly enhance conductivity and luminous efficiency, which, of course, cannot occur without prior knowledge of the chemical properties of the organic materials fused together as layers within the OLED.

6.2 CONSTRUCTION AND DEVICE STRUCTURES

Understanding the structure and components is essential to grasp complete knowledge of a working OLED system (Figure 6.2). Recent research and advances reveal that the OLED structure is critical and dominant in laying the foundation for a successful and efficient OLED design. The OLED structure comprises two electrodes and several organic layers. The following sections provide a comprehensive description. The OLED structure, and specifically the organic layers, can evaluate the capacity of electron-photon conversion, which often defines quantum efficiency (QE). Science, terminology, and measurement of light and optical instruments use the guidelines set by the Commission Internationale de l'Eclairage (CIE). Thus, proving

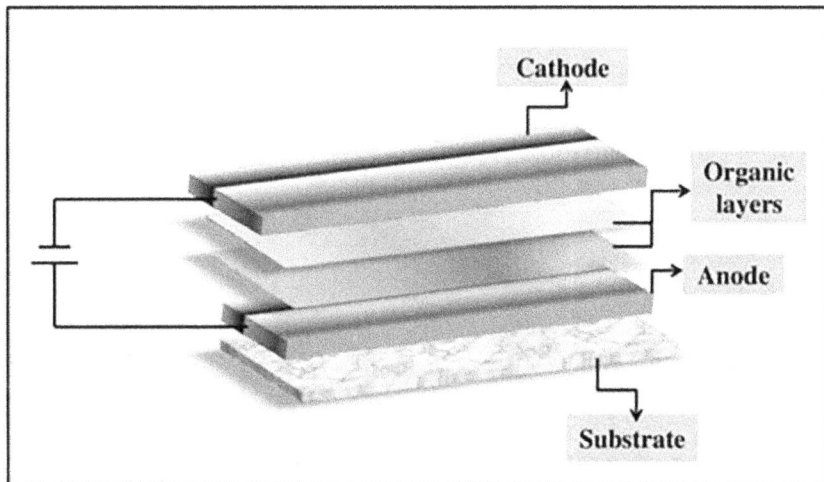

FIGURE 6.2　General structure of OLEDs.

to be of great importance in OLED devices.[1] The development of monolayer OLEDs sparked research interest in the 1960s. Numerous studies have demonstrated that many devices show commendable full-color and white display through the primary colors (red, green, and blue).[2]

OLEDs consist of a thin film, a monolithic semiconductor device that shows the emission of light on the applied voltage, displaying electroluminescence when the organic material is exposed to an electric field.[3] A generalized OLED structure consists of electrodes, substrate, and organic layers. Further, these organic layers may be charge transport, injection or blocking layers, and an EML. In a broader sense, the OLED classification based on the number of layers is single-layer, bilayer, three-layer, and multilayer OLEDs (Figure 6.3).[4]

Within a single-layer OLED, an EML is sandwiched by the electrodes (cathode and anode), exhibiting hole and electron transport properties.[5] In a bilayer OLED, two thin layers are present between the electrodes. The thin layers are the hole and electron transfer layers (HTL and ETL, respectively). They effectively cause recombination within the interfacial region, resulting in electroluminescence. A third additional layer (EML) is present between the HTL and ETL in a three-layer OLED. It effectively causes the charge recombination mechanism and acts as the site for electroluminescence.[6] An advanced, highly evolved structure of OLED is the multilayer OLED structure comprising two or more functional layers. Herein, the hole blocking and injection layers (HBL and HIL, respectively), help to eliminate charge leakage and exciton quenching. The following section provides a brief description of each of the components.

6.2.1　Substrate

The substrate may be clear plastic, metal foil, or glass (rigid and ultrathin flexible), given that they are transparent, exhibit conductivity, and have a high work function.

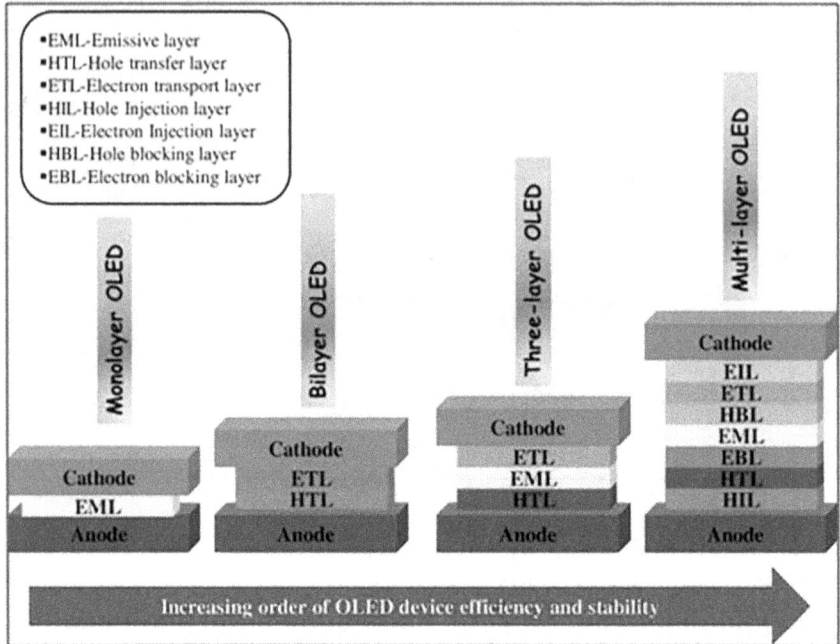

FIGURE 6.3 Diagrammatic representation of OLED structures based on organic layers. (Reconstructed from Bauri et al.[4])

The flexibility of OLEDs improves as a consequence of organic and organic-inorganic hybrid substrate materials. The properties that are essential in an efficient substrate are as follows.[7]

a. The transmission coefficient of the substrate must be >90% across the visible spectrum.
b. The substrate should exhibit properties like optical transparency, thermal and chemical stability, and compatible fabrication.
c. The substrate must have low roughness in the range of R_{RMS} <2 nm and R_{max} <20 nm.
d. Permeability towards air and moisture.
e. Work function in higher ranges of 4–5 eV.

6.2.2 Electrodes

The cathode and anode are the two electrodes that have crucial importance in the structure of OLEDs. This section provides a brief discussion of the individual properties of both electrodes. Usually, either of the electrodes is transparent or semitransparent to facilitate easy passage for light to escape. The anode tends to show work function in a higher range than the cathode, and is a transparent electrode that injects holes into the various organic layers. The working function also influences the operating voltage and device luminescence. Most often, indium tin oxide (ITO) is the de facto standard material used as an anode. Other anode materials can be a transparent conducting oxide (TCO) and carbon-based conductive thin films.[8] The properties that must be in the anode are:

a. Conducting materials exhibiting thermal stability, high work function, and low resistivity (in the range of or less than $4 \times 10^{-4} \Omega cm$).
b. The transmission coefficient of the material must be at least 85% or greater across the visible spectrum.
c. The anode surface roughness must show $Z_{RMS} < 4$ nm and $Z_{max} < 30$ nm.

In contrast, the cathode is often a material possessing a low work function that may be composed of elements containing rare earth elements, alkali metals, or alloys of magnesium and silver. The major properties that must be in the cathode are:[9]

a. Materials have an electron work function of 2–4 eV.
b. Materials exhibit chemical inertness in air and when in contact with adjacent organic layers.
c. For top emitting OLEDs, the cathode material must ensure superior transmission coefficient.

6.2.3 ORGANIC LAYERS

6.2.3.1 Charge Injection and Transport Layer

The OLED structure comprises many organic layers, two of which broadly are the charge injection and transport layers. The injection layer consists of the hole and electron injection layers (HIL and EIL, respectively), whereas the transportation layer consists of the HTL and ETL. The HIL is present between the anode and the HTL, enhancing OLED stability. It occurs by simultaneously blocking electrons from the anode and injecting holes into the HTL. The markers for efficient HIL and HTL materials are properties such as mobility and energy separations between their highest occupied molecular orbital (HOMO) and lowest unoccupied molecular orbital (LUMO) levels. For instance, compounds such as Alq_3 and CuPc have been integrated as efficient HIL materials in different OLED devices; their structures are depicted in Figure 6.4.[10] The HTL is a layer that ascends on the HIL and enhances OLED efficiency by providing a good supply of holes for charge recombination. Some of the primary distinctive qualities of HTL are high morphologically stable layer, high hole transportation but low electron mobility, and the HOMO level of HIL and the anode's Fermi level must have low barrier width.[10] HTL p-type materials such as N,N'-diphenyl-N,N'-bis(m-tolyl)-1,1'-biphenyl-4,4'-diamin (TPD) and 4,4'-bis[N-(1-napthyl)-N-phenylamino]biphenyl (NPB) have been investigated.[3]

The placing of the ETL between the HBL and the EIL helps in developing better lifetime and achieving higher efficiency of OLEDs.[11] Properties that contribute to the excellent performance of ETLs are outstanding conductivity, thermal and chemical stability, maximum affinity towards injection and mobility of electrons, strong

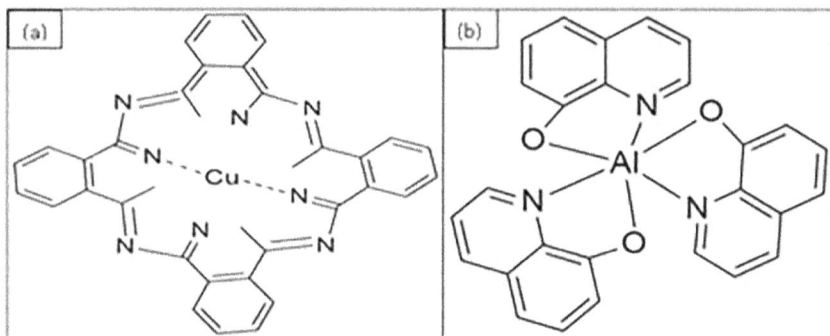

FIGURE 6.4 Chemical structures of materials used as the hole injection layer: (a) copper phthalocyanine and (b) Alq_3.

capacity to block holes, low barrier of energy separation, and minimum triplet energy for blocking of excitons. Kandulna and Choudhary and Mandal and Choudhary have demonstrated the utility of materials that were polymer-based nanocomposites, such as PPV/ZnO, PANI/GO/Mo_2S, etc.[12]

6.2.3.2   Charge Blocking Layers

OLEDs have two charge blocking layers (CBL), the hole and electron blocking layers, HBL and EBL, respectively. These blocking layers are present between the carrier transport and an emissive layer. They can remarkably enhance lifetime, luminescence, and internal efficiency. In the OLEDs framework, HBL is present between the ETL and EML. Similarly, the EBL is present between the HTL and EML. These hinder the transfer of charge and drastically decrease the quenching effect, thus preventing leakage of the charge. Materials such as Alq_3 and 2,9-dimethyl-5,7-diphenyl-1, 10-phenanthroline are good examples of blocking layers. The energy levels play a significant role in stimulating performance. HBL has HOMO at a higher and LUMO at a lower position, while EBL shows the opposite. Besides, the low barrier width of the HOMO/LUMO level of the EBL/HBL is roughly equivalent to that of the HTL/ETL.[13]

6.2.3.3   Emissive Layer/Active Layer

In single-layered OLEDs, the EML is present between the two electrodes. However, in multilayered OLEDs, the EML is present specifically between the CBLs. EML materials such as small organic molecules and polymers have a molecular mass of less than 100 AMU and more than 1000 AMU, respectively. They exhibit properties such as transparency and light emission in the visible light range and low bandwidth. Furthermore, EML materials for hole injection expect to have barriers between HOMO and the work function of the anode. However, materials for electron injection have barriers between LUMO and the work function of the anode.[14] Some studies reveal that small-molecule materials allow more advanced layer engineering and sophisticated architectures than polymer-based materials.[15]

6.3   WORKING PRINCIPLE OF OLEDs

In OLEDs, there are several organic layers between the anode and cathode. Depending on the number of organic layers, OLEDs can be single-, double-, three-, or multilayer. OLEDs are also known as dual injection LEDs as they contain HTLs and ETLs. OLEDs follow the principle of electroluminescence, that is, dropping electrons into holes in the EML (Figure 6.5).[4] On applying an external voltage, current starts flowing through electrodes. The anode is connected to the positive terminal, while the cathode is connected to the negatively charged terminal of the applied voltage. The current flows through the organic layers, originating from the cathode and in the direction of the anode. The role of the cathode is to make electrons available to the EML, while the anode ejects electrons from the conductive layer. The recombination of the electrons and holes (excitons) occurs in the EML. On deexcitation, these excitons emit photons, producing visible light. The color and the intensity of the emitted light are determined by the material used for the EML and the quantity of the applied electric current, respectively. OLEDs with transporting and blocking layers can reduce power consumption and achieve long life. In the construction of OLEDs, one of the electrodes is necessarily transparent. The total thickness of the device is 1.8 mm, of which the OLED layers constitute a thickness of less than 500 nm.

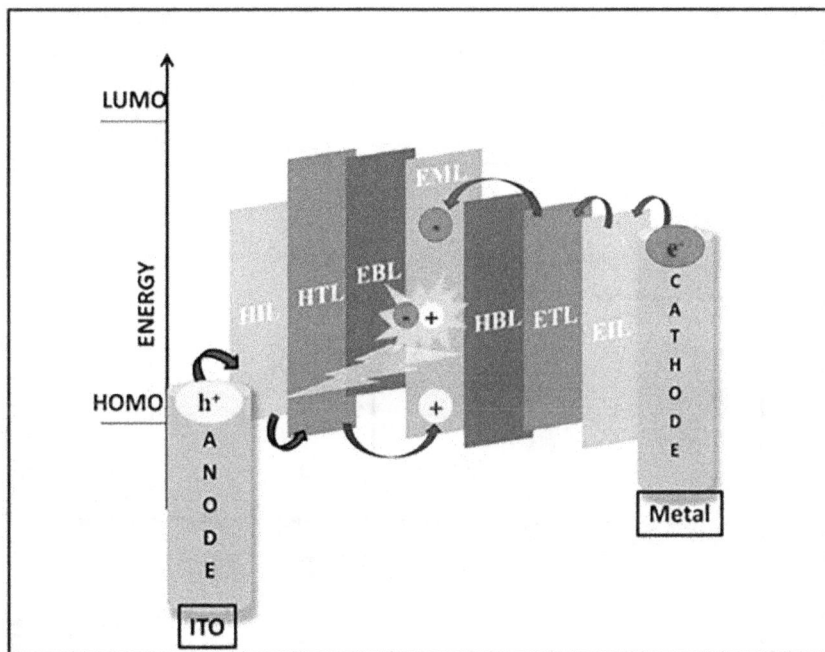

FIGURE 6.5   Schematic diagram showing the electroluminescence mechanism in an OLED. EBL = electron blocking layer; EIL = electron injection layer; EML = emissive layer; ETL = electron transfer layer; HBL = hole blocking layer; HIL = hole injection layer; HTL = hole transfer layer; ITO = indium tin oxide. (Reconstructed from Bauri et al.[4])

6.4 HOMO/LUMO-BASED LIGHT EMITTING MECHANISM

When two atomic orbitals (AOs) 2sp^2 of carbon directly overlap in C-C bonding, they give rise to molecular orbitals (MOs). These MOs formed from direct overlap produce one σ_1 (bonding MO) and one σ_2^* (antibonding MO). However, when two 2p$_z$ orbitals undergo side-to-side overlap, they produce two MOs (π_1 bonding and π_2^* antibonding). Studies on bond energy levels reveal that σ bonds are stronger than π bonds. Hence, electronic transitions are not carried out by the σ MOs. After the formation of MOs, the bonding MO is the HOMO, while the antibonding MO is the LUMO in the organic molecules. Each MO contains two electrons only, and follows Aufbau's principle. Thus, the electron transitions from the lower lying HOMO level to the higher LUMO level is called the $\pi \rightarrow \pi^*$ transition.

On light absorption, electron excitation takes place from HOMO to LUMO, producing a singlet (S; electrons having antiparallel spin) and a triplet (T; electrons having parallel spin) (Figure 6.6).[16] When electrons from the HOMO interact with that of LUMO, Coulomb and exchange interactions are produced. The energy level of the S or T state is reduced due to the Coulomb interactions. However, the exchange interactions reduce the energy of the T state while increasing the S state energy.[17] However, when there is a lack of exchange energy, the S state is destabilized as a result of the electronic repulsions observed. Consequently, this forces the S state to occupy a higher energy level and the T state to occupy a lower energy level. This occurs due to stabilization through more exchange of energy.[18]

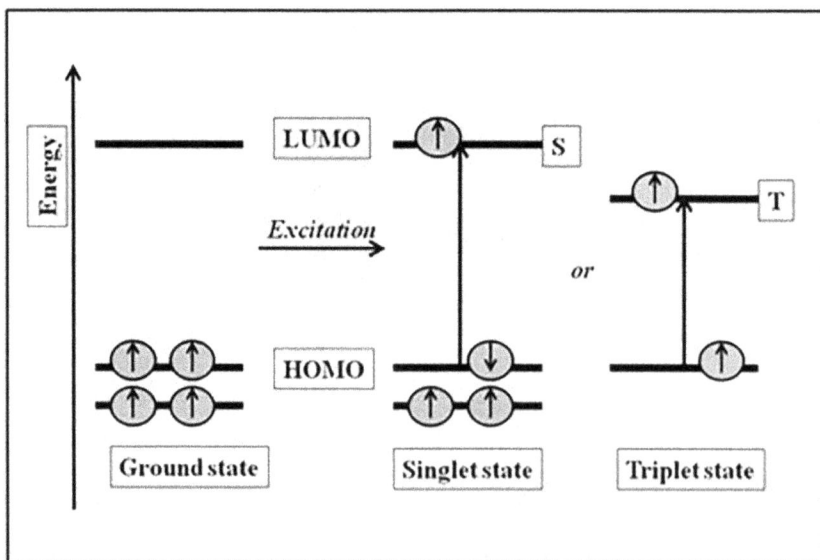

FIGURE 6.6 Electron excitation to show HOMO-LUMO transition in organic materials for generation of singlet or triplet excited states. (Reconstructed from Pode.[16])

6.5   MATERIALS USED IN OLEDs

Today, we live in an era driven by electronic devices. With the escalating demands, energy storage and optoelectronic devices are potential resources exhibiting paramount applications. OLEDs have emerged as an excellent technology.[4] Several materials have been synthesized in the last decade, which has facilitated the development of enhanced performance of OLEDs.[19] The type of material used to categorize OLEDs includes polymer- and small-molecule materials.[20] The following subsections provide a brief discussion of the materials.

The primary task in improving the performance of an OLED device is the appropriate selection of organic materials for the formulation of functional layers. Injection and charge transportation characterizes the efficiency of the device. The charge transfer occurs from the holes (anode surface), having a higher work function, to the electrons (cathode surface), having a lower work function. The balance of work functions between the electrodes minimizes the driving voltage. The energy barrier between the HOMO of the HTL and anode and the LUMO and cathode of the ETL is high (>0.6 eV). Then, there might be a significant drop in driving voltage.

6.5.1   HOLE TRANSPORT MATERIALS

The fabrication of the HTL uses hole transport materials (HTM). The HOMO and LUMO are critical in determining which HTM to use in an OLED device. For the transfer of electrons from the anode into the EML, HOMO has a lower energy barrier, while the LUMO has the potential to prevent further electron transfers from the EML to the HTL. Generally, the HTMs are electron-donating species owing to their lowered ionization potential and greater charge mobility.[19] Typical examples of such emitters containing electron-donating moieties are N,N'-diphenyl-N,N'-bis(m-tolyl)-1,1'-biphenyl-4,4'-diamin (TPD), m-methyl-tris(diphenylamine)-triphenylamine (MTDATA), 4,4'-bis[N-(1-naphthyl)-N-phenylamino]biphenyl (NPB), 4-(9H-carbazol-9-yl)triphenylamine carbazole derivatives, etc.[21] The chemical structures of some HTMs are depicted in Figure 6.7.

Because of their high visible light transparency and appropriate position of the HOMOs, aromatic amine derivatives are widely used as HTM in OLEDs. Several other HTM contain triarylamines and benzidines. Table 6.1 provides examples of some materials that use HTM in OLED devices.

6.5.2   ELECTRON TRANSPORT MATERIALS

The fabrication of the ETL can be achieved by using appropriate electron transport materials (ETM). The ETM improves electron mobility, electron affinity, charge recombination rate, thermal stability, and other electron-transporting characteristics.[19] Electron mobility and thermal stability are important ETM properties for successful development. However, controlled electron mobility has not yet been achieved with completely satisfactory results. Thermal stability is preferably high (>150°C), enhancing the device's durability and stability. In addition, factors such as layer thickness and shape influence the efficiency of ETM. Typical examples of emitters are Alq$_3$, 2-(4-biphenyl)-5-(4-tert-butylphenyl)-1,3,4-oxadiazol

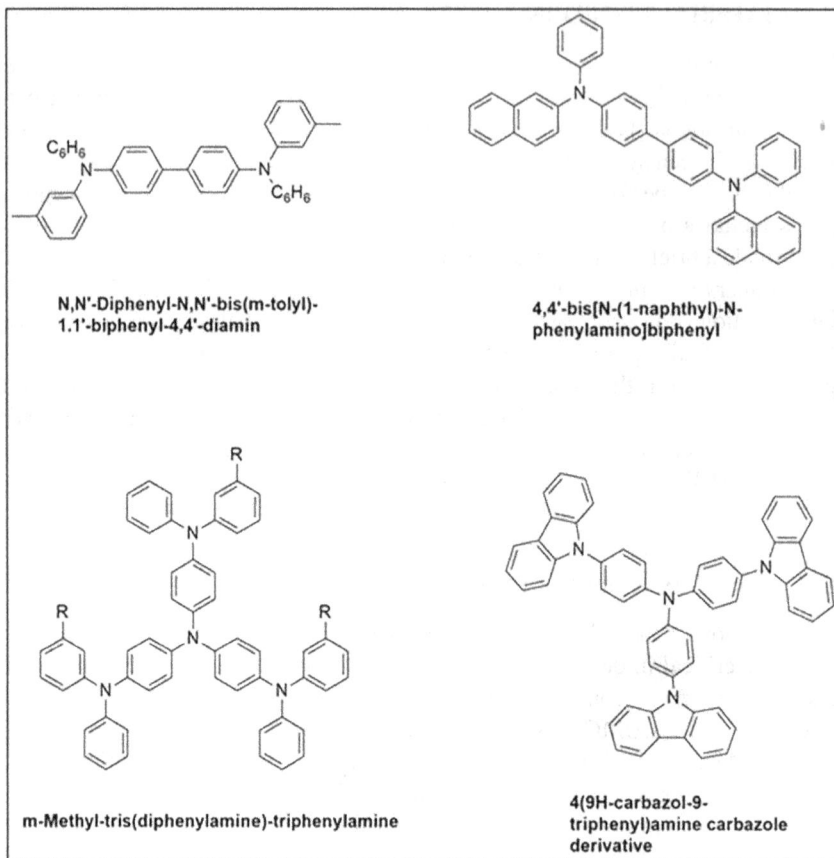

FIGURE 6.7 Chemical structures of example hole transport materials in OLEDs.

(PBD), poly(vinyl) alcohol (PVA)-ZnO nanocomposite, 1,3,5-tri(3-pyrid-3-yl-phenyl) benzene (TmPyPB), 4,7-diphenyl-1,10-phenanthroline (BPhen), 2,9-dimethyl-4,7-diphenyl-1,10-phenanthroline (BCP), etc.[21,26,27] The chemical structures of some ETMs are depicted in Figure 6.8.

Different ETMs can be selected based on the design of the OLED device. Properties such as their molecular structure, thermal stability, electron mobility and affinity, and various others help to develop more efficient ETL. Yet, it is often difficult to claim that a superior ETL has been designed that works without any constraints. Table 6.2 provides examples of some materials that are used as ETM in OLED devices.

6.5.3 POLYMER-BASED MATERIALS

Polymer-based materials are used extensively in OLEDs. They demonstrate a broad spectrum of utilization due to their flexibility, filter, reduced operating voltage, and many other characteristics.[4] Additionally, the manufacturing cost is low, and the probability of designing new polymers is endless compared to other materials.

TABLE 6.1

Properties of Some Hole Transport Materials Used in OLED Devices

Electron Transport Material	Chemical Name	Abbreviation	Glass Transition Temperature (°C)	Absorption Wavelength (nm)	Emission Wavelength (nm)	Ref.
Triphenylamine-based material	tris(3′-(1-phenyl-1H-benzimidazol-2-yl) biphenyl-4-yl)amine	TBBI	148	295.352	510	22
	tris(2-methyl-3′-(1-phenyl-1H-benzimidazol-2-yl)biphenyl-4-yl)amine	Me-TBBI	144	302	470	22
Phenothiazine and phenoxazine-substituted fluorene core based	10-hexyl-3-[2,7-di(naphthalen-1-yl)-fluoren-9-ylmethylene]phenoxazine	DNFPhe	91	404	527	23
	10-hexyl-3-[2,7-di(4-(diphenylamino)-phenyl)fluoren-9-ylmethylene]phenoxazine	DDPFPhe	93	386	513	23
Carbazole-based materials	3′,6′-bis(carbazol-9-yl)-bis[9-(2-ethylhexyl) carbazol-3,6-diyl]	GlCBC	138	294	384	24
	3′,6′-bis[3,6bis(carbazol-9-yl)carbazol-9-yl]-bis[9-(2-ethylhexyl)carbazol-3,6-diyl]	G2CBC	245	294	384	24
Spiro based	2,2′,7,7′-tetrakis(N,N′—di-p methylphenylamino)-9,9-spirobifluorene (Spiro-TTB)	Spiro-TTB	149	385	418	25

TABLE 6.2

Properties of Some Electron Transport Materials Used in OLED Devices

Electron Transport Material	Chemical Name	Abbreviation	Electron Mobility	Electron Affinity (eV)	Absorption Wavelength (nm)	Emission Wavelength (nm)	Ref.
Oxadiazole derivatives	2-(4-Biphenyl)-5-(4-tert-butylphenyl)-1,3,4-oxadiazole	PBD	$\sim 2 \times 10^{-5}$ cm^2/Vs at high electric fields ($\sim 1 \times 10^6$ V/cm)	2.16	309	369	28
	2,5-bis(4-Naphthyl)-1,3,4-oxadiazole	BND	$\sim 2 \times 10^{-5}$ cm^2/Vs at high electric fields ($\sim 1 \times 10^6$ V/cm)	–	–	–	29
Metal chelates	tris(8-Hydroxyquinoline) aluminium	Alq$_3$	Of the order of 1×10^{-5} cm^2/Vs at electric fields of $0.39–1.3 \times 10^6$ V/cm	3.0	378, 360 (in a 1:1 mixed solvent of methanol and ethanol (ME) at 77 K), 334, 316, 300, 263, 255.8, and 233 in ethanol	567 and 478	30
	tris(8-Hydroxyquinoline) indium	Inq$_3$	–	3.4	390	540	31
Azole-based materials	4-Naphthalen-1-yl-3,5-diphenyl-4-[1,2,4-triazole	TAZ-1	–	~2.3	264	367 in dichloromethane-d$_2$	18
	3-(4-Biphenyl)-4-phenyl-5-tert-butylphenyl-1,2,4-triazole	TAZ	–	~2.3	290	370	32
Dendritic molecule	1,3,5-tris(N-phenylbenzimidizol-2-yl)benzene	TPBi	3.3×10^{-5} cm^2/Vs	2.7	305	359, 370	33

Polybenzobisazoles	poly(p-Phenylenebenzo[1,2-d:4,5-d']bis-thiazole-2,6-diyl)	PBZT	$\sim 2 \times 10^{-7}$ cm^2	2.4–3	—	—	34
Quinoxaline based	1,3,5-tris[3-Phenyl-6-trifluoromethyl) quinoxaline-2-yl]benzene	TPQ	$\sim 10^{-4}$ cm^2/Vs (at $\sim 10^6$ V/cm)	3.6	364	—	35
	4,7-Diphenyl-1,10-phenanthroline	Bphen	10^{-4} cm^2/Vs (at 10^5 V/cm)	—	272	379	36
	2,9-Dimethyl-4,7-diphenyl-1,10-phenanthroline	BCP	5.0×10^{-6} cm^2 V^{-1} s^{-1}	—	277	386	37
	8-Hydroxyquinolinolato-lithium	Liq	—	—	260	485	38

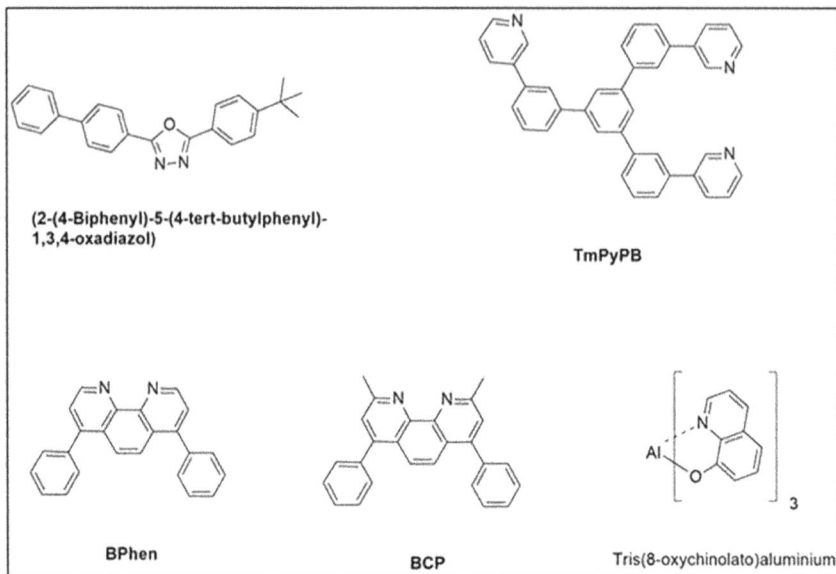

FIGURE 6.8　Chemical structures of several electron transport materials in OLEDs.

Typical examples of emitters are poly(fluorene), poly(p-phenylenvinylene) (PPV), poly(2-(2-ethylhexyloxy)-5-methoxy-p-phenylen)vinylene) (DPVBI), rubrene, tris[2-phenylpyridine]iridium(III) (Ir(ppy)$_3$), bis[2-(4,6-difluorophenyl)Pyridinato-FIrpic, tris(4-carbazol-9-phenyl)amine (TCTA), etc.[4,21] Figure 6.9 depicts the chemical structures of polymer-based materials. These polymer-based materials exhibit a wide range of innovations in designing OLED devices. Zhong et al. innovate a thermally activated delayed fluorescent (TADF) OLED device via a novel strategy. In their fabrication, they use a ternary polymer complex, that is, poly(9-vinylcarbazole)-(1,3,5-triazine-2,4,6-triyl)-tris(benzene-3,1-diyl)-tris(diphenylphosphineoxide) (PVK@PO-T2T@mCP). The TADF OLED exhibits better quantum yield for photoluminescence.[39]

6.5.4　SMALL-MOLECULE-BASED MATERIALS

Small-molecule-based materials include metal oxides, carbazole, cross-linkable compounds, etc. They are low molecular weight compounds of less than 900 Da. The vacuum deposition method is widely used for the synthesis of these materials. Additionally, their thickness ranges up to 1 μm.[4,19] They exhibit excellent properties and cost-effective manufacturing, forming pure materials for better OLED performance.[19] Typical examples of emitters are tris(8-oxychinolato)aluminium (Alq3), 4,4′-bis(2,2-diphenylethene-1-yl)diphenyl (DPVBI), 4-(dicyanomethylene)-2-methyl-6-(p-dimethylaminostyryl)-4H-pyran (DCM), etc.[4,21] Figure 6.10 illustrates several small-molecule-based materials and their chemical structures. These materials also offer many opportunities for designing OLED devices.

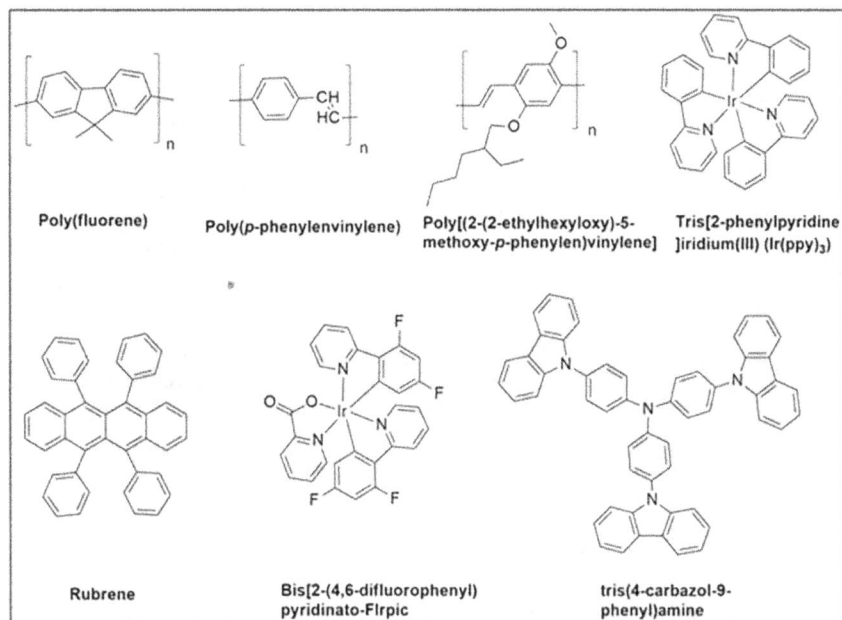

FIGURE 6.9 Chemical structures of some emitting materials based on polymers in OLEDs.

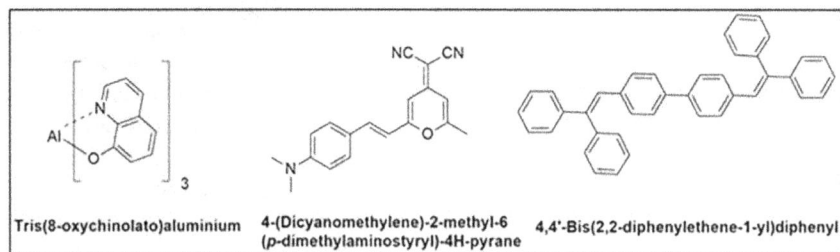

FIGURE 6.10 Chemical structures of some emitting materials based on small molecules in OLEDs.

Furthermore, the classification of these materials is also based on their transport, that is, HTM and ETM.[20,21] These materials use charge injection property optimization to enhance OLED performance.[20,40] An efficient material for charge injection reflects glaring characteristics. These properties are charge distribution, recombination rate, balanced injection, etc. These characteristics are responsible for innovating high-performance OLEDs.[20]

However, that's not all. The preceding sections discussed the various layers making up OLEDs, and the numerous materials that have been developed and shown to enhance OLED performance.[19,40] After this brief glance at the different materials used in OLEDs, the following subsections briefly discuss the methods for the synthesis of the materials used in OLEDs.

6.6   METHODS FOR SYNTHESIS OF MATERIALS USED IN OLEDs

6.6.1   SOLUTION CASTING METHOD

A solution casting method is a facile approach in which the polymer and pre-polymer molecules are mixed in an equimolar ratio and are soluble in an appropriate solution. Lim et al. used the solution casting method for the fabrication of an ethynylated Schiff's base (ESB) along with ITO and NO_2/Au. Their work demonstrates that a single layer of the ESB/NO_2 exhibited better electroluminescence for yellow emission. Therefore, it shows a promising OLED material tailored to nematogenic dyes.[41] Kandulna et al. fabricated a PVA nanocomposite with ZnO via a solution casting approach. Their results demonstrated that the synthesized composite has promising application in the OLED electron transport layer owing to the desired band gap, visible range current density, and charge carrier recombination rate.[26] Mai et al. have designed undoped OLEDs using a multipurpose material, (E)-1,2-bis(5,11-bis(2-(2-ethoxyethoxy)ethyl)-5,11-dihydroindolo[3,2-b]carbazole-2-yl)ethane (BEDCE), which is synthesized via the solution casting method. Furthermore, the designed material has demonstrated promising application in perovskite solar cells (PSCs).[42]

6.6.2   SPIN COATING METHOD

The spin coating method is a type of wet deposition strategy reliable for the synthesis of OLED materials. Typically, the material solution is applied dropwise onto the substrate, that is, while rotating at a range of 1000 to 3000 rpm. The material exerts centrifugal force and spreads onto the substrate surface up to a few nanometers thick. Therdkatanyuphong et al. demonstrated the promising fluorophore OLED application in biomedicine and optoelectronics. For that, they designed two isomeric forms of bis(n-hexylthienyl)thiadiazole[3,4-c]pyridines; namely, C3HTP and C4HTP with tetracarbazole (an electron-rich) moiety. The synthesized isomers demonstrated an enhanced emission peak at 725 nm. Of the two isomers, C3HTP showed better photoluminescence of 34%. Therefore, C3HTP could be a potential candidate material for near-infra-red OLED applications.[43] Feng et al. have designed tetraphenylbenzosilole (TPBS) along with silyl radicals. They developed an undoped spin-coated OLED with the emission of deep blue light.[44] Uddin and Teo designed a hybrid OLED with potential application in display technology. For the synthesis, they used cadmium selenite (CdSe) and quantum dots (QDs) via the spin coating method. Therefore, the intervention of QDs in this hybrid OLED has enhanced the emission threefold.[45] Kuznetsov et al. has synthesized a europium-(2-thenoyltrifluoroacetone) ligand complex with dipyrido[3,2-a:2′c,3′c-c]phenazine (Eu(tta)$_3$DPPZ) via the spin coating method. Their results demonstrated enhanced brightness of OLED, although the operating voltage is lower than the same OLED material synthesized via the vacuum deposition technique.[46]

6.6.3   FREE RADICAL POLYMERIZATION

Free radical polymerization starts with the initiation of free radicals, which have the potential to break into two fragments. The generated radical species tend to bind to

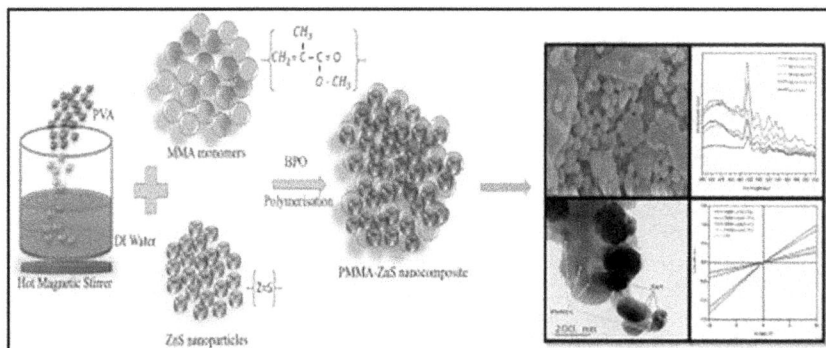

FIGURE 6.11   Schematic of PMMA-ZnS nanocomposite synthesis. (Reprinted from Nayak and Choudhary.[47] Copyright © 2019 Elsevier B.V. All rights reserved.)

the monomer one after the other to form the polymer chain and finally terminate the chain, hence, forming the desired polymer. Nayak and Choudhary fabricated a nano-composite using poly(methyl methacrylate) (PMMA)-ZnS (Figure 6.11). The synthesized nanocomposite exhibits desired band gap and recombination rate of charges and therefore has a promising role in the emission layer of OLEDs.[47]

Kandulna et al. fabricated an Ag-doped ZnO@PMMA nanocomposite. Their results demonstrated that reinforcing doping would improve the conduction and absorption properties of the OLED and has potential application in the ETL of OLEDs.[48] Another study by Kandulna et al. designed a PMMA and TiO_2 nanocomposite. They demonstrated that the nanocomposite exhibited four band attributes: blue-violet, blue, green, and yellow. Additionally, it exhibited better transport properties, making them a suitable candidate, therefore, for the electron transport layer of OLEDs.[49]

6.6.4   INKJET PRINTING PROCESS

Inkjet printing is the most often used approach for fabricating multilayer OLED devices. The OLED materials are placed onto the substrate and the ink dropped from the nozzle with digital control. It can be accomplished in two modes (Figure 6.12): (1) continuous injection (CIJ) and (2) drop on demand (DOD).[50] Hu et al. fabricated phosphorescent iridium (III) OLED devices to exhibit enhanced efficiency.[51]

6.7   APPLICATIONS OF OLEDs

OLEDs exhibit characteristics such as thinness, wide viewing angles, mechanical flexibility, lightweight, solution processability, low operating voltage, high power efficiency, and low fabrication cost. These remarkable parameters are for portable applications, particularly wearable electronics, pulse oximetry sensors, patches for advanced wound care, the display screen on TVs, laptops, smartphones, watches, etc.[52] A thorough discussion of the smart lighting and display technology of OLEDs follows.

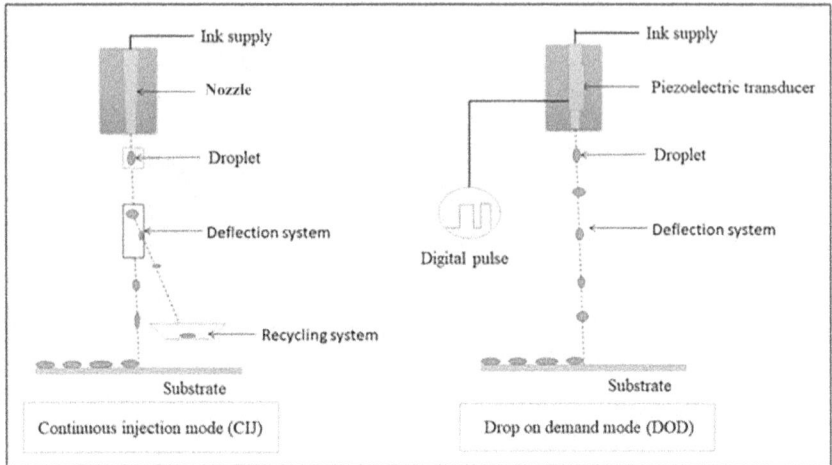

FIGURE 6.12 Schematic representation of the two modes of inkjet printing: (left) continuous injection and (right) drop on demand. (Reprinted with permission from Lan et al.[50] Copyright ©2017, Higher Education Press and Springer-Verlag GmbH Germany.)

6.7.1 OLEDs in Display Technology

As development progressed from the 1980s, OLED displays stood as great competitors against traditional cathode-ray tubes (CRTs) and liquid-crystal displays (LCDs) in the flat-panel display market.[53] The key distinction between OLED and LCDs is that the OLED displays can precisely control brightness variation and emit colors red, green, and blue inside subpixels within a panel, but LCDs are transmissive devices that cannot completely block the backlight. OLED display screens have high color contrast, fast response speed, low cost, less power consumption, and a thinner body.[54] Moreover, OLED display screens are more flexible. However, there are still several critical issues with OLEDs that must be resolved. For instance, as in large-size displays, there is significantly faster fading of commercial blue emitters as compared to red and green emitters during operation, which leads to color deviations. Although certain solutions have been suggested to overcome the lack of blue emission, the commercial production of highly stable and reliable blue emitters is still at its inception point.[55]

At present, OLEDs have significant benefits in small-size display screens such as smartphones and smartwatches. This is why OLED display screens are used in digital cameras, mobile phones, OLED TVs, laptops, and audio and video players. Several companies run OLED display screens in their audio and camera products, such as Sony, Fujifilm, Panasonic, and Nikon. Furthermore, the global market for wearable devices is expected to expand in the near future. Several elegant flexible and wearable devices, such as head-mounted displays and fitness bands, have emerged to use OLED displays for better color quality. In addition, OLED displays have also found applications in virtual reality (VR) screens for three-dimensional (3D) viewing due to their excellent resolution and comfort. Today, OLED technology can be found in a wide range of smartphones, TVs, and computers. For example, Apple, Redmi, Samsung, LG, Dell, HP, Lenovo, etc., all use OLED displays in their products.[56]

Display technology is of two types: passive matrix display (PMD) and active matrix display (AMD). In PMD, the cross-over region of the linear electrodes is determined by each pixel on all the sides of the emissive or liquid crystal material. These linear electrodes are positioned such that they are 90° apart from each other.[3] Each line has a T/N period, where N denotes the total count of lines on the screen, and T is the frame duration. Throughout, all of the required pixels are on, following the image content. After that, the succeeding line is chosen. To avoid any flickering effect while using a transducer without memory, the image is required to be refreshed at a rate of at least 50 Hz. The electrodes of each pixel contribute to other pixels in a similar line and column, and regulation of the voltage supply to the electrode is continuous throughout the line selection time. However, parasitic signals may affect the vulnerable pixels for the remaining frame time. An increase in the current value as per the number of lines to attain the effective luminous effect in the T/N span and across a specific pixel column is important. As a result, the columns consume a lot of power and have a lot of resistive losses, causing stress and damage to the pixels.[15]

The AMD is an alternative to PMD in which the pixels are determined by their electrodes and powered by circuitry consisting of a capacitor and a thin-film transistor (TFT). The "active" OLED material is put on the active matrix circuitry, with the opposite electrode acting as a ground electrode because it is not patterned. The capacitor in such a device is designed to keep the information for a while. Despite the challenges of fabricating these matrixes, this approach is so far widely utilized in flat-panel screens and confirms the demand for massive, high-resolution devices.[15] Figure 6.13 depicts a full-color active matrix OLED.[57]

Various research advances have occurred to fabricate high-performance OLED substrates. Tao et al. fabricated a thermally stable and transparent cellulose nanofibril/polyarylate (CNF/PAR) hybrid polymer substrate for OLED devices, as shown in Figure 6.14. The mechanical and thermal ability of CNFs/PAR has improved due to their specific morphology and efficient interfacial interaction with PAR.[7] Moreover, fabricated hybrid polymer substrates of nanocellulose/polyimide have excellent mechanical, optical, and thermal properties introduced in flexible OLED displays.[58]

6.7.2 SMART LIGHTING

Smart lighting systems (SLSs) are important in communication, health care, and gardening. Nowdays, the discovery of OLED illuminants for use in SLSs is inadequate to meet the demand. The main challenge is fabrication of an OLED illuminant with a tunable spectrum and intensity.[59] SLSs for practical or aesthetic applications, such as environmental changes and color temperature matching, are fast emerging because of the diverse uses.[60]Additional features critical for SLSs include altering the intensity and spectrum of the illumination source. Human-centric lighting is an essential SLS area that demands emission intensity and color temperature to respond to environmental changes to benefit humans.[61]

Zhang et al. fabricated an in-planar-electrodes (IPE) OLED on ITO post-coated glass substrates that work as bottom IPE. The color of the emission light by IPE

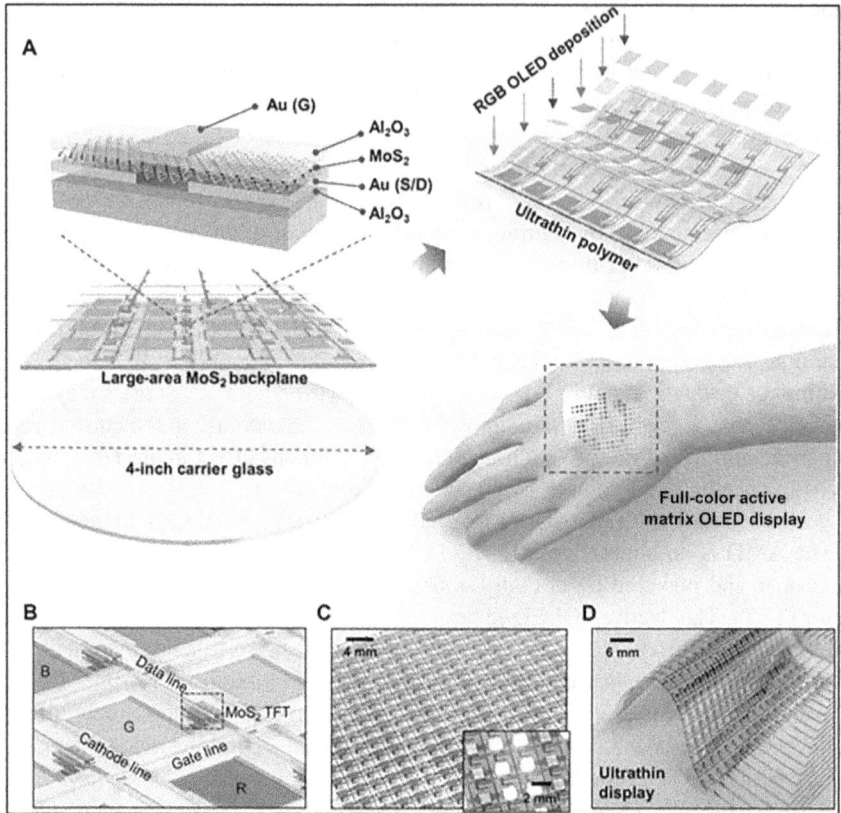

FIGURE 6.13 Illustration of a full-color MoS_2-based, backplane active matrix OLED. (A) Al_2O_3 doping effects on MoS_2 film applied on an ultrathin polymer substrate and tested on a human hand; (B) integration of a pixel array with MoS_2 transistors interconnected via gate, data, and scanning lines; (C) a digitally obtained photograph of an active matrix display; (D) demonstrates the flexibility of the ultrathin polymer substrate due to low bending stiffness. (Reprinted with permission from Choi et al.[57] Copyright © 2020, Science Advances.)

OLEDs transitions from the color blue to white and then finally to a yellow color. As a result, IPE-OLED-based SLSs are used not only to provide human-centric lighting but also to alert workers to toxic chemical leakage. With increasing modulation frequency, IPE-OLEDs is also a light source utilized in visible light communication. The result demonstrates an SLS with a brightness and color temperature-controllable illuminant.[59]

6.8   CHALLENGES

- Multilayers in OLEDs are important to optimize device performance and directly impact the cost. Hence, device engineering is necessary to reduce the cost.

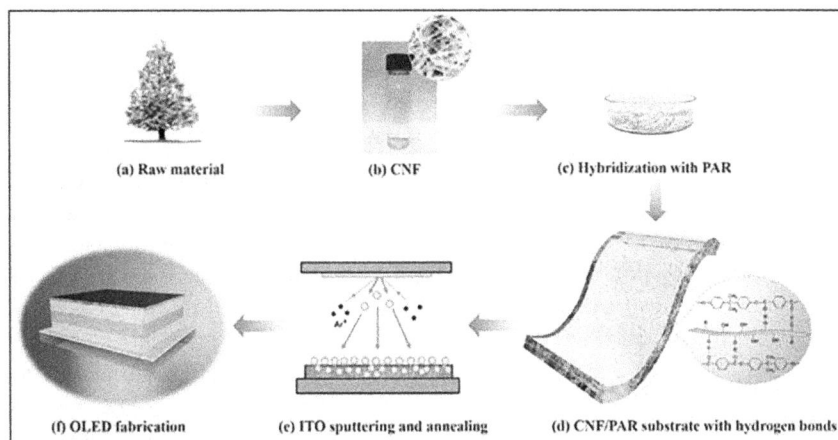

(a) Raw material   (b) CNF   (c) Hybridization with PAR

(f) OLED fabrication   (e) ITO sputtering and annealing   (d) CNF/PAR substrate with hydrogen bonds

FIGURE 6.14 Schematic diagram of fabrication of cellulose nanofibril/polyarylate hybrid polymer. CNF = cellulose nanofibril; ITO = indium tin oxide; PAR = polyarylate. (Reprinted with permission from Tao et al.[7] Copyright © 2020 ACS Publications.)

- In the past two decades, phosphorescent OLED materials have used iridium-based complexes, enhancing full-color display and illumination. Iridium is present in very low abundance in nature. Thus, alternative phosphorescent dopants containing metal centers are essential.
- Sensitivity to moisture, heat, and direct current persist as major hurdles for the organic materials incorporated in OLEDs. Hence, a protective film is essential to protect it from any environmental damage. Thus, increasing the lifetime of OLED devices remains a challenge to overcome the aging of the organic layer.
- The extraction of trapped light within the device is a challenging task for scientific communities.
- The fabrication of multilayer OLEDs does not involve simple processes.

6.9 FUTURE SCOPE

OLEDs rose to the top of the competition and continued to be a strong contender against other LEDs or lighting devices in the last three decades. However, additional advancements need further documentation for better OLED performance. Lifetime, chemical stability, energy, and quantum efficiency, emission color, selection and optimization of organic materials incorporated, and different optical properties require the largest section of attention from researchers for the construction of a high-performance OLED. Addressing the economic factor of cost reduction is necessary for the production of OLEDs, which can include various segments such as organic and inorganic materials, depreciation, and labor. Among small-molecule-based and polymer-based OLEDs, the former provides a greater scope of commercialization, exhibiting maximum efficiency and excellent performance metrics. Phosphorescent metal complex-based OLEDs present opportunities for possible future studies and developments, although high cost and low stability are limiting factors.

The automotive lighting sector is a market that could flourish with the flexible OLED technology, leading to global revolutionary changes in the economic and commercial status. Exploration in the automatic sector could occur through numerous displays and lighting applications of OLEDs. The struggle with current technological limitations cannot envisage many possible ways in the automotive sector, including Head-Up Display (HUD), dashboard displays, internal lighting, etc., which are made possible by OLED technology. Another emerging futuristic platform with exciting outcomes is the digital compatibility of SSL with modern-day electronics and sensors, paving the way for the convergence of SSL with the Internet of Things (IoT). Thus, OLEDs are distinctly grabbing the attention of the global lighting market while also finding meticulous use in medical applications. Further improvements in OLED technology and manufacturing will take us a step closer to commercially viable OLEDs that can fulfill the current global demand for better life quality.

REFERENCES

(1) Sliney, D. International Commission on Illumination. Radiometric quantities and units used in photobiology and photochemistry: recommendations of the commission internationale de l'Eclairage (International Commission on Illumination). *Photochemistry and Photobiology* **2007**, *83*, 425–432.
(2) Spindler, J. P.; Hatwar, T. K.; Miller, M. E.; Arnold, A. D.; Murdoch, M. J.; Kane, P. J.; Ludwicki, J. E.; Alessi, P. J.; Van Slyke, S. A. System considerations for RGBW OLED displays. *Journal of the Society for Information Display* **2006**, *14* (1), 37–48.
(3) Sain, N.; Sharma, D.; Choudhary, P. A review paper on: organic light-emitting diode (OLED) technology and applications. *International Journal of Applied Science and Technology* **2020**, *4* (11), 587–591.
(4) Bauri, J.; Choudhary, R. B.; Mandal, G. Recent advances in efficient emissive materials-based OLED applications: a review. *Journal of Materials Science* **2021**, *56* (34), 18837–18866.
(5) Jeon, W. S.; Park, T. J.; Kim, K. H.; Pode, R.; Jang, J.; Kwon, J. H. High efficiency red phosphorescent organic light-emitting diodes with single layer structure. *Organic Electronics* **2010**, *11* (2), 179–183.
(6) Rothe, C. Organic light emitting diodes: Energy saving lighting technology—A review. *Laser & Photonics Reviews* **2007**, 1, 303–306; (d) Kalyani, N. T. and Dhoble, S. J. *Renewable and Sustainable Energy Reviews* **2012**, *16*, 2696–2723.
(7) Tao, J.; Wang, R.; Yu, H.; Chen, L.; Fang, D.; Tian, Y.; Xie, J.; Jia, D.; Liu, H.; Wang, J. Highly transparent, highly thermally stable nanocellulose/polymer hybrid substrates for flexible OLED devices. *ACS Applied Materials Interfaces* **2020**, *12* (8), 9701–9709.
(8) Wang, Z.; Helander, M.; Lu, Z. Transparent conducting thin films for OLEDs. In *Organic Light-Emitting Diodes (OLEDs)*, Elsevier, 2013; pp 49–76.
(9) Jin, Y.; Ding, X.; Reynaert, J.; Arkhipov, V.; Borghs, G.; Heremans, P.; Van der Auweraer, M. Role of LiF in polymer light-emitting diodes with LiF-modified cathodes. *Organic Electronic* **2004**, *5* (6), 271–281. Yang, H. I.; Jeong, S. H.; Cho, S. M.; Lampande, R.; Lee, K.-M.; Hong, J.-A.; Choi, J.-W.; Kim, B.-s.; Park, Y.; Pode, R. Efficient cathode contacts through Ag-doping in multifunctional strong nucleophilic electron transport layer for high performance inverted OLEDs. *Organic Electronic* **2021**, *89*, 106031.
(10) Han, T.-H.; Song, W.; Lee, T.-W. Elucidating the crucial role of hole injection layer in degradation of organic light-emitting diodes. *ACS Applied Materials Interfaces* **2015**, *7* (5), 3117–3125.

(11) Sasabe, H.; Kido, J. Development of high performance OLEDs for general lighting. *Journal of Materials Chemistry C* **2013**, *1* (9), 1699–1707.

(12) Kandulna, R.; Choudhary, R. Robust electron transport properties of PANI/PPY/ZnO polymeric nanocomposites for OLED applications. *Optik* **2017**, *144*, 40–48; Mandal, G.; Choudhary, R. rGO–Y2O3 intercalated PANI matrix (PANI–rGO–Y2O3) based polymeric nanohybrid material as electron transport layer for OLED application. *Research on Chemical Intermediates* **2019**, *45* (7), 3755–3775.

(13) Hu, B.; Ci, Z.; Liang, L.; Li, C.; Huang, W.; Ichikawa, M. Spiro derivatives as electron-blocking materials for highly stable OLEDs. *Organic Electronic* **2020**, *86*, 105879. Hagen, J. A.; Li, W.; Steckl, A.; Grote, J. Enhanced emission efficiency in organic light-emitting diodes using deoxyribonucleic acid complex as an electron blocking layer. *Applied Physics Letters* **2006**, *88* (17), 171109. Negi, S.; Mittal, P.; Kumar, B. Impact of different layers on performance of OLED. *Microsystem Technologies* **2018**, *24* (12), 4981–4989.

(14) Erickson, N. C.; Holmes, R. J. Investigating the Role of Emissive Layer Architecture on the Exciton Recombination Zone in Organic Light-Emitting Devices. *Advanced Functional Materials* **2013**, *23* (41), 5190–5198. Li, X.; Cui, J.; Ba, Q.; Zhang, Z.; Chen, S.; Yin, G.; Wang, Y.; Li, B.; Xiang, G.; Kim, K. S. Multiphotoluminescence from a Triphenylamine Derivative and Its Application in White Organic Light-Emitting Diodes Based on a Single Emissive Layer. *Advanced Materials* **2019**, *31* (23), 1900613.

(15) Geffroy, B.; Le Roy, P.; Prat, C. Organic light-emitting diode (OLED) technology: materials, devices and display technologies. *Polymer International* **2006**, *55* (6), 572–582.

(16) Pode, R. Organic light emitting diode devices: an energy efficient solid state lighting for applications. *Renewable and Sustainable Energy* **2020**, *133*, 110043.

(17) Turro, N. J. *Modern Molecular Photochemistry*, University Science Books, 1991.

(18) Pode, R. Organic light emitting diode devices: an energy efficient solid state lighting for applications. *Renewable Sustainable Energy Reviews* **2020**, *133*, 110043.

(19) Kalyani, N. T.; Dhoble, S. Novel materials for fabrication and encapsulation of OLEDs. *Renewable and Sustainable Energy Reviews* **2015**, *44*, 319–347.

(20) Swayamprabha, S. S.; Nagar, M. R.; Yadav, R. A. K.; Gull, S.; Dubey, D. K.; Jou, J.-H. Hole-transporting materials for organic light-emitting diodes: an overview. *Journal of Materials Chemistry C* **2019**, *7* (24), 7144–7158.

(21) Nuyken, O.; Jungermann, S.; Wiederhirn, V.; Bacher, E.; Meerholz, K. Modern trends in organic light-emitting devices (OLEDs). *Monatshefte für Chemie/Chemical Monthly* **2006**, *137* (7), 811–824.

(22) Ge, Z.; Hayakawa, T.; Ando, S.; Ueda, M.; Akiike, T.; Miyamoto, H.; Kajita, T.; Kakimoto, M.-A. Solution-processible bipolar triphenylamine-benzimidazole derivatives for highly efficient single-layer organic light-emitting diodes. *Chemistry of Materials* **2008**, *20* (7), 2532–2537.

(23) Nagar, M. R.; Choudhury, A.; Tavgeniene, D.; Beresneviciute, R.; Blazevicius, D.; Jankauskas, V.; Kumar, K.; Banik, S.; Ghosh, S.; Grigalevicius, S. Solution-processable phenothiazine and phenoxazine substituted fluorene cored nanotextured hole transporting materials for achieving high-efficiency OLEDs. *Journal of Materials Chemistry C* **2022**, *10* (9), 3593–3608.

(24) Thaengthong, A.-M.; Saengsuwan, S.; Jungsuttiwong, S.; Keawin, T.; Sudyoadsuk, T.; Promarak, V. Synthesis and characterization of high Tg carbazole-based amorphous hole-transporting materials for organic light-emitting devices. *Tetrahedron Letters* **2011**, *52* (37), 4749–4752.

(25) Murawski, C.; Fuchs, C.; Hofmann, S.; Leo, K.; Gather, M. C. Alternative p-doped hole transport material for low operating voltage and high efficiency organic light-emitting diodes. *Applied Physics Letters* **2014**, *105* (11), 145_141.

(26) Kandulna, R.; Choudhary, R. Concentration-dependent behaviors of ZnO-reinforced PVA–ZnO nanocomposites as electron transport materials for OLED application. *Polymer Bulletin* **2018**, *75* (7), 3089–3107.

(27) Earmme, T.; Jenekhe, S. A. High-performance multilayered phosphorescent OLEDs by solution-processed commercial electron-transport materials. *Journal of Materials Chemistry* **2012**, *22* (11), 4660–4668.

(28) Sarojini, B. K.; Hegde, S. The structural studies of 2-[(4-tert-butylphenyl)-5-(4-biphenyl)]-1, 3, 4-oxadiazole (PBD) incorporated γ ray irradiated chitosan films for optoelectronic applications. *Journal of Molecular Structure* **2021**, *1242*, 130699.

(29) Wu, C.-C.; Sturm, J. C.; Register, R. A.; Tian, J.; Dana, E. P.; Tnompson, M. Efficient organic electroluminescent devices using single-layer doped polymer thin films with bipolar carrier transport abilities. *IEEE Transactions on Electron Devices* **1997**, *44* (8), 1269–1281.

(30) Hoshi, T.; Kumagai, K.-I.; Inoue, K.; Enomoto, S.; Nobe, Y.; Kobayashi, M. Electronic absorption and emission spectra of Alq3 in solution with special attention to a delayed fluorescence. *Journal of Luminescence* **2008**, *128* (8), 1353–1358.

(31) Thangaraju, K.; Kumaran, R.; Ramamoorthy, P.; Narayanan, V.; Kumar, J. Study on photoluminescence from tris-(8-hydroxyquinoline) indium thin films and influence of light. *Optik* **2012**, *123* (15), 1393–1396.

(32) Pfeiffer, S.; Hoerhold, H.-H.; Boerner, H.; Nikol, H.; Busselt, W. Vinylene-bridged triphenylamine dimers as an emitting material in trilayer organic electroluminescent devices. In *Organic Light-Emitting Materials and Devices II*, SPIE, 1998: Vol. 3476; pp 258–266.

(33) Yeh, H.-C.; Meng, H.-F.; Lin, H.-W.; Chao, T.-C.; Tseng, M.-R.; Zan, H.-W. All-small-molecule efficient white organic light-emitting diodes by multi-layer blade coating. *Organic Electronic* **2012**, *13* (5), 914–918.

(34) Osaheni, J. A.; Jenekhe, S. A. Synthesis and processing of heterocyclic polymers as electronic, optoelectronic, and nonlinear optical materials. 1. New conjugated rigid-rod benzobisthiazole polymers. *Chemistry of Materials* **1992**, *4* (6), 1282–1290.

(35) Redecker, M.; Bradley, D.; Jandke, M.; Strohriegl, P. Electron transport in starburst phenylquinoxalines. *Applied Physics Letters* **1999**, *75* (1), 109–111.

(36) Choo, D. C.; Lee, K. S.; Kim, T. W.; Seo, J. H.; Park, J. H.; Kim, Y. K. Luminence efficiency and color stabilization of blue organic light-emitting devices fabricated utilizing a 4, 7-diphenyl-1, 10-phenanthroline electron transport layer containing an ultra thin trap layer. *Molecular Crystals Liquid Crystals* **2010**, *530* (1), 137/[293]–143/[299].

(37) Wang, X.; Yu, J.; Zhao, J.; Lei, X. Comparison of electron transporting layer in white OLED with a double emissive layer structure. *Displays* **2012**, *33* (4–5), 191–194.

(38) Schmitz, C.; Schmidt, H.-W.; Thelakkat, M. Synthesis of lithium-quinolate complexes and their use as emitter and interface materials in OLEDs. In *Organic Light-Emitting Materials and Devices IV*; SPIE, 2001: Vol. 4105; pp 183–193.

(39) Zhong, P.-L.; Zheng, C.-J.; Zhang, M.; Zhao, J.-W.; Yang, H.-Y.; He, Z.-Y.; Lin, H.; Tao, S.-L.; Zhang, X.-H. Highly efficient ternary polymer-based solution-processable exciplex with over 20% external quantum efficiency in organic light-emitting diode. *Organic Electronic* **2020**, *76*, 105449.

(40) Zhang, Y.; Song, J.; Qu, J.; Qian, P.-C.; Wong, W.-Y. Recent progress of electronic materials based on 2, 1, 3-benzothiadiazole and its derivatives: synthesis and their application in organic light-emitting diodes. *Science China Chemistry* **2021**, *64* (3), 341–357.

(41) Lim, S. K. J.; Rahamathullah, R.; Sarih, N. M.; Khairul, W. M. Tailoring tail-free nematogen of ethynylated-schiff base and its evaluation as solution-processable OLED emitting material. *Journal of Luminescence* **2018**, *201*, 397–401.

(42) Mai, R.; Wu, X.; Jiang, Y.; Meng, Y.; Liu, B.; Hu, X.; Roncali, J.; Zhou, G.; Liu, J.-M.; Kempa, K. An efficient multi-functional material based on polyether-substituted indolocarbazole for perovskite solar cells and solution-processed non-doped OLEDs. *Journal of Materials Chemistry A* **2019**, *7* (4), 1539–1547.

(43) Therdkatanyuphong, P.; Chasing, P.; Kaiyasuan, C.; Boonnab, S.; Sudyoadsuk, T.; Promarak, V. High solid-state near infrared emissive organic fluorophores from thiadiazole [3, 4-c] pyridine derivatives for efficient simple solution-processed nondoped near infrared OLEDs. *Advanced Functional Materials* **2020**, *30* (31), 2002481.

(44) Feng, W.; Su, Q.; Ma, Y.; Džolić, Z.; Huang, F.; Wang, Z.; Chen, S.; Tang, B. Z. Tetraphenylbenzosilole: an AIE building block for deep-blue emitters with high performance in nondoped spin-coating OLEDs. *The Journal of Organic Chemistry* **2019**, *85* (1), 158–167.

(45) Uddin, A.; Teo, C. Fabrication of high efficient organic/CdSe quantum dots hybrid OLEDs by spin-coating method. In *Organic Photonic Materials and Devices XV*, SPIE, 2013: Vol. 8622; pp 140–148.

(46) Kuznetsov, K. M.; Kozlov, M. I.; Aslandukov, A. N.; Vashchenko, A. A.; Medved'ko, A. V.; Latipov, E. V.; Goloveshkin, A. S.; Tsymbarenko, D. M.; Utochnikova, V. V. Eu (tta) 3 DPPZ-based organic light-emitting diodes: spin-coating vs. vacuum-deposition. *Dalton Transactions* **2021**, *50* (28), 9685–9689.

(47) Nayak, D.; Choudhary, R. B. Augmented optical and electrical properties of PMMA-ZnS nanocomposites as emissive layer for OLED applications. *Optical Materials* **2019**, *91*, 470–481.

(48) Kandulna, R.; Choudhary, R.; Maji, P. Ag-doped ZnO reinforced polymeric Ag: ZnO/ PMMA nanocomposites as electron transporting layer for OLED application. *Journal of Inorganic and Organometallic Polymers and Materials* **2017**, *27* (6), 1760–1769.

(49) Kandulna, R.; Choudhary, R.; Singh, R.; Purty, B. PMMA–TiO2 based polymeric nanocomposite material for electron transport layer in OLED application. *Journal of Materials Science: Materials in Electronics* **2018**, *29* (7), 5893–5907.

(50) Lan, L.; Zou, J.; Jiang, C.; Liu, B.; Wang, L.; Peng, J. Inkjet printing for electroluminescent devices: emissive materials, film formation, and display prototypes. *Frontiers of Optoelectronics* **2017**, *10* (4), 329–352.

(51) Hu, Y.-X.; Lin, T.; Xia, X.; Mu, W.-Y.; Sun, Y.-L.; He, W.-Z.; Wei, C.-T.; Zhang, D.-Y.; Li, X.; Cui, Z. Novel phosphorescent iridium (iii) emitters for both vacuum-deposition and inkjet-printing of OLEDs with exceptionally high efficiency. *Journal of Materials Chemistry C* **2019**, *7* (14), 4178–4184.

(52) Yokota, T.; Zalar, P.; Kaltenbrunner, M.; Jinno, H.; Matsuhisa, N.; Kitanosako, H.; Tachibana, Y.; Yukita, W.; Koizumi, M.; Someya, T. Ultraflexible organic photonic skin. *Science Advances* **2016**, *2* (4), e1501856. Choi, S.; Kwon, S.; Kim, H.; Kim, W.; Kwon, J. H.; Lim, M. S.; Lee, H. S.; Choi, K. C. Highly flexible and efficient fabric-based organic light-emitting devices for clothing-shaped wearable displays. *Scientific Reports* **2017**, *7* (1), 1–8. Kwon, S.; Hwang, Y. H.; Nam, M.; Chae, H.; Lee, H. S.; Jeon, Y.; Lee, S.; Kim, C. Y.; Choi, S.; Jeong, E. G. Recent progress of fiber shaped lighting devices for smart display applications—a fibertronic perspective. *Advanced Materials* **2020**, *32* (5), 1903488.

(53) Adachi, C.; Baldo, M. A.; Forrest, S. R.; Thompson, M. E. High-efficiency organic electrophosphorescent devices with tris (2-phenylpyridine) iridium doped into electron-transporting materials. *Applied Physics Letters* **2000**, *77* (6), 904–906.

(54) Jia, H. Who will win the future of display technologies? *National Science Review* **2018**, *5* (3), 427–431.

(55) Jung, Y. K.; Choi, H. S.; Ahn, S. Y.; Kim, S.; Choi, H.; Han, C. W.; Kim, B.; Kim, S. J.; Kim, J.; Choi, J. H. 52-3: Distinguished paper: 3 stacked top emitting white OLED for high resolution OLED TV. In *SID Symposium Digest of Technical Papers*, Wiley Online Library, 2016: Vol. 47; pp 707–710.

(56) Shin, H. J.; Takasugi, S.; Park, K. M.; Choi, S. H.; Jeong, Y. S.; Kim, H. S.; Oh, C. H.; Ahn, B. C. 50.1: invited paper: technological progress of panel design and compensation methods for large-size UHD OLED TVs. In *SID Symposium Digest of Technical*

Papers, Wiley Online Library, 2014: Vol. 45; pp 720–723. Zou, S.-J.; Shen, Y.; Xie, F.-M.; Chen, J.-D.; Li, Y.-Q.; Tang, J.-X. Recent advances in organic light-emitting diodes: toward smart lighting and displays. *Materials Chemistry Frontiers* **2020**, *4* (3), 788–820. Han, C.-W.; Han, M.-Y.; Joung, S.-R.; Park, J.-S.; Jung, Y.-K.; Lee, J.-M.; Choi, H.-S.; Cho, G.-J.; Kim, D.-H.; Yee, M.-K. 3-1: invited Paper: 3 Stack-3 Color White OLEDs for 4K Premium OLED TV. In *SID Symposium Digest of Technical Papers*, Wiley Online Library, 2017: Vol. 48; pp 1–4. Shin, H. J.; Takasugi, S.; Park, K. M.; Choi, S. H.; Jeong, Y. S.; Song, B. C.; Kim, H. S.; Oh, C. H.; Ahn, B. C. 7.1: invited Paper: novel OLED display technologies for large-size UHD OLED TVs. In *SID Symposium Digest of Technical Papers*, Wiley Online Library, 2015: Vol. 46; pp 53–56. Chen, C. Y.; Lin, L. F.; Lee, J. Y.; Wu, W. H.; Wang, S. C.; Chiang, Y. M.; Chen, Y. H.; Chen, C. C.; Chen, Y. H.; Chen, C. L. 21.3: a 65-inch amorphous oxide thin film transistors active-matrix organic light-emitting diode television using side by side and fine metal mask technology. In *SID Symposium Digest of Technical Papers*, Wiley Online Library, 2013: Vol. 44; pp 247–250.

(57) Choi, M.; Bae, S.-R.; Hu, L.; Hoang, A. T.; Kim, S. Y.; Ahn, J.-H. Full-color active-matrix organic light-emitting diode display on human skin based on a large-area MoS2 backplane. *Science Advances* **2020**, *6* (28), eabb5898.

(58) Chen, L.; Yu, H.; Dirican, M.; Fang, D.; Tian, Y.; Yan, C.; Xie, J.; Jia, D.; Liu, H.; Wang, J. Highly transparent and colorless nanocellulose/polyimide substrates with enhanced thermal and mechanical properties for flexible OLED displays. *Advanced Materials Interfaces* **2020**, *7* (20), 2000928.

(59) Zhang, X.; Liu, S.; Zhang, L.; Xie, W. In-planar-electrodes organic light-emitting devices for smart lighting applications. *Advanced Optical Materials* **2019**, *7* (3), 1800857.

(60) Tsao, J. Y.; Crawford, M. H.; Coltrin, M. E.; Fischer, A. J.; Koleske, D. D.; Subramania, G. S.; Wang, G. T.; Wierer, J. J.; Karlicek Jr, R. F. Toward smart and ultra-efficient solid-state lighting. *Advanced Optical Materials* **2014**, *2* (9), 809–836. Schubert, E. F.; Kim, J. K. Solid-state light sources getting smart. *Science* **2005**, *308* (5726), 1274–1278.

(61) Hye Oh, J.; Ji Yang, S.; Rag Do, Y. Healthy, natural, efficient and tunable lighting: four-package white LEDs for optimizing the circadian effect, color quality and vision performance. *Light: Science Applications* **2014**, *3* (2), e141–e141.

7 A New Generation of Organic Materials
Photophysical Approach and OLEDs Applications

Mushraf Hussain and Syed S. Razi

7.1 INTRODUCTION

An organic/inorganic family of semiconductors, namely carbon (C), silicon (Si), nitrogen (N), boron (B) and germanium (Ge), has changed all parts of our life through electronics over the past 70 years.[1] From personal cellular phones and computers to satellites, solar energy panels, and the worldwide internet, products from a multi-billion-dollar sector have pervaded our daily lives. The transmission of electronic information to human visual cognition via "display technology" is a critical aspect of this sector. Because it is used in so many electronic gadgets, this area of technology is anticipated to be the fastest-growing market in the future.[2] The enormous technological potential of organic light emitting diode (OLEDs) has been recognized in the display and lighting fields since C. W. Tang and S. A. Van Slyke of Eastman Kodak initially disclosed a brilliant OLED device in 1987.[3] Various scientific breakthroughs, technological improvements, prototypes, and commercial items have proven and demonstrated OLEDs' technological potential.[4] Colored or white self-emission, planar and solid devices, fast reaction speed, thin and light weight, and applicability to flexible applications are all appealing qualities of OLEDs.[4] As a result, it should be clear that not only are OLEDs a fascinating scientific topic but they also have enormous market potential. Due to the rapid performance development and cost reduction of LCDs and LEDs over the past 10 years, OLEDs have faced a substantial and challenging era from liquid crystal displays (LCDs) to light emitting diodes (LEDs).[5] However, new opportunities for OLED displays and lighting devices appear to be the enormous potential of flexible OLEDs, despite the fact that LCDs and LEDs remain the dominant display and lighting devices, respectively. Although organic light emitting devices have made their way into the display sector, improved device efficiencies, lower-cost ingredients, and easier, scalable production processes are still needed.[6] This necessitates a better understanding of organic materials, which can help achieve these objectives. As a result, there is an obvious demand for new organic materials, ideally with a high photoluminescence quantum yield, balanced charge transport, and high light out-coupling via dipole engineering.[7] The ramifications of these discoveries are that device structures should be much more simplified, with enhanced overall

DOI: 10.1201/9781003260417-7

efficiency. These results can be achieved by using highly luminous organic materials and improving device out-coupling, which implies that light can be emitted in a direction that is preferably perpendicular to the device output plane.[8] The fundamental investigations also established certain ground principles for engineering high radiative efficiencies in light-emitting field effect transistors, which should benefit future device iterations in terms of both material and architecture design.[9] Future research could focus on chemically altering material structures to improve characteristics and control dipole orientation. The design and production of sophisticated organic and tiny molecular emitting materials with adequate electrical and optical properties to fully utilize the singlet and triplet excitons for energy conversion is one of the most difficult aspects of creating high-performance OLEDs.[10]

In previous strategies, only 20%–30% of the energy flowing through the device can escape, limiting the development of extremely efficient OLEDs.[11] As a result, one of the primary problems for producing high-performance devices is the development of solutions for the efficient extraction of emitted light. Furthermore, because OLEDs have much potential in flexible displays and lighting, a lot of work has gone into developing flexible OLEDs, particularly flexible transparent conducting electrodes (TCEs).[12] For a better understanding of the backdrop of device structure design, a brief description of the emission mechanisms (optoelectronic) of organic electroluminescent materials and devices is presented first in this chapter.

7.2   PHOTOPHYSICAL MECHANISM AND DESIGN STRATEGY

Carbon chains serve as the foundation for organic semiconductors. Isolated carbon atoms have the electronic configuration $1s^2 2s^2 2p^2$ in their ground state, and they can act as conductive materials through orbital hybridization in the bonding configuration. The conjugated π-electron system in organic semiconductors is produced by overlapping P orbitals of sp^2-hybridized carbon atoms within the molecules.[13] Each carbon atom has three sp^2 orbitals in this arrangement, two of which are bound to neighboring carbon atoms in the carbon chain and the third to a hydrogen atom or another carbon atom. The fourth valence electron is housed in a p-orbital, which can be delocalized along the carbon chain by forming a chain of π-orbitals by overlapping p-orbitals. Bonding (π) and antibonding (π^*) molecular orbitals are formed when the p-orbitals overlap in phase or out of phase (Figure 7.1).[14] Due to a lack of orbital overlap, the π-bonding framework is weaker than the σ-bonding framework in forming the molecule's backbone. As a result, in conjugated molecules, the π-π^* transitions are often the lowest-energy electronic excitations, with energy gaps ranging from 1.5 to 3 eV.[15] The energy gap can be adjusted by the degree of conjugation in the organic semiconductor, allowing absorption and emission of light in the visible spectrum range. As a result, molecular engineering can be used to tune the optoelectronic properties.[15]

When one electron is taken from a molecule, it results in the formation of an ionized molecule with new energy levels. The ionization potential is the energy difference between the original molecule and its ionized state (IP). When an electron is added, however, a new molecule with an additional electron is formed. The electron affinity is the energy differential between the original molecule and the one with an

(a)

(b)

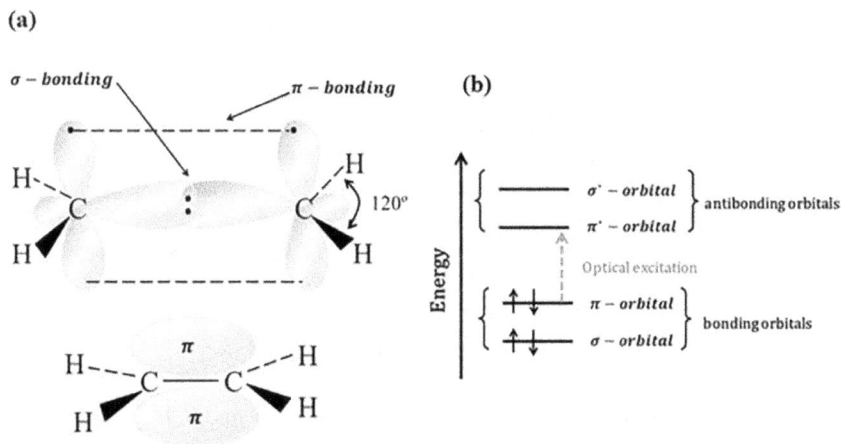

FIGURE 7.1 (a) The σ- and π-bonds, (b) energy level diagram of ethene. The electronic excitation shown is between the bonding π-orbital and the antibonding π*-orbital. (Reprinted with permission from Brütting.[15])

additional electron (EA). The terms IP and EA are frequently used interchangeably in the organic semiconductor research field to refer to highest occupied molecular orbitals (HOMO) and lowest unoccupied molecular orbitals (LUMO), respectively. The electrons are arranged in the molecular orbitals in ascending order of energy. The valence band and conduction band in inorganic semiconductors are equivalent to the HOMO and LUMO, respectively. The electron density distribution in these orbitals determines the charge transport properties of a neutral molecule. To achieve effective p-type charge transport, holes are injected into the HOMO, and the HOMO should be delocalized. In n-type charge transport, electrons are injected into the LUMO in a similar way. Nonetheless, the lack of orbital overlap with neighboring molecules creates a potential barrier between molecules, resulting in charge carriers being stuck on a molecule and poorer overall charge transmission.

7.2.1 SINGLET EMISSION

Singlet excitons are responsible for the emissions of both traditional fluorescent emitters and upconversion fluorescent emitters.[16] Upconversion fluorescence emitters have a higher exciton consumption ratio than traditional fluorescent emitters, which is the difference between the two. This is due to the process of upconversion from triplet to singlet state, which harvests 75% of triplet excitons for light.[16] Developing a high-efficiency, low-cost OLED requires the development of an efficient upconversion fluorescence emitter. Triplet–triplet annihilation (TTA), hybrid local and charge transfer (HLCT), and thermally activated delayed fluorescence (TADF) are the three types of upconversion emitters that is new-generation strategy for photophysics. Because of their exceptional performance, TADF emitters have gotten a lot of attention in recent years.[17]

7.2.2 PREDICTABLE FLUORESCENCE

As charges are injected according to the spin statistical limit, recombination of electrons and holes generates singlet and triplet excitons with a ratio of 1:3 in organic semiconductors.[17] Because of the radiative decay of singlet excitons and the nonradiative decay of triplet excitons, the upper limit of the internal quantum efficiency (IQE, the ratio of the number of generated photons to the number of injected charges) of OLEDs employing traditional fluorescent emitters is around 25%. And maximal external quantum efficiency (EQE) of such devices is only about 5%, as represented by the equation:[1,17]

$$EQE_{max} = \eta_r \times \eta_{ST} \times \Phi_{PL} \times \eta_{out} = IQE_{max} \times \eta_{out}$$

where EQE_{max} is the maximum exterior quantum efficiency and IQE_{max} is the highest internal quantum efficiency, respectively. η_r is the fraction of electron–hole recombination, which in the ideal scenario should be unity. The light out-coupling efficiency of typical OLEDs with an indium tin oxide (ITO)-based flat thin-film design is around 20%. η_{ST} is the fraction of radiative excitons, Φ_{PL} is the photoluminescence quantum yield (PLQY) of the emitting layer, and η_{out} is the light out-coupling efficiency. Fluorescent materials, despite their efficiency limitations, benefit from long lifetimes, especially for steady blue emitters.[18]

7.2.3 TRIPLET–TRIPLET ANNIHILATION

A Jablonski diagram can be used to show the photophysics of TTA upconversion (Figure 7.2).[19] The sensitizer is first stimulated by photo-irradiation (Figure 7.2). The singlet thrilled mood will be filled with excitation of electrons (S0→S1). The sensitizer's triplet excited state will then be populated using intersystem crossing (ISC, for example, S1→T1), which frequently requires the heavy atom effect of transition metal atoms. It should be emphasized that direct T1 excitation is not permitted (S0→T1 is usually a forbidden process).[19] The energy may be transmitted from the triplet sensitizer to the triplet acceptor via the triplet-triplet energy transfer (TTET) process because the lifespan of the triplet excited state is substantially longer than that of the singlet excited state. The upconverted fluorescence is produced by radiative decay from the singlet excited state of the acceptor, which has higher energy than the excitation light. TTA, also known as P-type delayed fluorescence (DF), is a process that can fuse two triplet excitons into a singlet excited state, increasing the IQE_{max} to 62.5%.[20] Through ISC, the generated high-lying triplet exciton jumps to the singlet excited state and emits delayed fluorescence. The TTA emitter's exciton lifespan has both prompt and delayed components, with the delayed part being substantially longer than the prompt part. TTA molecules are challenging to design because the energy difference between 2T1 and Sn must be minimal, and 2T1 must be greater than Sn. The TTA process in the doping system could potentially come from the host material molecules.[21] Finally, the energy is transferred to the guest molecules using a Forster energy transfer mechanism, as shown schematically in Figure 7.2.

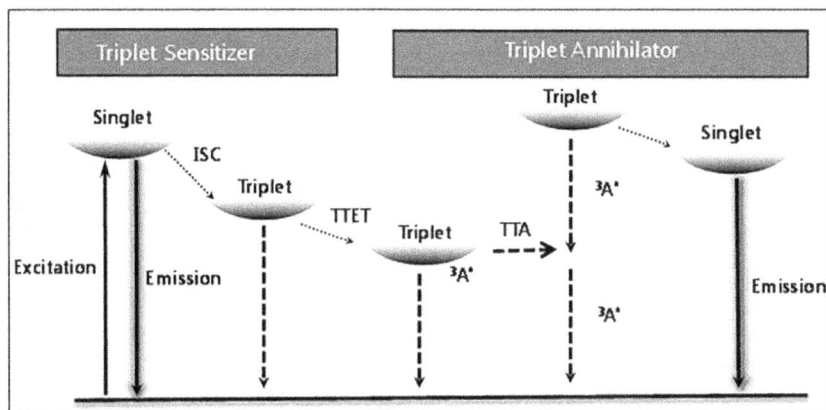

FIGURE 7.2 Schematic Jablonski diagram of triplet–triplet annihilation (TTA) upconversion process. ISC = intersystem crossing; TTET = triplet–triplet energy transfer. (Reprinted with permission from Zhao et al.[19])

7.2.4 THERMALLY ACTIVATED DELAYED FLUORESCENCE

Due to the effective endothermic upconversion, which can use all triplet excitons for emission, thermally activated delayed fluorescence (TADF) or E-type delayed fluorescence is regarded as a potential fluorescent emitter.[22] The excitons on the first triplet excited state in TADF molecules can move to the first singlet state through upconversion. ΔEST lower than 0.2 eV is required to start the reverse intersystem crossing (RISC) process at ambient temperature, which requires a small molecule's orbital overlap between the HOMO and LUMO. Adachi and colleagues published the first pure organic TADF molecule (PIC-TRZ) in 2011.[23] They discovered that PIC-twisted TRZ's donor–acceptor structure can result in less orbital overlap between the HOMO and LUMO. The ΔEST estimated is only 0.11 eV. Using PIC-TRZ as an emitter resulted in a high EQE of 5.3%, whereas the theoretical EQE for a traditional fluorescent emitter with a PLQY of 39% was just 2%. They attributed the results to the RISC process's efficient upconversion.[24] Adachi and coworkers then reported a series of pure organic TADF emitters with a modest ΔEST and a high PLQY.[25] The EQE of a green TADF emitter (4CzIPN) OLED was 19.3%, while the values for orange and sky-blue OLEDs were 11.2% and 8.0%, respectively.[26] These findings suggest that TADF could be a viable technique for producing high-efficiency OLEDs. Using various TADF-based emitters, the highest EQE values of ~38% for blue, ~37% for green, and ~28% for red OLEDs have been achieved up to the present, which are considerably beyond the device efficiency of typical fluorescent OLEDs and even similar to phosphorescent OLEDs.[27–29] However, the stability of TADF-OLEDs, particularly blue TADF devices, cannot match the demands of mass production. In 2018, Adachi and colleagues created $3Ph_2CzCzBN$,[30] a steady sky-blue TADF emitter. The gadget had the longest T80 of 118 hours at an initial brightness of 1000 cd m^{-2}. As a result, the low stability of blue TADF-OLEDs is still a concern that needs to be addressed.

7.2.5 PHOSPHORESCENT EMITTERS

In 1998, Ma and Forrest detected electroluminescence from the triplet excited states of phosphorescent emitters based on Os(II) and Pt(II) complexes.[31] Because of the high spin–orbit coupling effect, not only can the introduction of heavy metal atoms to biological skeletons break through the prohibited transition from T1 to S0, it can also promote the ISC process from S1 to T1.[32] With full usage of excitons, the IQE of phosphorescent OLEDs can reach unity. The fast development of phosphorescent emitters in the ensuing decades ushered in a revolution in the display and lighting industries. Several phosphorescent transition metal complexes, including as Ir(III),[33] Pt(II),[34] Au(III),[35] Os(II),[36] Re(I),[37] Ru(II),[38] and Cu(I),[39] have been employed to make highly efficient OLEDs thus far. Because of their high emission efficiency, short triplet exciton lifetime, and wide color range, Ir(III) complexes, such as Ir(ppy)3 and Ir(piq)3, are the most extensively employed.[40]

7.3   DEVICE STRATEGY AND ORGANIC MATERIALS

7.3.1 ORGANIC LIGHT EMITTING DIODES

An OLED is a light-emitting electrical device that responds to an applied voltage.[41] OLEDs are made up of a substrate and either an emissive layer (EML) or a more sophisticated layered stack of organic materials sandwiched between two electrodes (electron-transporting and hole-transporting layers; Figure 7.3a). When a potential

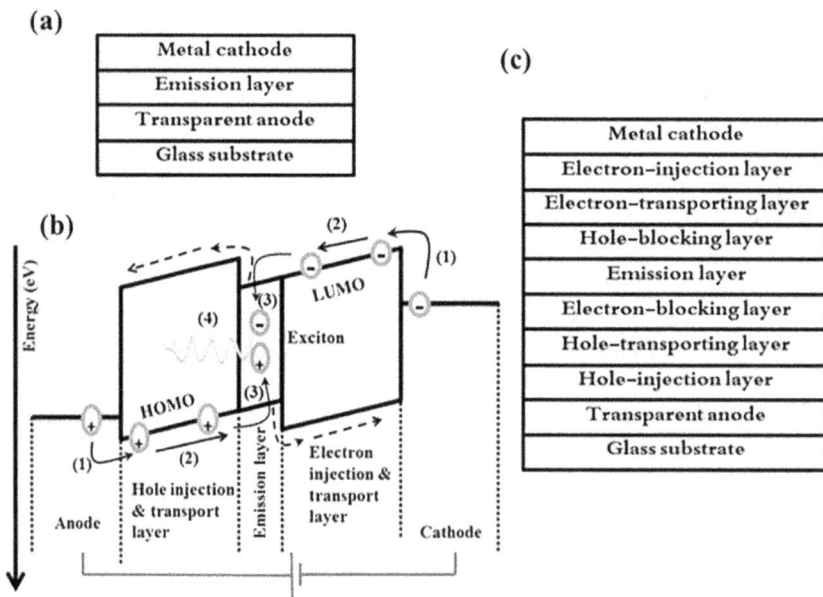

FIGURE 7.3   OLED device structure: (a) an OLED comprising a single active layer, (b) mechanistic processes of electroluminescence, (c) a multilayer OLED. (Reprinted with permission from So.[41])

is provided, electrons and holes are pumped into the organic layer from the cathode and anode, respectively.[41] Under the effect of the applied electric field, opposing charges might migrate toward each other, potentially resulting in recombination (Figure 7.3b), as described by:

$$A^+ + A^- = A_0 + A^*$$

A_0 is the ground state, and A^* is a molecule in an excited state, where A^+ and A^- represent holes (cations) and electrons (anions), respectively. While A^+ and A^- are made up of distinct materials, A^* is made up of a Coulombically bound electron and a hole that can radiatively decay back to the ground state by producing light or nonradiatively cross to another excited state.[41] As a result, the four fundamental processes in the light emission process in OLEDs are charge injection, charge transport, recombination, and radiative decay.

Ionization potentials (IPs—the energy required to remove an electron from the HOMO) of hole-transporting materials (HTMs) are low. Photoelectron spectroscopy or electrochemical oxidation potentials in solution can be used to calculate IPs.[6] Furthermore, HTMs benefit from having a substantial amount of hole drift mobility. Similarly, electron-transporting materials (ETMs) should have high electron drift mobility as well as an appropriate EA. The EA is the inverse of the energy produced when an atom or molecule (X) gains an electron and forms the negative ion (X–). The connection of laser photo-electron spectroscopy data with the LUMO energy levels can yield EAs. Four criteria are commonly used to describe the performance of OLED devices: drive voltage, efficiency, longevity, and color,[6] which are defined in the following sections.

7.3.2 DRIVE VOLTAGE

Drive voltage is the voltage that must be delivered to achieve 1 cd/m² emission intensity, also known as the turn-on voltage. A number of factors influence this voltage, including the built-in potential between electrodes, energy barriers to charge injection, and the electric field required for charge carrier migration through the device.[6] As a result, the switch-on voltage in a well-optimized device will approach the energy of the emitted photons. Furthermore, by carefully selecting the EML, electron-transporting layer (ETL), and hole-transporting layer (HTL) materials, the turn-on voltage can be decreased, lowering the energy barriers for charge injection from the transport layer into the EML.[6]

7.3.3 EFFICIENCY

The ratio of output light energy to energy intake or electrical energy expended is defined as external quantum efficiency (EQE), current efficiency (cd/A), or power efficiency (lm/W).[6]

7.3.4 LIFETIME

The number of hours necessary for the device's photo-intensity to decline to half of its starting value at a certain current density is referred to as potential stability.

The degradation of the interface between the metallic electrodes and the organic layers, chemical reactions from current flow, internal heating of the device at high current densities, and changes in the film morphology of the organic layers are all factors that contribute to device failure.[6] Typical testing current densities are roughly 80 mA/cm^2, or the current necessary to produce 1000 cd/m^2 photo-intensity. Due to the inverse exponential relationship between lifespan and current density, doubling the current density results in a decrease factor of 3–4 in a lifetime.[6]

7.3.5 COLOR

CIEx,y values were defined by the International Commission for Illumination (Commission Internationale de l'Eclairage, CIE) in 1931.[42] The color of a device's whole emission spectrum is reduced to two numbers that define the color as seen by the human eye. The CIE coordinates are (0.14, 0.08) for blue, (0.21, 0.71) for green, (0.67, 0.33) for red, and about (0.33, 0.33) for white, according to the National Television Standard Committee (NTSC). Other performance measures, such as growing drive voltage, power consumption, and changing emission color with device age, can affect one or more of the characteristics listed above. Various strategies for optimizing device performance metrics include: additional functional layers, such as hole-injection layers (HILs), electron-injection layers (EILs), hole-blocking layers (HBLs), and electron-blocking layers (EBLs), are added to the device (see Figure 7.3c). The injection of holes and electrons into the transporting layers is improved by HIL and EIL, respectively. By restricting electrons and holes in the EML, the blocking layers can also improve recombination efficiency.

7.3.6 OLEDs-BASED ORGANIC MATERIALS

Due to their broad emission spectra resulting from intrinsic vibronic coupling and structural relaxation of the S1 state, conjugated organic fluorophores have low color purity.[43] Color filters and optical microcavities have been studied in the manufacturing of fluorescent OLEDs to generate narrow EL spectra with a short full width at half maximum (FWHM).[43] However, developing efficient organic fluorescent emitters with narrow emission bandwidths for good color purity is still a priority. The Huang–Rhys factor (S)[44] determines the relative intensity ($I0–1/I0–0$) of the 0–0 (between the $v = 0$ vibrational levels of S0 and S1) and 0–1 (from $v = 0$ of S1 to $v = 1$ of S0) vibronic transitions.45 A significant orbital overlap results in a dominant 0–0 vibronic transition, converging the S value to zero and producing a sharp single emission peak when conjugated organic molecules have a locally excited (LE) state with a similar equilibrium geometry to the ground state (structural distortion, $\Delta Q \approx 0$).[45] The development of the CT state, on the other hand, might cause a substantial contribution from 0 to n ($n = 1, 2, 3, \ldots$) vibronic transitions, resulting in a broader emission peak.[43,45] To reduce the S value and generate organic fluorescent materials with narrow band emission, stiff structures with no CT character in the excited state must be designed.

In this chapter, we categorize the molecules into the following categories: (i) twisted structures with bulky substituents and suppressed intermolecular aggregation and (ii) rigid/fused aromatic molecules without CT character.[46] Much of the prior research on narrow-emission fluorescent singlet emitters focused on blue-emitting materials and associated EL devices. Green and red emission spectra with exceptionally modest FWHM values have been observed in several anthracene-, pyrene-, and BODIPY-based fluorophores (Figure 7.4).[46] The enhanced LE character and insignificant CT interaction in stiff and symmetrical molecule structures can be attributed to the narrow emission with a dominating 0–0 vibronic transition. The light emitting qualities of representative singlet emitters with narrow band emission are summarized, as well as their device features.[46]

Recently Richard et al. reported organic LEDs based on solution-processed polymers using the phenomenon of TTA upconversion with a number of triplet acceptors such as perylene, 9,10-diphenylanthracene (DPA), TIPS-pentacene, and rubrene that were selected based on their high fluorescence quantum yield and short singlet lifetime (Figure 7.5).[47] These highly emissive molecules were doped in a broad-bandgap poly(9-vinylcarbazole) polymer by solution processing to fabricate LEDs. These emitter-doped polymer LEDs achieved the EQE of 5% and exceeded the IQE by 25%.[47]

In another study styryl pyrene appended with electron donor groups (carbazole, diphenyl amine, difluoro triphenyl amine) were synthesized and demonstrated an

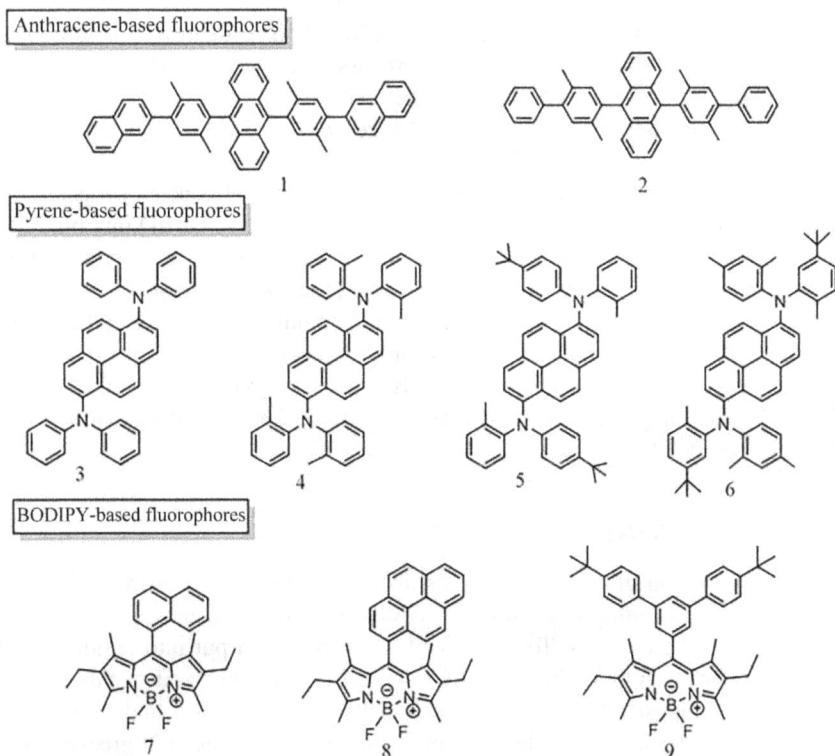

FIGURE 7.4 OLED-based emissive fluorophores. (Reprinted with permission from Zhou et al.[46])

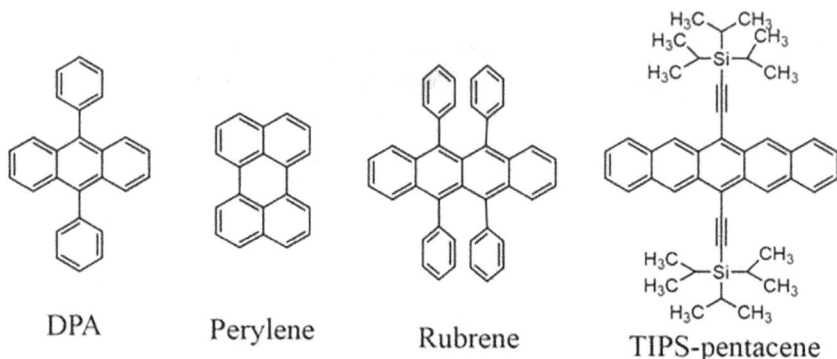

FIGURE 7.5 The molecular structures of triplet acceptors. (Reprinted with permission from Di et al.[47])

outstanding thermal stability along with high fluorescence quantum yield.[48] These molecules were doped in CBP (4,40 -bis(N-carbazolyl)-1,10-biphenyl host to construct blue-light emitting devices that demonstrated excellent efficiency. The maximum efficiency was observed with a DPASP device (EQE = 12%; Figure 7.6) with CBP and DMPPP hosts.[48] The detailed photophysical investigation was carried out in this study and it was revealed that TTA process is responsible for the efficient EL output of the fabricated devices. EL studies with different doping concentrations further confirmed that DPASP (Figure 7.6) is responsible for the TTA-type delayed fluorescence observed in the device.[48]

A team of German and Australian scientists used a novel concept of using TTA upconversion in OLEDs by converting green light emitted by OLED into blue light by using a solid layer of upconverting molecules.[49] Palladium tetraphenylporphryrin (PdTPP) was used as a triplet sensitizer capable of absorbing the green light of the visible spectrum and DPA was used as a triplet annihilator and blue light emitter. The molecular structures of the sensitizer and acceptor are shown in Figure 7.7.[49] The commercially available Ir(ppy)3 was also used as a green emitter in the tandem OLED for high brightness, and EQE reached 19 %. The anti-Stokes shift for the emission was found to be 65 nm.[49]

7.4 APPLICATIONS

The new generation of organic materials and their unique optoelectronic behavior can be utilized in multiple useful applications, ranging from molecular engineering to medical theranostics. To date, much work has been put into building high-performance white-light emitting OLEDs (WOLEDs) for solid-state lighting (SSL) applications in industry.[50] SSL, which is defined as the use of semiconductor electroluminescence to generate visible light for illumination, will result in greener buildings that consume less electricity, reducing our reliance on fossil fuels. SSL will play a critical role in the construction of low-cost, high-performance buildings that use less energy and emit fewer greenhouse gases than their counterparts in the future

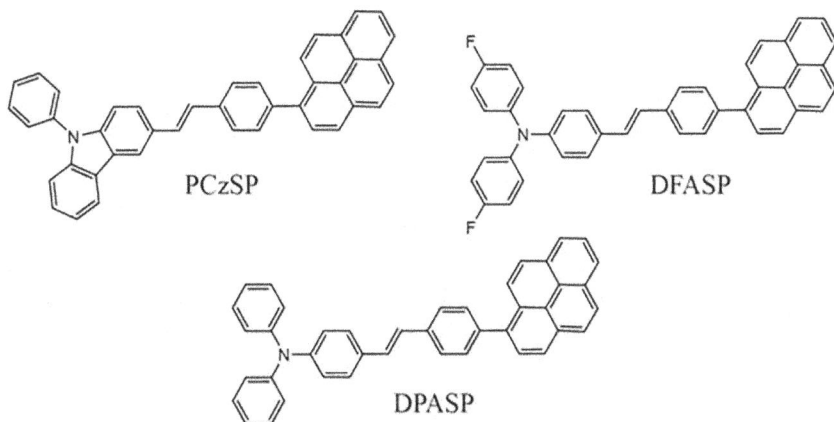

FIGURE 7.6 Pyrene-based triplet acceptors. (Reproduced with permission from Chen et al.[48])

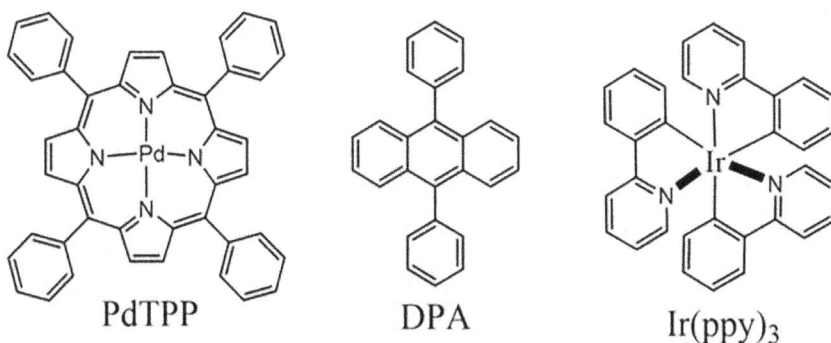

FIGURE 7.7 TTA-UCbased palladium and iridium sensitizers and DPA acceptor. (Reprinted with permission from Reventlow et al.[49])

decade.[51] Since Kido's first revelation of a WOLED in 1994, SSL sources based on WOLEDs have gotten a lot of attention from academic and industrial research communities as the next-generation lighting sources.[52] In terms of color quality, a limited CIE chromaticity coordinate close to the black-body curve with a correlated color temperature (CCT) between 2500 K and 6500 K, as well as a CRI above 80, are necessary for a high-quality WOLED.[53] In 2005, GE Global Research released a 2 ft × 2 ft white OLED panel with a power efficiency (PE) of 15 lm W^{-1}, a CCT of 4400 K, and a CRI of 88.[54] In 2008, the Japanese Research Institute of Organic Electronics developed 14 cm × 14 cm prototype WOLED panels with a PE of 20 lm W^{-1} (@5000 cd m^{-2}).[55] In 2010, Panasonic revealed high-quality WOLED panels with PE >30 lm W^{-1}, CRI >90, and a half-decay lifespan of >40,000 h (@1000 cd m^{-2}).[56]

Owing to the widespread range of applications, devices using organic materials include drivers for neuronal edges (all optical images), pulse oximeters (optical devices to monitor pulse level), chemotherapy (triggers for light-activated devices),

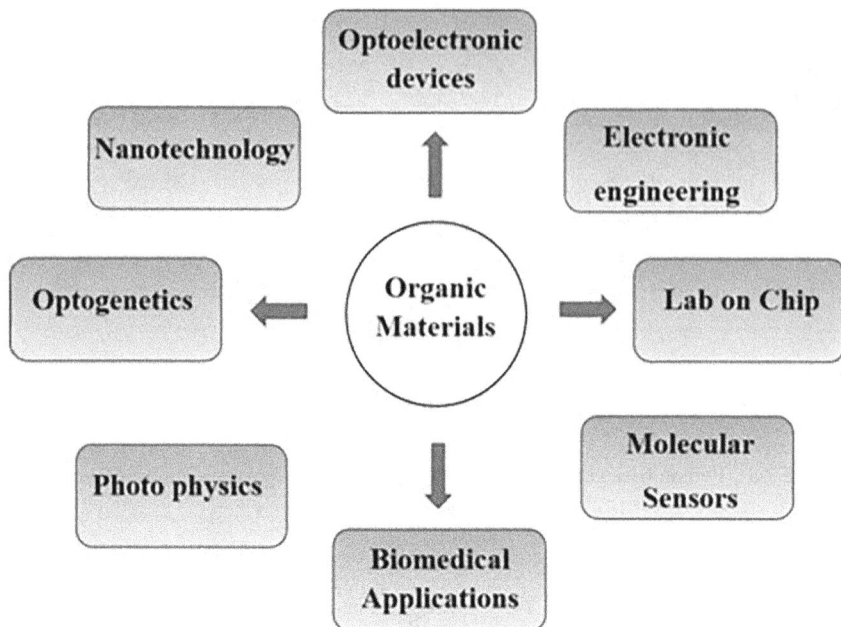

FIGURE 7.8 Schematic of organic materials used in different fields.

biomedical theranostics, and nano devices for ultracompact spectrometers (probe devices) that can perform quantitatively (Figure 7.8).

In 2018, the OLED Works company used thin glass substrates to develop a flexible OLED lighting screen.[57] In 2018, Kim et al. published a paper on kirigami-based 3D OLED lighting in addition to the planar lighting panel.[58] OLED lighting panels have a number of advantages, including their exceptional lightness, thin profile, glare-free display, flexibility, and light homogeneity, which allow for a wide range of appealing designs and novel applications. OLED displays are now widely employed in audio and video players, digital cameras, wearable devices, smartphones, laptops, and OLED television screens.

7.5 CONCLUSIONS AND FUTURE PERSPECTIVES

In this chapter, recent improvements in OLED materials, device topologies, light extraction methodologies, fabrication procedures, and breakthroughs in organic SSL and displays have all been thoroughly explored. First, various materials and their relevant emission principles were summarized. A number of light extraction strategies, including the use of micro-nano structures, have been shown to aid in the release of trapped photons in devices. Future flexible or foldable OLED-based applications will benefit from the rapid development strategies of flexible and stretchable OLEDs. Meanwhile, there are several methods, such as solution-processed fabrication, including spin coating and inkjet printing, that have the potential to lower manufacturing costs significantly. The performance of OLEDs has improved as a result of these scientific endeavors, and their commercial application has accelerated. However, huge technological hurdles must be overcome before OLEDs can become a commercial success in the future. Despite the enormous problems that OLEDs face, fast progress in their development in recent years has given rise to significant optimism regarding this potential technology for future lighting and displays. We believe that scientific research on OLEDs is thriving and flourishing, and that OLEDs will soon infiltrate our daily lives in exciting new forms.

ACKNOWLEDGMENTS

SSR acknowledges the support received through a UGC-BSR Research Start-Up-Grant award from the University Grant Commission, New Delhi (F.No. 30440/2018(BSR)) and Gaya College Gaya, Magadh University, Bodh Gaya, Bihar, India.

REFERENCES

1. Pope, M., Swenberg, C. E. 1999. Electronic processes in organic crystals and polymers, Oxford University Press.
2. Reineke, S., Lindner, F., Schwartz, G., Seidler, N., Walzer, K., Lüssem, B., Leo, K. 2009. White organic light-emitting diodes with fluorescent tube efficiency. *Nature*, 459: 234–238.
3. Tang, C. W., VanSlyke, S. A. 1987. Organic electroluminescent diodes. *Appl. Phys. Lett.*, 51: 913.
4. Schwartz, G. Reineke, S., Rosenow, T. C., Walzer, K., Leo, K. 2009.Triplet harvesting in hybrid white organic light-emitting diodes. *Adv. Funct. Mater.*, 19: 1319–1333.
5. Kuei, C. Y., Tsai, W. L., Tong, B., Jiao, M., Lee, W. K., Chi, Y., Chou, P. T. 2016. Bis-Tridentate Ir(III) complexes with nearly unitary RGB phosphorescence and organic light-emitting diodes with external quantum efficiency exceeding 31%. *Adv. Mater.*, 28: 2795–2800.
6. Shirota, Y., Kageyama, H. 2007. Charge carrier transporting molecular materials and their applications in devices. *Chem. Rev.* 107: 953–1010.
7. Hong, G., Gan, X., Leonhardt, C., Zhang, Z., Seibert, J., Busch, J. M., Bräse, S. 2021. A brief history of OLEDs-emitter development and industry milestones. *Adv. Mater.*, 33: 2005630.
8. Qian, G., Dai, B., Luo, M., Yu, D., Zhan, J., Zhang, Z., Ma, D., Yang, Z. Y. 2008. Band gap tunable, donor-acceptor-donor charge-transfer heteroquinoid-based chromophores: near infrared photoluminescence and electroluminescence. *Chem. Mater.*, 20: 6208–6216.
9. Gierschner, J., Mack, H. G., Luer, L., Oelkrug, D. 2002. Fluorescence and absorption spectra of oligophenylenevinylenes: vibronic coupling, band shapes, and solvatochromism. *J. Chem. Phys.*, 116: 8596–8609.
10. Cui, L. S., Ruan, S. B., Bencheikh, F., Nagata, R., Zhang, L., Inada, K., Nakanotani, H., Liao, L. S., Adachi, C. 2017. Long-lived efficient delayed fluorescence organic light-emitting diodes using n-type hosts. *Nat. Commun.*, 8: 2250.
11. Hecht, D. S., Hu, L., Irvin, G. 2011.Emerging transparent electrodes based on thin films of carbon nanotubes, graphene, and metallic nanostructures. *Adv. Mater.*, 23: 1482–1513.

12. Zhang, D., Ryu, K., Liu, X., Polikarpov, E., Ly, J., Tompson, M. E., Zhou, C. 2006. Transparent, conductive, and flexible carbon nanotube films and their application in organic light-emitting diodes. *Nano Lett.*, 6: 1880–1886.

13. Heeger, A. J., Sariciftci, N. S., Namdas, E. B. 2010. Semiconducting and metallic polymers, Oxford University Press.

14. Baldo, M., Thompson, M., Forrest, S. 2000. High-efficiency fluorescent organic light-emitting devices using a phosphorescent sensitizer. *Nature*, 403: 750–753.

15. Brütting, W. 2006. Physics of organic semiconductors, John Wiley & Sons.

16. Baldo, M., O'Brien, D., Thompson, M., Forrest, S. 1999. Excitonic singlet-triplet ratio in a semiconducting organic thin film. *Phys Rev B*, 60: 14422.

17. Singh-Rachford, T. N., Castellano, F. N. 2010. Photon up conversion based on sensitized triplet–triplet annihilation. *Coord. Chem. Rev.*, 254: 2560–2573.

18. Chiang, C.-J., Kimyonok, A., Etherington, M. K., Griffiths, G. C., Jankus, V., Turksoy, F., Monkman, A. P. 2013. Ultrahigh efficiency fluorescent single and bi-layer organic light emitting diodes: the key role of triplet fusion. *Adv. Funct. Mater.*, 23: 739–746.

19. Zhao, J., Ji, S., Guo, H. 2011. Triplet–triplet annihilation based up conversion: from triplet sensitizers and triplet acceptors to up conversion quantum yields. *RSC Adv*, 1: 937–950.

20. Kondakov, D. Y. 2015.Triplet–triplet annihilation in highly efficient fluorescent organic light-emitting diodes: current state and future outlook. *Philos. Trans. R. Soc.*, A, 373: 2014032.

21. Zhou, J., Chen, P., Wang, X., Wang, Y., Wang, Y., Li, F., Yang, M., Huang, Y., Yu, J., Lu, Z. 2014. Charge-transfer-featured materials-promising hosts for fabrication of efficient OLEDs through triplet harvesting via triplet fusion. *Chem. Commun.*, 50: 7586–7589.

22. Uoyama, H., Goushi, K., Shizu, K., Nomura, H., Adachi, C. 2012. Highly efficient organic light-emitting diodes from delayed fluorescence. *Nature*, 492: 234–238.

23. Endo, A., Sato, K., Yoshimura, K., Kai, T., Kawada, A., Miyazaki, H., Adachi, C. 2011. Efficient up-conversion of triplet excitons into a singlet state and its application for organic light emitting diodes. *Appl. Phys. Lett.*, 98: 083302.

24. Li, W., Pan, Y., Xiao, R., Peng, Q., Zhang, S., Ma, D., Li, F., Shen, F., Wang, Y., Yang, B., Ma, Y. 2014. Employing ~100% excitons in OLEDs by utilizing a fluorescent molecule with hybridized local and charge-transfer excited state. *Adv. Funct. Mater.*, 24: 1609–1614.

25. Chen, X.-K., Kim, D., Brédas, J.-L. 2018. Thermally activated delayed fluorescence (TADF) path toward efficient electroluminescence in purely organic materials: molecular level insight. *Acc. Chem. Res.*, 51: 2215–2224.

26. Shu, Y., Levine, B. G. 2015. Simulated evolution of fluorophores for light emitting diodes. *J. Chem. Phys.*, 142: 104104.

27. Dae Hyun, A., Si Woo, K., Hyuna, L., Ik Jang, K., Durai, K., Ju Young, L., Jang Hyuk, K. 2019. Highly efficient blue thermally activated delayed fluorescence emitters based on symmetrical and rigid oxygen-bridged boron acceptors. *Nat. Photon*, 13: 540–546.

28. Tien-Lin, W., Min-Jie, H., Chih-Chun, L., Pie-Yun, H., Tsu-Yu, C., Ren-Wu, C.-C., Hao-Wu, L., Rai-Shung, L., Chien-Hong, C. 2018. Diboron compound-based organic light-emitting diodes with high efficiency and reduced efficiency roll-off. *Nat. Photon*, 12: 235–240.

29. Zhang, Y. L., Ran, Q., Wang, Q., Liu, Y., Hanisch, C., Reineke, S., Fan, J., Liao, L. S. 2019. High-efficiency red organic light-emitting diodes with external quantum efficiency close to 30% based on a novel thermally activated delayed fluorescence emitter. *Adv. Mater.*, 31: 1902368.

30. Chan, C. Y., Tanaka, M., Nakanotani, H., Adachi, C. 2018. Efficient and stable sky-blue delayed fluorescence organic light-emitting diodes with CIE y below 0.4. *Nat. Commun.*, 9: 5036.
31. Fan, C., Yang, C. 2014. Yellow/orange emissive heavy-metal complexes as phosphors in monochromatic and white organic light-emitting devices. *Chem. Soc. Rev.*, 43: 6439–6469.
32. Yang, X., Zhou, G., Wong, W. Y. 2015. Functionalization of phosphorescent emitters and their host materials by main-group elements for phosphorescent organic light-emitting devices. *Chem. Soc. Rev.*, 44: 8484–8575.
33. Adachi, C., Baldo, M. A., Thompson, M. E., Forrest, S. R. 2001. Nearly 100% internal phosphorescence efficiency in an organic light-emitting device. *J. Appl. Phys.*, 90: 5048.
34. Li, K., Tong, G. S. M., Wan, Q., Cheng, G., Tong, W. Y., Ang, W. H., Kwong, W. L., Che, C. M. 2016. Highly phosphorescent platinum(ii) emitters: photophysics, materials and biological applications. *Chem. Sci.*, 7: 1653–1673.
35. Tang, M. C., Chan, A. K. W., Chan, M. Y., Yam, V. W. W. 2016. Platinum and Gold complexes for OLEDs. *Top. Curr. Chem.*, 374: 46.
36. Liao, J.-L., Chi, Y., Yeh, C.-C., Kao, H.-C., Chang, C.-H., Fox, M. A., Low, P. J., Lee, G.-H. 2015. Near infrared-emitting tris-bidentate Os(II) phosphors: control of excited state characteristics and fabrication of OLEDs. *J. Mater. Chem. C*, 3: 4910–4920.
37. Yu, T., Tsang, D. P. K., Au, V. K. M., Lam, W. H., Chan, M. Y., Yam, V. W. W. 2013. Deep red to near-infrared emitting rhenium(I) complexes: synthesis, characterization, electrochemistry, photophysics, and electroluminescence studies. *Chem. Eur. J.*, 19: 13418–13427.
38. Tung, Y. L., Lee, S. W., Chi, Y., Chen, L. S., Shu, C. F., Wu, F. I., Carty, A. J., Chou, P. T., Peng, S. M., Lee, G. H. 2005. Organic light-emitting diodes based on charge-neutral Ru^{II} phosphorescent emitters. *Adv. Mater.*, 17: 1059–1064.
39. Wu, F., Li, J., Tong, H., Li, Z., Adachi, C., Langlois, A., Harvey, P. D., Liu, L., Wong, W.-Y., Wong, W.-K., Zhu, X. 2015. Phosphorescent Cu(i) complexes based on bis(pyrazol-1-yl- methyl)-pyridine derivatives for organic light-emitting diodes. *J. Mater. Chem. C*, 3: 138–146.
40. Wei, Q., Fei, N., Islam, A., Lei, T., Hong, L., Peng, R., Fan, X., Chen, L., Gao, P., Ge, Z. 2018. Small-molecule emitters with high quantum efficiency: mechanisms, structures, and applications in OLED devices. *Adv. Opt. Mater.*, 6: 1800512.
41. So, F. 2009. Organic electronics: materials, processing, devices and applications, CRC Press.
42. Smith, T., Guild, J. 1931.The CIE colorimetric standards and their use. *Trans. Opt. Soc.*, 33: 73.
43. Frobel, M., Fries, F., Schwab, T., Lenk, S., Leo, K., Gather, M. C., Reineke, S. 2018. Three-terminal RGB full-color OLED pixels for ultrahigh density displays. *Sci. Rep.*, 8: 9684.
44. de Jong, M., Seijo, L., Meijerink, A., Rabouw, F. T. 2015. Resolving the ambiguity in the relation between Stokes shift and Huang-Rhys parameter. *Phys. Chem. Chem. Phys.*, 17: 16959–16969.
45. Kalinowski, J., Fattori, V., Cocchi, M., Williams, J. A. G. 2011. Light-emitting devices based on organometallic platinum complexes as emitters. *Coord. Chem. Rev.*, 255: 2401–2425.
46. Zou, S.-J., Shen, Y., Xie, F.-M., Chen, J.-D., Li, Y.-Q., Tang, J.-X. 2020. Recent advances in organic light-emitting diodes: toward smart lighting and displays. *Mater. Chem. Front.*, 4: 788–820.
47. Di, D., Yang, L., Richter, J. M., Meraldi, L., Altamimi, R. M., Alyamani, A. Y., Credgington, D., Musselman, K. P., MacManus-Driscoll, J. L., Friend, R. H. 2017. Efficient triplet exciton fusion in molecularly doped polymer light-emitting diodes. *Adv. Mater.*, 29: 1605987.

48. Chen, Y.-H., Lin, C.-C., Huang, M.-J., Hung, K., Wu, Y.-C., Lin, W.-C., Chen-Cheng, R.-W., Lin, H.-W., Cheng, C.-H. 2016. Superior up conversion fluorescence dopants for highly efficient deep-blue electroluminescent devices. *Chem. Sci.*, 7: 4044–4051.

49. Reventlow, L. G. V., Bremer, M., Ebenhoch, B., Gerken, M., Schmidt, T. W., Colsmann, A. 2018. An add-on organic green-to-blue photon-up conversion layer for organic light emitting diodes *J. Mater. Chem. C*, 6: 3845–3848.

50. Reineke, S. 2015. Complementary LED technologies. *Nat. Mater.*, 14: 459–462.

51. Lee, S. Y., Yasuda, T., Komiyama, H., Lee, J., Adachi, C. 2016. Thermally activated delayed fluorescence polymers for efficient solution-processed organic light-emitting diodes. *Adv. Mater.*, 28: 4019–4024.

52. Kido, J., Hongawa, K., Okuyama K., Nagai, K. 1994. White light-emitting organic electroluminescent devices using the poly(*N*-vinylcarbazole) emitter layer doped with three fluorescent dyes. *Appl. Phys. Lett.*, 64: 815.

53. D'Andrade, B. W., Forrest, S. R. 2004. White organic light-emitting devices for solid-state lighting. *Adv. Mater.*, 16: 1585–1595.

54. Duggal, A. R., Shiang, J. J., Foust, D. F., Turner, L. G., Nealon, W. F., Bortscheller, J. C. 2005. *Invited Paper:* large area white OLEDs. *SID Symp. Dig. Tech. Pap.*, 2005, 36: 28–31.

55. So, F., Kido, J., Burrows, P. 2008. Organic light-emitting devices for solid-state lighting. *MRS Bull.*, 33: 663–669.

56. Komoda, T., Tsuji, H., Ito, N., Nishimori, T., Ide, N. 2010. *Invited Paper*: high-quality white OLEDs and resource saving fabrication processes for lighting application. *SID Symp. Dig. Tech. Pap.*, 41: 993–996.

57. Spindler, J., Kondakova, M., Boroson, M., Buchel, M., Eser, J., Knipping, J. 2018. *Invited Paper:* advances in high efficacy and flexible OLED lighting. *SID Symp. Dig. Tech. Pap.*, 49: 1135–1138.

58. Kim, T., Price, J. S., Grede, A., Lee, S., Choi, G., Guan, W., Jackson, T. N., Giebink, N. C. 2018. Kirigami-inspired 3D organic light-emitting diode (OLED) lighting concepts. *Adv. Mater. Technol.*, 3: 1800067.

8 Mixed Valence π-Conjugated Coordination Polymers for OLEDs

Bhanu Pratap Singh Gautam, Girijesh Kumar Verma, Nidhi Sharma, Himanshu Pandey, Virendra Kumar, and Manjul Gondwal

8.1 INTRODUCTION

The development of new molecular electronic materials is of recent interest all over the world. Building blocks of suitable molecules assembled into structural architectures by controlling molecular orientation lead to material for specific applications.[1] The solid-state orbital connectivity derived from controlled assembly determines electronic band structure, local magnetic interactions, polarizability, and/or the extent of orbital overlap. These fundamental properties ceate the electrical conductivity, bulk magnetism, and optical nonlinearities found in numerous organic, inorganic, and polymeric materials. The development of solid-state materials with interesting electrical conductivity depends upon synthesis of its building blocks with appropriate electronic structures, followed by controlled assembling of the building blocks in the solid state and their interconnectivity. The design and development of building blocks of organic and organometallic conducting materials that have been reported so far have achieved tremendous success in the formation of semiconductor and superconducting materials.[2–10]

Taking an original chain of conjugated polyene with easily polarizable side groups, W. A. Little explained a theoretical model assuming delocalized electrons of polyene moving in a narrow one-dimensional (1D) band along a chain and predicted that such systems may exhibit high T_c superconductivity.[11] Similarly, Collman suggested chains of metal atoms with polarizable ligands as candidates for the low-dimensional conductors.[12]

In 1D stacks of coordination polymers/charge transfer complexes, the dz^2 orbital of metal or ligand π orbitals of the adjacent chain overlap and form a 1D highly conducting band in the solid state. These interactions are inherently weak and, therefore, are prone to Mott transition (metal–insulator transition).[2] This interaction can be improved by close stacking of perfectly planar building blocks so that overlap of the dz^2 orbital of metal atoms or ligand π orbitals of the adjacent chain is significantly large.[2,3] The stacks of charge transfer complexes as well as covalently bonded

TABLE 8.1

Various Kinds of Modern Light Sources and the Technology behind Them

S. No.	Light Source	Technology Used
1	Arc lamp	Charcoal strips as electrodes
2	Edison's light bulb	Carbon filament
3	Incandescent lamps	Tungsten filament
4	Compact fluorescent lamps	Mercury vapour
5	Light emitting diodes	Semiconductors between electrodes (electrons and holes)
6	Organic light emitting diodes	Organic molecules sandwiched between electrodes
7	Polymer light emitting diodes	Polymers between electrodes

π-conjugated oligomers/polymers undergo Peierls transition (metal–insulator transition).[2,9,13,14] Peierls transitions are suppressed by adding bulky side groups perpendicular to the conduction chain and increasing the dimensionality from one to two or quasi-2D in the formation of mixed metal complexes.[2,15–17]

The stacks of charge transfer complexes and covalently bonded π-conjugated oligomers/polymers are insulators or semiconductors with large band gaps. Both exhibit good electrical conductivity after redox treatment.[2,3,4,7,16] When the level of redox reagents is high or the polymer is exposed for a long time to redox reagents or redox reagents are in homogeneously dispersed throughout the polymer chain, the side reactions on the polymer backbone occur as a result, saturated sites are produced on the polymer backbone along the conducting direction, and the polymer's ability to achieve high conductivity is inhibited.[2,7]

From the early human to modern civilization, illuminating nearby areas to remove darkness is one of the most basic requirements of human civilization. This requirement was accomplished by flambeau in early times, to various kinds of bulbs in this modern era. In modern times, the evolution of light kept on changing with the chain technology age, beginning with Humphrey Davy's arc lamp, Edison's light bulb, to incandescent lamps, compact fluorescent lamps (CFL) and light emitting diodes (LEDs; Table 8.1. All of these have their merits and demerits concerning their technology, lumen efficiency, life span, cost effectiveness, and compatibility. OLEDs and polymer light emitting diodes (PLEDs) are the latest in this chain of development. OLEDs are very thin and flexible; they vary in shape, size, and color, with a pleasant glow.[18–21]

8.2 SOURCES OF LIGHT

Depending upon the source, light can be obtained in two ways:

(a) Incandescence light (hot source), that is, when something is heated to high enough temperature to produce light or, in other words, light is coming out from a hot material/source. For example, light produced by the sun, stars, gas mantle, filament-based bulbs, arc lamp, etc.

(b) Luminescence (cool source), that is, all sources other than a hot source, such as
   1) Chemiluminescence, that is, light produced by a chemical reaction.
   2) Bioluminescence, that is light produced by a living organism, such as a firefly, jelly fish, glowworm, etc.
   3) Electroluminescence, that is, light produced by a material under the influence of an electric current or a strong electric field.
   4) Mechanoluminescence, that is, light is produced by mechanical action on a solid.
   5) Fluorescence, that is, light produced by an excited molecule undergoing a relaxation process from excited singlet state to ground singlet state.
   6) Phosphorescence, that is, light produced by an excited molecule while undergoing a relaxation process from triplet excited state to lower triplet excited state to singlet ground state.

Crystalloluminescence, radioluminescence, thermoluminescence, and cryoluminescence are some other examples of luminescence where light is produced by the material through various processes.

8.3   HISTORY OF LEDs AND OLEDs

The optoelectrical incident in which light is produced by the combination of electrons and holes in a semiconducting material, that is, a diode, is known as electroluminescence and the devices as light emitting diodes (LEDs). This phenomenon was first noticed by Henry Joseph Round, a British researcher, in 1907. He published a 24-line note to the editors of the journal *Electrical World* with the title "A Note on Carborundum", where he mentioned, "On applying a potential of 10 volts between two points on a crystal of carborundum, the crystal gave a yellowish light". Later, a Russian scientist Oleg Vladimirovich Losev independently invented an LED in the late 1920s and early 1930s. His first work was published by the journal *TelegrafiyaiTelefoniya bez Provodov* (Wireless Telegraphy and Telephony) in 1927, with the title, "Luminous Carborundum [Silicon Carbide] Detector and Detection with Crystals".[22] The legacy of LED was carried forward by James R. Biard and Gary Pittman, who patented the technology of infrared radiation from diodes.[23] An American engineer and educator Nick Holonyak, Jr., while working at General Electric, invented LEDs with red light for the first time in 1962.[24] He was known as the "father of the light emitting diode" for his continuous efforts and discoveries in this field. Efforts in this field continue.[25]

Electroluminence was first studied in organic molecules by A. Bernanose in the 1950s, followed by P. Magnante and coworkers.[26–28] Researchers Tang and Van Slyke at Eastman Kodak's research laboratories were working on OLED technology in the 1970s and demonstrated the first viable OLED device in 1987.[29] Since then OLEDs have been explored extensively by researchers in academia and industries. Thousands of patents, journal publications, reviews, and several books and book chapters have been released on this topic.[29,30]

The invention of OLEDs by Tang and Van Slyke in 1987 brought a revolution in display and lighting technologies.[31] OLEDs are monolithic, thin-film, semiconductive devices that emit light when an external voltage is applied. OLEDs use organic fluorescent dyes, and have turned out to be one of the best candidates for the next generation of flat panel displays (FPD) and optoelectronic devices (OEDs).OLEDs are very light, thin like paper, cheap, and flexible enough that they can be fabricated on cloth and produce colored and brighter images. Due to their luminance properties, these sources of energy can be used as energy-saving, eco-friendly, and solid-state lighting agents.[32–36] OLEDs do not require the use of back-lighting and, thus, have much lower power consumption. Due to these exemplary qualities electroluminescent devices based on organic materials have engrossed both researchers and technologists[37,38] Studies show that the number of applications of OLEDs in technology has increased and is expected to pass 20 billion by 2030.[33]

8.4 STRUCTURE AND WORKING OF OLEDs

OLEDs are solidstate area light sources that contain a thin and flexible layer of organic electroluminescent material sandwiched between two electrodes of different work function (Figure 8.1). One of the electrodes (cathode) is metallic while the other (anode) is transparent. Materials generally used for the cathode are of low work function, such as Al, Mg:Al, L:Al, Ca, thin insulator, LiF, and MgO_x. Typically, materials used in the anode are of high work function, such as indium tin oxide (ITO), IZO, TCP (PANI, PEDOT), Au, Pt, Ni, p-Si, thin insulator AlO_x, and SiO_x. The substrate on which these materials are placed is generally made of glass, clear plastic, or metal foil of high work function.

The conductive layer and emissive layer, which transport holes and electrons, respectively, are made up of organic molecules and are 100–500 nanometers thick.

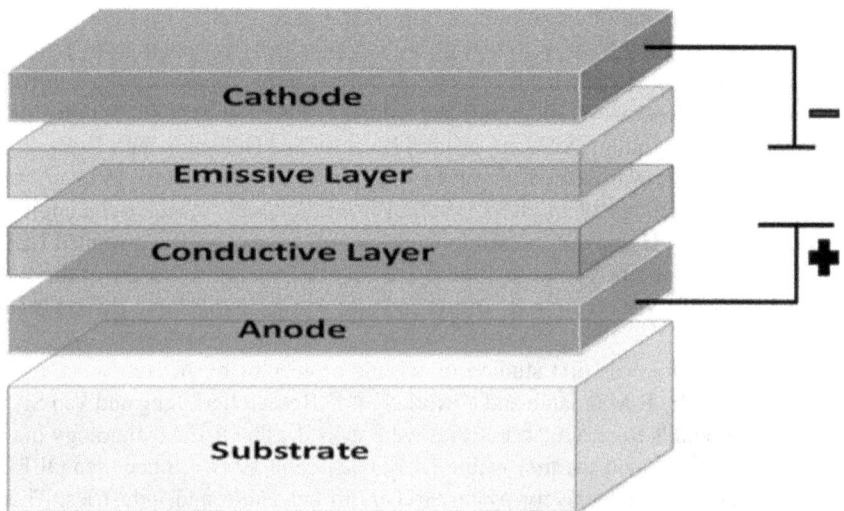

FIGURE 8.1 Typical OLED structure. (Reprinted with permission from Hong et al.[40])

These organic layers placed between electrodes, are carefully deposited on a transparent substrate. The conjugation of –π-electrons present in part of or the entire organic molecule provides a platform for the conduction of electrons and holes in OLEDs. These materials have a conductivity level ranging from insulators to conductors and are therefore considered organic semiconductors. The highest occupied molecular orbital (HOMO) and the lowest unoccupied molecular orbital (LUMO) serve as the valence band and conduction band of the inorganic semiconductors. In a circuit, electrons flow from cathode to anode. When potential difference is applied between the electrodes, electrons are injected from the cathode side to the LUMOs of the organic material, which requires a low barrier of 2–3 eV. On the other hand, for an efficient hole injection from the anode to the HOMOs of the conduction layer a low barrier of 5–6 eV is required. Removal of electrons from the conductive layer, having HOMOs, leaves this layer with holes. These holes need to be filled by the electrons of the emissive layer having LUMOs. The holes, which are more mobile than electrons, jump to the emissive layer and combine with the electrons provided by the cathode. As the electrons fill these holes, they form excitons (the electro-hole pair). The decay of this exciton releases energy in the form of light (photon). The colour of this light photon is a function of difference of the energy gap between HOMOs and LUMOs.[21] The colour of the photon has been tuned by engineering the HOMO-LUMO gap of the organic material, as well as the materials being used in the cathode and anode.[30]

The structure of OLED devices has constantly been improved with time. This may be single layer, double layer, triple layer, or multilayer, depending upon the number of organic layers placed in between the layers of the anode and cathode (Figure 8.2).[20–21,39] In a single-layer OLED device, the organic layer, that is, the emitting layer must have the ability to transport hole and electrons as well as high quantum efficiency for photoluminescence. In a two-layer system, one layer should have good quality to transport holes, while the other layer transports electrons. The electron-hole recombination takes place at the interface of these two layers to generate electroluminescence. In a three-layer system, a specific emitting layer has been sandwiched in between the hole-transport layer and electron-transport layer for the recombination of hole and electron and formation of excitons and corresponding electroluminescence. In a multilayer device, the hole injection layer is placed in between the anode and the hole-transport layer. A hole-blocking layer is also placed between the emitting layer and the electron transport layer to limit the holes in the emitting layer. A multilayer system prevents charge carrier leakage and exciton quenching. In fact, charge injection and mobility are the factors that limit the operating voltage and luminescence efficiency. Radiative decay of excitons, charge balance, and light extraction are the factors that determine the efficiency of an OLED device.

8.5 TYPES OF OLEDs

The types of OLEDs are as follows:[22]

(a) **The active-matrix OLED** (AMOLED) is a multilayer OLED in which the anode has a thin-film transistor (TFT) plane parallel to it to form a matrix.

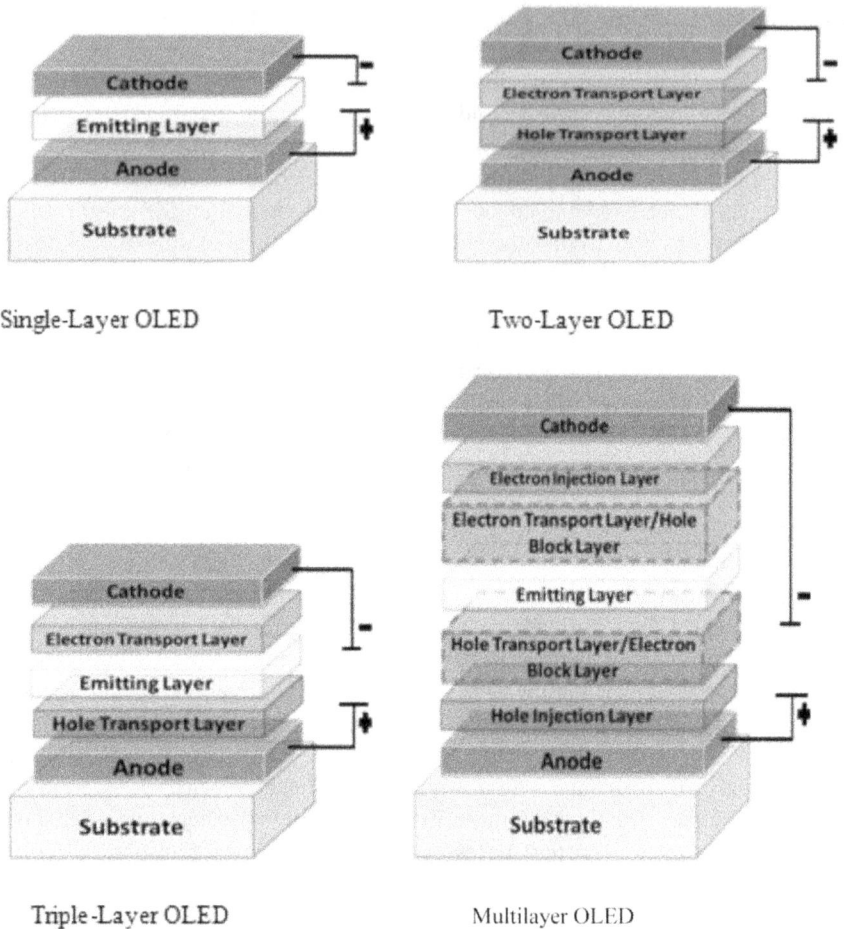

FIGURE 8.2   Evolution in the structure of OLED devices.[18,39]

This gives a better control on each pixel to turn off and on and provides better picture display.

(b) **The passive-matrix OLED** (PMOLED) is based on a multilayer architecture, with anode and cathode strips perpendicular to each other, the intersection of which makes up the pixels. Light emits from these points and the brightness of these pixels are proportional to the current applied.

(c) **Transparent OLEDs** (TOLEDs) are devices that have all transparent components, which allows light to pass in both directions.

(d) **Top-emitting OLEDs** are devices in which the substrate layer is either opaque or reflective. These are better suited for AMOLEDs.

(e) **Foldable OLEDs** have substrates of very flexible metallic foils or plastics.

(f) **White OLEDs** emit white light, which is brighter, more efficient, and a better replacement for CFLs.

8.6   OLED GENERATIONS

Two types of excitons are generated in OLED devices: 1) singlet excitons, 25%, and 2) triplet excitons, 75%. A better, more efficient OLED device needs to harvest both singlet and triplet excitons, that is, materials that may harvest both kind of excitons to produce light are of utmost importance. To meet such a material requirement, OLEDs have gone through three generations, in which various materials have been thoroughly studied. However, next-generation materials for OLEDs are under parallel investigation.[40]

8.6.1   First Generation

OLEDs employ organic dyes and complexes of aluminium (8-hydroxyquinoline aluminium: Alq_3). The organic dyes used were fluorescent molecules. So, these OLEDs were also known as fluorescent OLEDs. These were able to harvest only about 25% of the singlet excitons to produce light. With these materials, green, blue, and orange-red lightemitting OLED devices were developed.

8.6.2   Second Generation

OLEDs relied heavily on phosphorescent complexes based on Ir, Pt, and Au metals. These phosphorescent molecules were able to harvest the remaining 75% of triplet excitons. Second-generation OLEDs performed well and were highly commercialized, but the use of heavy metals raised environmental concerns, limiting the widespread use of second-generation OLEDs.

8.6.3   Third Generation

OLEDS use molecules based on TADF (thermally activated delayed fluorescence) technology. TADF molecules are able to up-convert the triplet excitons to the singlet excitons.[40,41]

After fluorescence, phosphorescence, and TADF emitters, a **next-generation** class of materials are being investigated for high performance of OLEDs. They are showing submicron emission caused by rotationally accessed spin-state inversion (RASI).[42] The use of surface-mounted metal organic frameworks (SURMOFs) in OLED technology will tune the electron-hole mobility and thus will improve the light output.[43]

8.7   ROLE OF METALS IN OLEDs

As described earlier, the anode and cathode are made up of metals. Typically, a transparent substrate of glass or flexible plastic supports a transparent anode. Here the challenge is the anode, which should be a conductor of electricity, have high work function, and be transparent, as well. The anode is typically prepared from the oxides of indium and tin, that is, ITO $[(In_2O_3)_{0.9}(SnO_2)_{0.1}]$. In between these two, different layers are sandwiched, which may inject and transport or block electrons and holes, along with an emitting layer. This is clear from a typical structure of a multilayer OLED device.

Initially some organic dyes and aluminium-based complexes have been used in OLED devices to produce light by hunting excitons. But, with the buildup of knowledge and understanding of OLED devices, their architecture has been updated and numerous materials have been discovered with different characteristics. Metal complexes played a major role in the development of different layers of OLED devices, especially iridium and platinum. However, these were very costly and were also rare earth metals. To address the environmental concerns, sustainability, and commercial aspects, industry and academia focused on developing metal complexes of copper, silver, gold, zinc, tin, aluminium, iron, nickel, manganese, titanium, cobalt, magnesium, etc.[44] Aluminium- and zinc-based complexes are fluorescent emitters, whereas copper- and gold-based complexes are phosphorescent. However, TADF-based emitters are mostly copper, tin, zinc, silver, and gold complexes. RASI-based emitters are carbene-metal-amide complexes of gold (I) and copper (I).

After the utilization of tris(8-hydroxyquinolinato)aluminum (Alq3) by Tang and Van Slyke for OLEDS, Forrest and coworkers developed platinum(II) complex, PtOEP, as a phosphorescent emitter with 23% internal quantum efficiencies.[31,45] This advancement promoted research based on the design and synthesis of metal-based emitters having phosphorescent properties for OLEDs.[46–55] Coordination complexes based on metals can be classified either according to the luminescence processes that they undergo or based on the coordination ligands used and central metals, since same metals can present different luminescence mechanisms in different ligand environments. Mainly platinum- and iridium-based OLED emitters have attracted major development efforts. The availability of rare metals, especially iridium, is a limiting factor for the development of OLEDs. Iridium is one of the rarest elements on the planet, with an abundance of only 0.0007 ppm in the earth's crust.[46–51] These concerns will become the most dominant in the future. To solve this sustainability concern, researchers are trying to find new OLED emitters based on abundant metal complexes.[56] Sustainable metal complexes have been employed and fluorescent emitters developed based mainly on Al(III) and Zn(II).[57–63] Thermally activated delayed fluorescence (TADF)-based emitters, a promising class of photo-functional materials are mostly based on Cu(I), but some examples can be found also for Sn(IV), Ag(I), Au(III), and Zn(II).[64] The rotationally accessed spin-state inversion (RASI)-based emitters having sub-microsecond emission are mainly based on carbene–metal–amides linear complexes, with Au(I) and Cu(I) as metal core.[58] In 2017, Bizzarri et al. concluded that copper(I) compounds seem to be the most relevant candidates, followed by zinc compounds.

8.8 BASIS OF APPROACH TO ACHIEVE MIXED METAL COMPLEXES

The concept of partially filled bands of electronic structure arises from careful analysis of reports in the literature, described in the following sections.

8.8.1 Mixed Valence Metal Oxides

The mixed valence systems where a metal ion in different oxidation states occupying sites A and B bridged by a ligand has been described by Robin and Day,[65] such as:

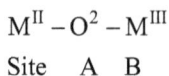

$$M^{II} - O^2 - M^{III}$$

$$\text{Site} \quad A \quad B$$

The properties of a mixed valency system are not equal to the sum of the properties of the two metal ions taken separately, that is, the property of $M^{II} - O^{2-} - M^{III} \neq$ the property of $M^{II} - O^{2-} - M^{II}$ + the property of $M^{III} - O^{2-} - M^{III}$. It was proposed that mixed valence compounds exhibit resonating structures X and Y similar to the resonance structures of organic molecules.

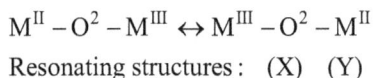

$$M^{II} - O^2 - M^{III} \leftrightarrow M^{III} - O^2 - M^{II}$$

Resonating structures : (X) (Y)

There is interaction between metal ions at site A to B, which results in changes in the physical properties of the system. For example, $W^{VI}O_3$ and LiW^VO_3 are insulators, whereas a mixed oxide of these, $Li_xW^V_xW^{VI}_{1-x}O_3$, is a conductor, possibly due to the resonance structures of the mixed oxide, as shown by X and Y bridged by oxide ion. Thus, each tungsten metal is expected to possess a fractional oxidation state and one electron is delocalized over both tungsten metal ions.

8.8.2 FACTORS CONTROLLING THE INTERACTION BETWEEN TWO METAL IONS (RESONANCE BETWEEN TWO STRUCTURES X AND Y) IN MIXED VALENCE SYSTEMS

On the two different metal sites of different valences A and B the electrons will only be delocalized when mixing coefficient α has a finite large value. This is possible when molecular orbitals on both the sites A and B have the same point group symmetry as Oh, Td, or square planar, etc., and the energy difference between resonating structures X and Y is not large. This depends upon the nature of the metal ion at two different sites A and B, and ligand field strength around the respective metal ions have been shown by the quantum mechanical calculations of Robin and Day.[65]

A system where metal ions at sites A and B are in different symmetry and ligand field strength, the mixing coefficient $\alpha = 0$ and the valences are trapped at their respective sites and delocalization is restricted. So, there are no interactions between two valences site A and B. They are classified as class I.[66] Such systems are expected to be insulators, such as in the case of $Co^{II}Co^{IV}_2O_4$ (Co_3O_4), where Co(II) and Co(III) occupy tetrahedral and octahedral sites, respectively.

The interaction between two structures where metal ions at site A and B possess the same symmetry but different ligand field strength (i.e., bond lengths between metal and ligand at site A differ from those at site B) because of different oxidation states of the same metal ion or the different nature of the metal ions, will possess a finite value of mixing coefficient α and will involve the following: (a) reorganizational energy that must be supplied to stretch or compress the bonds to make the entire sites equivalent, and (b) resonance stabilization energy, which accrues from the oscillation of valences. If reorganizational energy > resonance stabilization energy, the magnitude of α is very small and the two sites are distinguishable; an electron does not spend equal time on both metal ions, that is, slight delocalization occurs between the two structures X and Y. Such systems belong to Class II and behave as semiconductors. For example, $Ti^{IV}_{n-2}Ti^{III}_2O_{2n-1}$ (Ti_nO_{2n-1}) or TiO_x (x = 1.95–2.00) where both cations Ti(IV) and Ti(III) occupy Oh sites.

TABLE 8.2

Characteristic of Various Classes of Mixed Valence Systems

	Class I	Class II	Class III A	Class III B
1	Metal ions in very different symmetry and ligand field strength (Td vs Oh).	Metal ions in nearly identical ligand fields and symmetry differing from one another by distortion of only a few tenths of Å.	Metal ions indistinguishable but grouped into polynuclear clusters. There is no interaction among different clusters.	All metal ions are indistinguishable.
2	Valences of metal ions are very firmly trapped at respective sites.	Valences distinguishable, but with slight delocalization.	Delocalization of valences within the clusters.	Complete delocalization of valences over the cation sublattice.
3	Insulator (resistivity of 10^{10} ohm cm or more).	Semiconductor (resistivity in the range 10–10^7 ohm cm).	Probably insulating	Metallic conductivity (resistivity in the range 10^{-2}–10^{-6} ohm cm).
4	Magnetically dilute (paramagnetic or diamagnetic).	Magnetically dilute (ferro and antiferromagnetic at low temperature).	Magnetically dilute	Ferromagnetic with high curie temperature or diamagnetic.

The system has a maximum value of mixing coefficient α and the two sites A and B are indistinguishable if the resonance stabilization energy > reorganizational energy. Such mixed valence systems exhibit metallic conductivity and belong to Class III A or III B. For example, magnetite Fe_3O_4, which possesses an inverse spinel structure where 8 Td holes are occupied by Fe^{III} ions, 8 out of 16 Oh holes are occupied by Fe^{III}, and the remaining 8 Oh holes are occupied by Fe^{II} ions. Hence, there is strong interaction between the Fe^{II} and Fe^{III} ions in the Oh sites and the system behaves as Class III B as its electrical resistivity at room temperature is 4×10^{-3} ohm-cm (electrical conductivity $\sigma = 2.5 \times 10^3$ S cm^{-1}). Thus, Robin and Day classified the mixed valence systems into three groups. The salient features of mixed valence systems are summarized in Table 8.2.

A superconducting mixed oxide of composition $YBa_2Cu_3O_{7-x}$ ($x \leq 0.5$) having T_c 95 K has been prepared by K. M. K. Wu et al.,[67] which was found to be structurally similar to the perovskite structure (such as $CaTiO_3$, which has a cubic unit cell with a calcium atom at the centre, a titanium atom at each corner, and an oxygen at the middle of each edge. If all the positions of oxygen were occupied the formula would have been $YBa_2Cu_3O_9$ and the average oxidation state of copper equal to $11/3 = 3.67$. Such a composition would be unstable because of the strong oxidizing power of +4 copper and thus one can rationalize oxygen vacancies in the lattice. The $YBa_2Cu_3O_{7-x}$ unit cell is essentially a group of three adjacent perovskite unit cells, with an yttrium and two barium atoms replacing calcium atoms and copper atoms replacing titanium atoms. The top and bottom planes of the cell contain only two oxygen atoms each, and there are no oxygen atoms in the horizontal plane passing through the yttrium atom.

In $YBa_2Cu_3O_7$ there are two structurally distinct sites for the copper atoms: Cu^{III} has a square planar coordination of oxide ions, whereas Cu^{II} is located near the base of the square pyramid of oxide ions. Five coordinate Cu^{II} is displaced about 0.3 Å from the plane of oxide ions. The distortion gives a dimpled CuO_2 plane, which may be of importance to the superconducting properties of this material.[68] The superconducting $YBa_2Cu_3O_7$ is an example of a mixed valence compound of Class III B and may be written as $YBa_2Cu^{III}Cu^{II}_2O_7$.

The well-known high Tc ceramic-based superconductors,[69] for example, $YBa_2Cu_3O_7$, can be considered examples of 3D networked coordination polymers. If reorganizational energy is of the order of thermal energy, then the valence may oscillate via a hopping process. It is to be noted that in this system, symmetry and ligand field strength of both sites are different and the system still behaves like Class III B.

Additionally, an attempt to predict in which class a pair of mixed valence metal ions might fit is aggravated by the fact that unusual coordination can result in mixed valence compounds. For example, one would ordinarily guess a Cu^I, Cu^{II} system to be class I or II because these ions in general have very different coordination symmetries. However, in isostructural $KCu^I_3Cu^{II}S_3$ (KCu_4S_3) and $RbCu^I_3Cu^{II}S_3$ ($RbCu_4S_3$) all the copper ions are in identical regular tetrahedral sites and behave as good conductors and so are classified as class III B. It is to be noted that a tetrahedral geometry has never been found for Cu^{II} salt/complexes.

8.8.3 CREUTZ-TAUBE ION: MIXED VALENCE COORDINATION COMPLEX

In the Creutz-Taube ion[70] both ruthenium ions are expected to exhibit identical redox potential because the two ruthenium metal ions are apparently in identical chemical environments (Figure 8.3). However, in the Creutz-Taube ion the two successive $Ru^{II}Ru^{III}$ couples occur at +0.37 and +0.76V versus the standard hydrogen electrode, a separation of 0.39V. Within the potential domain between these two potentials the complex is therefore in a mixed valence state (denoted {2, 3}), containing one Ru^{II} ion and another Ru^{III} ion. This result indicates that oxidation of one metal center results in a change of electron density at one metal center which is communicated to the other across the bridging ligand and the second metal ion feels the additional positive charge and is therefore more difficult to oxidize than the first. The most apparent demonstration of this interaction is a separation of the two metal-centred redox potentials for metals that are apparently in chemically identical environments.

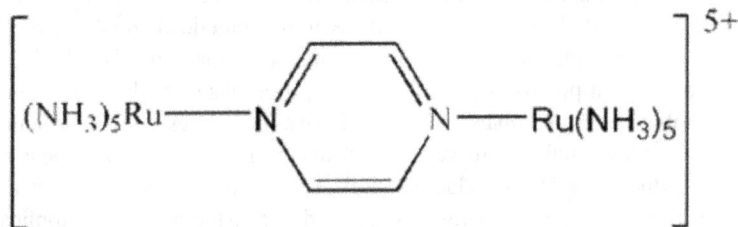

FIGURE 8.3 Creutz-Taube ion.

Creutz and Taube concluded that electronic interaction between two metal ions occurs because the d electrons of the metal ion are in the d(π) orbital, which can effectively overlap with the π–acceptor ligand and are, therefore, delocalized to a certain extent between both metals across a conjugated bridge.[71,72] The complex is therefore in a mixed valence state (denoted by {2,3}) containing one Ru^{2+} and one Ru^{3+} ion.

In very strongly interacting class III complexes according to the Robin and Day classification, the odd electron of the mixed valence state is evenly delocalized between both metals, and a description {2.½, 2.½} is more appropriate than {2,3} for the oxidation states. In such cases the odd electron may be promoted from one orbital delocalized over the whole metal–bridge–metal system to a higher-energy orbital which is likewise delocalized: this process is effectively a π–π^* transition.

Since the preparation of the electronically communicated ruthenium metal ions in the Cruetz-Taube ion, several inorganic chemists studied mixed valence coordination compounds and reported electronic communication between two metal centers bridged by π-conjugated ligands.[73–79] The molecule exhibits strong electronic communication between the two metal centers and shows good electrical conductivity.[65,73,74] Thus, a system with mixed valence but with no interaction between two adjacent metal ions is not a good conductor, whereas a mixed valence system with strong interaction (exhibiting strong electronic communication) between two adjacent metal ions is a good conductor. In a noninteracting mixed valence compound, each metal ion possesses integer oxidation states, whereas in systems with strong interaction between adjacent metal ions, each metal ion possesses noninteger oxidation states.

8.9 MIXED-METAL COMPLEXES

Making use of their coordinating properties, metal-ligand complexes have been successfully incorporated into OLEDs for various applications. A wide variety of ligands, such as bipyridyls, phenanthrolines, and related aromatic nitrogen-containing heterocycles, alkynes, phosphines, porphyrines, and phthalocyanines have emerged as suitable candidates with metal partners for luminescence applications. Although a large number of bimetallic and trimetallic metal-ligand complexes are known for their enhanced luminance, only a few complexes of copper-silver, copper-silver alkenyl complexes, silver-gold and silver-copper complexes, etc., have been studied for their luminance properties. Dinuclear complexes have been studied thoroughly and confirmed through experiments.[80–86]

Mixed-metal complexes bound collectively in a wide variety of structural dispositions have been of great interest to chemists, material scientists, and physicists, as well. Their concerted efforts, backed by the skill of synthetic chemists, have yielded multitudes of interesting and novel polymeric materials that have found wide application. Coordination polymers have the mixed properties of both pure organic and inorganic polymers. This imparts unique and novel properties to the molecular systems formed, which make them very useful materials for use in nonlinear optical materials, synthetic metallic conductors, molecular ferromagnets, ferroelectrics, and one-dimensional conductors. In this chapter we describe the properties, applications, and significance of p-conjugated ligands, which play an important role in the synthesis of LEDs.

8.9.1 TYPES OF COORDINATION POLYMERS

The low-dimensional coordination polymers are broadly classified into two major categories: stack of charge transfer/metal complexes and covalently bonded π-conjugated polymers/oligomers.

8.9.1.1 Stack of Charge Transfer/Metal Complexes

Several efforts have been made in the solid state to form segregated stacks of donor and accepter molecules. For example, tetrathiafulvalene-tetracyanoquinodimethane (TTF-TCNQ),[87] $K_2[Pt(CN)_4]Br_{0.3} \cdot 3H_2O$ (KCP)[2] (Figure 8.4), superconducting $(TMTSF)_2PF_6$ (TMTSF = tetramethyltetraselena-fulvalene)[88] and the first superconducting transition metal complexes of 1,3-dithiole-2-thione-4,5-dithiolate

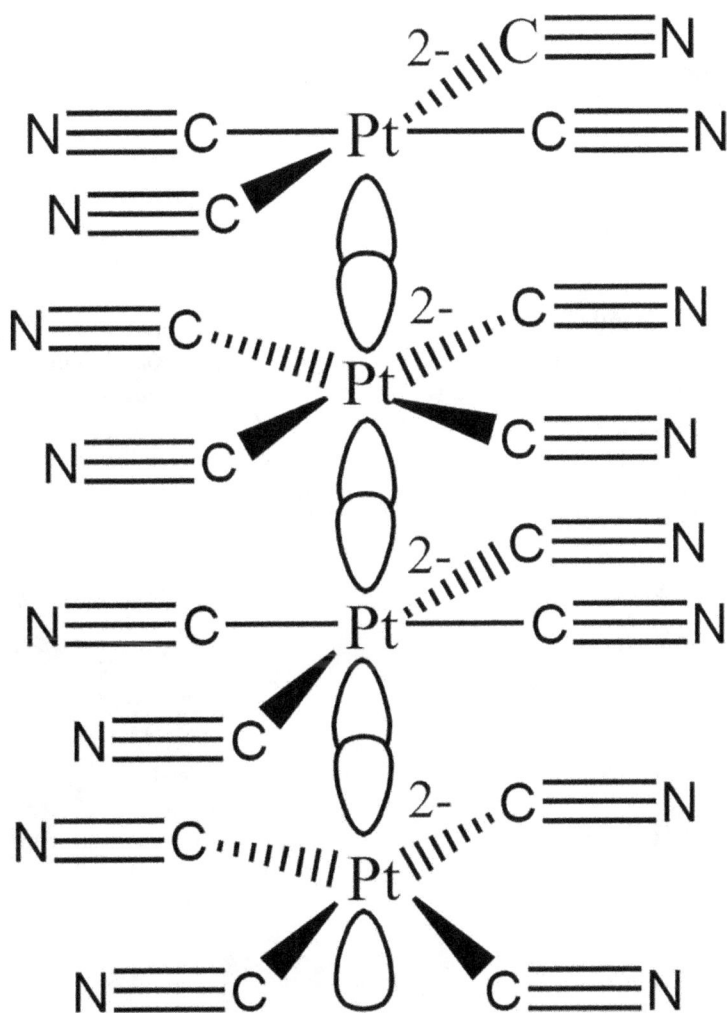

FIGURE 8.4 One-dimensional segregated stacking of $K_2[Pt(CN)_4]Br_{0.3} \cdot 3H_2O$ (KCP).

(DMIT^{2-})[89,90] of the type TTF[M(DMIT)$_2$]$_2$. Also, there are reports of metal coordinated TTF systems showing good conducting properties when doped with iodine compared with their undoped analogues (5–7 orders of magnitude higher).

The tendency for 1D solid interaction owing to the particular combination of (i) constituent molecular units having directional character of orbital and (ii) checked structures in solid imposed by molecular and/or crystal structure. As in the case of KCP, the cyano groups surrounding Pt are staggered with respect to those coordinating the adjacent Pt atoms. This material is a 1D conductor with conductivity greater than 1 Ω^{-1} cm^{-1}. The close stacking is obtained with perfect square planer complexes of elements of d^8 electronic configuration possessing a completely filled band and a completely vacant conduction band. It requires partial oxidation to get a partially filled conduction band. An α-edit-TTF[Ni(DMIT)$_2$]$_2$ transforms to a superconducting state[3,6] at 7 Kbar and at 1.62 K. In solid state, this polymer contains segregated stacks of the donor TTF and acceptor M(dmit)$_2$ molecules. A number of short inter-stack S . . . S contacts are found between M(dmit)$_2$ units in adjacent stacks, and between M(dmit)$_2$ and TTF molecular units. Magnetic interactions due to metal-metal bonding and/or the super exchange phenomenon through the ligand also contribute to the conducting behavior of transition metal complexes. Several metal (Ni, Pd, Pt, and Rh) complexes with dithio-ligands, such as 1,3-dithiol-2-thione-4,5-dithiolate (DMIT^{2-}), exhibit very high conductivity due to 2D molecular interactions of the DMIT^{2-} ligand.

8.9.1.2 Covalently Bonded π-Conjugated Polymers/Oligomers

The π–conjugated polymers, such as polyacetylene, polypyrrole (Figure 8.5), poly (p–phenylene), poly(p–phenylene)sulphide, etc., are semiconductors in the neutral state with a large band gap in excess of ~1.5 eV.[7,73,74,91]

π-Conjugated polymers can be made highly electrical conductors following a redox chemical treatment (Figure 8.6), first shown by Alan Heeger et al.[17] in 1976, for which he was awarded a Nobel Prize in 2000.

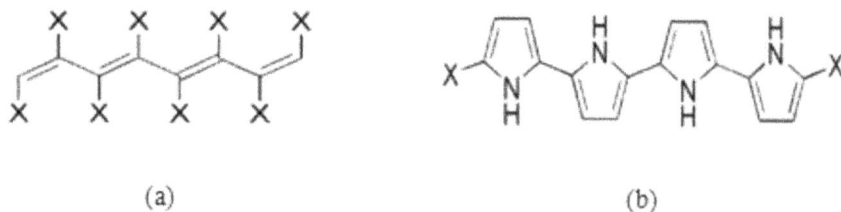

(a)   (b)

FIGURE 8.5 One-dimensional chain of π-conjugated polymers: (a) polyacetylene and (b) polypyrrole.

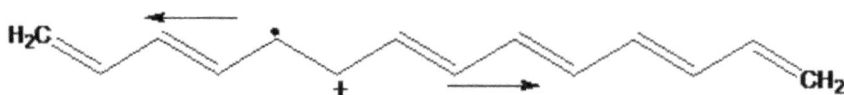

FIGURE 8.6 Partial oxidation of polyacetylene.

Conjugated organic polymers show high conductivity on partial oxidation/reduction due to the creation of polarons whose energy lies between the valence and conduction bands of the neutral chain (Figure 8.7), as in the case of polypyrrole/polyacetylene.[7]

Upon partial oxidation with chemical treatment, semiconducting π-conjugated polymers can be made highly conducting. In the Creutz-Taube ion, two chemically equivalent ruthenium ions coordinated simultaneously to conjugated pyrazine exhibit electronic communication with each other. Inhomogeneous distribution of the reagents during redox treatment yields side reactions on the polymeric backbone and produces saturated sites along the conducting chain direction throughout the polymer chain. As a result, conductivity of the polymer chain is inhibited. Thus, it is essential to create partial filled bands by avoiding the use of redox reagents.

The electrical conductivity of the polymers increases on addition of copper ions along the conjugated polymer chain. Probably the copper (II) ions inject unpaired electrons into the conjugated chain through delocalization of injected unpaired electrons on the donor atoms of the ligand by forming π-back bonds of metal to the coordinated donor atoms of the ligand. Thus, by employing metal ions with unpaired electrons, the use of redox reagents can be avoided to create partial filled bands to form 1D coordination polymers. Kitagawa et al. reported the presence of antiferromagnetic exchange interaction in the 1D coordination polymer $\{[Cu(CA)(H_2O)](H_2O)\}_n$ (CA = chloranilic acid). The antiferromagnetic interaction localizes the unpaired electrons along the conjugated chain direction, thereby preventing the availability of

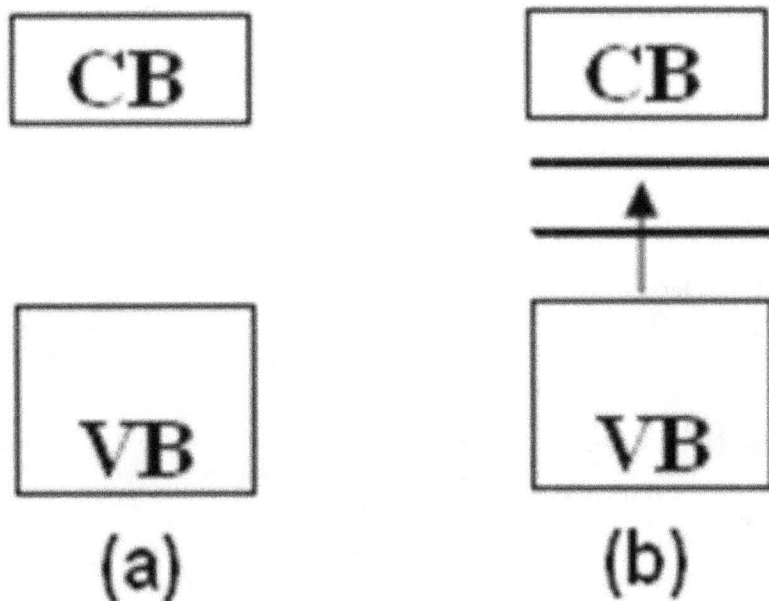

FIGURE 8.7 (a) Electronic band energy of neutral polypyrrole/polyacetylene; (b) creation of polarons on partial oxidation of polypyrrole/polyacetylene chains whose energy lies between the conduction band (CV) and the valence band (VB).

unpaired electrons to form conduction bands. Thus, designing 1D coordination polymers using π-conjugated ligands and metal ions having unpaired electrons diluted with other metal ions to avoid antiferromagnetic interactions is desirable. Organic ligands such as 2,5-dichloro-3,6-bis-(methylamino)1,4-benzoquinone, 2-chloro-5-ethoxy-3,6-bis (methylamino)1,4-benzoquinone and other 1,4-benzoquinones, which possess two donor atoms at each end, are capable of forming mixed metal complexes. In the heterobimetallic complexes of these types of ligands, two adjacent metal ions are different; therefore, antiferromagnetic interactions are expected to be minimized.

8.9.2 Applications of Mixed Metal Complexes

Rapid growth in the field of mixed metal complexes in recent times can be attributed to their expected role as new materials such as optical (luminescence/SHG) materials, molecular magnets, metallic and superconducting polymers, ion exchangers, and microporous solids. A brief discussion of the applications of mixed metal complexes follows.

8.9.2.1 Metallic and Superconducting Mixed Metal Complexes

Chen and Suslick reviewed various coordination polymeric systems that show conductivity similar to that of metals.[92] These include stacked structures having $[Pt(CN)_4]^{2-}$ stacks; planar metal complexes such as $Ni(Pc)(I_3)_{0.33}$; macrocyclic complexes linked by bridging ligands, such as [M(Pc)(L-L)] where (L-L = pyrazine, bipyridine, CN⁻, N_3^-, etc.); iodine-doped polysiloxane polymers of phthalocyanine [M(Pc)X] where (M = Si, Ge, Sn, Al, Ga, Cr, X = O, F) (Pc = the dianion of phthalocyanine). Extremely high metallic conductivity of the order 500,000 S cm⁻¹ at 3.5 K and decreasing steadily with temperature was found in the radical anion salt 2,5-dimethyl-N,N'-dicyanoquinonediimine and Cu(I).[93] An iodine-doped Ag(I) coordination polymer of 2,5-bis(4,5-bis(methylthio)-1,3-dithiol-2-ylidene)-1,3,4,6-tetrathia-pentalene (TTM-TTP) exhibited good conductivity (room temperature [RT] conductivity 0.85 S cm⁻¹)[94] (Figure 8.8).

8.9.2.2 Molecular Magnets

Molecular magnetism deals with the synthesis and characterization of new materials with expected magnetic properties. It plays an important role in the

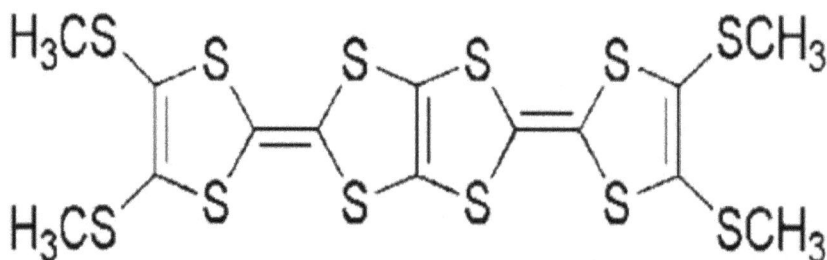

FIGURE 8.8 Structure of the TTM-TTP molecule.

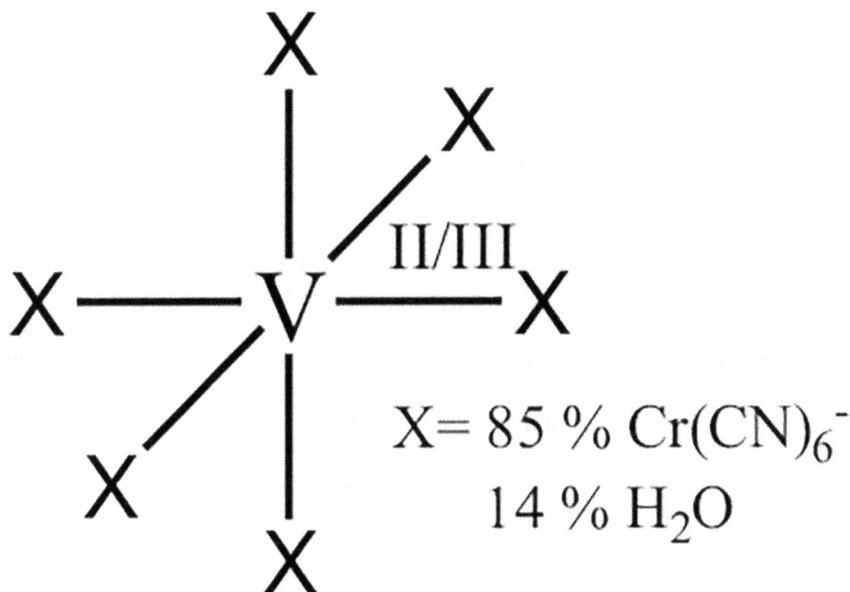

FIGURE 8.9 The first high Tc molecular magnet, $V[Cr(CN)_6]_{0.86} \cdot 2.8H_2O$, showing nonstoichiometry as well as a mixed oxidation state.

emerging field of molecular electronics, where molecular magnetic systems are used in digital display systems and in devising novel magnetic systems useful for information storage and processing.[95] The goal of all studies in this field is to have ferromagnetic (FM) materials that exhibit magnetic ordering at RT. With an aim to develop materials exhibiting magnetism at RT, several systems have been designed and their magnetic behaviour studied; these systems often showed good FM or antiferromagnetic (AFM) interactions. A high Tc (315 K) molecular magnet was first realized in the Prussian blue analogue $V[Cr(CN)_6]_{0.86} \cdot 2.8H_2O$ (Figure 8.9).[96] This system is structurally similar to Prussian blue, except for the small variation in the geometry of V(II). While the above system is a high Tc molecular magnet, it is a nonstoichiometric as well as a mixed-oxidation-state material.

8.10   CRITERIA FOR A GOOD CONDUCTING MOLECULE

To have good conducting properties, a molecule should meet the following criteria:[65,71] (i) symmetry and ligand field strength around both elements with non-integer oxidation states (metal ions) should be as close as possible; (ii) some elements of the molecule, preferably metal ions, should possess noninteger oxidation states; (iii) elements with noninteger oxidation states (metal ions) should be bridged by π-conjugated ligands; (iv) resonance energy should be greater than reorganizational energy.

8.10.1   DESIGN OF A NEW MOLECULAR CONDUCTOR

One-dimensional stacks arise in the charge transfer complexes due to noncovalent interactions in the solid state, which vanish when dissolved in suitable solvents. These 1D interactions are inherently weak in nature; therefore, on increasing the temperature and/or when interstack distance is not so close, Mott transition (metal-insulator transition) appeared. In the case of a π-conjugated oligomer/polymer without redox treatment, creation of a partial filled band (mixed valence state) is not possible. Therefore, stacks of charge transfer complexes as well as π-conjugated oligomer/ polymers independently are not a better design for molecular conductor.

However, a hybrid design of the above two types of conducting materials that have mixed metal complexes with π-conjugated ligands will be a better design for good conducting material. All the bonds are covalent in this proposed design; therefore, the limitation of Mott transition will be avoided. Mixed metal ions present in the system are capable of introducing noninteger oxidation states provided the standard electrode potential difference between them is suitable and bridging π-conjugated ligands satisfy the second criteria of good molecular conductance and conditions for electronic communications between the two metal centers.

Other criteria for a good molecular conductor are dependent on the selection of metal ions used in coordination polymers. The selection of metal ions should be such that metal ions at sites A and B possess similar symmetry, ligand field strengths, and M-L bond lengths based on known literature reports. Metal ions in mixed valence complexes sometimes possess unusual symmetry as discussed above;[65] consequently, it is very difficult to predict decisively the metal ion pair suitable for a good molecular conductor and, hence, all possible metal ion pairs need to be screened.

8.10.2   SIGNIFICANCE OF P–CONJUGATED LIGANDS AND THEIR MIXED METAL COMPLEXES/COORDINATION POLYMERS

The π-conjugated ligands (oligomers/polymers) are redox active[98] and exhibit semiconductor, conductor, and nonlinear optical properties.[7,91,17,98] These properties of π-conjugated ligands make plastic electronics and photonics possible.[7,17] The π-conjugated oligomers/polymers have been used to design a variety of devices such as LEDS,[98–100] electrophosphorescent diodes,[101] photodiodes, dye-sensitized solar cells,[102–105] chemical sensors,[106] transistors,[106–110] and batteries.[111] However, such types of work with π-conjugated oligomers/polymers have been explored with inorganic π-conjugated mixed metal complexes. The interest in these materials is vibrant due to benefit from a unique set of characteristics combining the electrical properties of these semiconductor materials with properties typical of the plastic, that is, low cost, the versatility of chemical synthesis, ease of processing, and flexibility,[7,98] etc.

Cameron and Pickup[81] have shown the electronic communication between two metal ions in mixed metal complexes bridged by π-conjugated ligands. Transition metal ions interact electronically with each other through π-conjugated ligands in the formation of metal complexes and behave as catalysts,[82] as multi-redox systems,[65,83] and as molecular metals,[84] and exhibit much higher photoconductivity.[85]

8.11 FUTURE PERSPECTIVE OF OLED DEVICES

With various types of OLEDs in hand, the future of our society seems to be changing to a great extent. OLEDs may provide options for display technology, including for newspapers, automotive dashboards, billboards, new-era transparent mobiles, and curved televisions. White light emitter OLEDs provide solutions for lighting in home and offices, as well as wearable devices such as smartwatches. White OLEDs are perfectly suited for car lighting because they can display very deep black as well as light so that the displays can be crisp and easy to use while also showing a higher contrast than LCD and LED backlights. Modern fashionable clothing also uses OLEDs for various colors. Research and development in the field of OLEDs are proceeding rapidly and may lead to future real-time applications in heads-up displays, automotive dashboards, billboard-type displays, home and office lighting, and flexible displays.

REFERENCES

1. Etter, M. C. Encoding and decoding hydrogen bond patterns of organic compounds. Acc. Chem. Res., 1990, 23, 120–126.
2. Miller, J. S.; Epstein, A. J. One-dimensional inorganic complexes. Prog. Inorg. Chem., 1976, 20, 1–151, and references therein.
3. Cassoux, P.; Valade, L. 'Molecular inorganic superconductors' in Inorganic materials. Bruce D. W.; O' Hare D. Eds. John Wiley and Sons Ltd, London, 1991.
4. Cassoux, P.; Valade, L.; Kobayashi, H.; Kobayashi, A.; Clark, R. A.; Underhill, A. E. Molecular metals and superconductors derived from metal complexes of 1,3-dithiol-2-thione-4,5-dithiolate (dmit). Coord. Chem., Rev., 1991, 110, 115–160.
5. Ouahab, L. Coordination complexes in conducting and magnetic molecular materials. Coord. Chem. Rev., 1998, 1501–1531.
6. Canadell, E. Electronic structure of transition metal complex based molecular metals and superconductors. Coord. Chem. Rev., 1999, 185, 629–651.
7. Rubner, M. F. 'Conjugated polymeric conductor' in Molecular electronics. Ashwell, G. J. Ed. John Wiley and Sons Inc, New York, 1992, 65–116.
8. Ferraro, J.; Williams, J. M. Introduction to synthetic electrical conductor. Academic Press, Inc., Orlando, FL, 1987.
9. Kobayashi, A.; Fujiwara, E.; Kobayashi, H. Single component molecular metals with extended TTF–dithiolate ligands. Chem. Rev., 2004, 104, 5243–5264, and references therein.
10. Yamada, J.; Akutsu, H.; Nishikawa, H.; Kikuchi, K. New trend in synthesis of π-electron donors for molecular conductors and superconductors. Chem. Rev., 2004, 104, 5057–5083.
11. Little, W. A. Possibility of synthesizing an organic superconductor. Phys. Rev. A, 1964, 134, 1416–1424.
12. Fried, J.; Collman, J.; Doering, W.; Ginzburg, V. L.; Hoffman, R.; West R. Panel discussion: the challenges of superconductivity to organic chemistry. J. Polym. Sci. C, 1970, 29, 133–155.
13. Peierls, R. Quantum theory of solids. Oxford University Press, Oxford, 1955.
14. Bredas, J. L.; Street, G. B. Polarons, bipolarons, and solitons in conducting polymers. Acc, Chem. Res., 1985, 18, 309–315.
15. Reefman, D.; Carnelissen, J. P.; Haasnoot, J. G.; Graff, R. A.G. de; Reedjik, J., Single – crystal structure and electrical conductivity of 1,2,3–trimethylimidazoliu bis – (4,5–dimercapto–1,3–dithiole–2–thionato) nickelate(III). Inorg. Chem., 1990, 29, 3933–3935.

16. Underhill, A. E.; Watkins, D. M., One dimensional metallic complexes. Chem. Soc. Rev., 1980, 9, 429–448.

17. Chiang, C. K.; Fincher, C. R.; Park, Jr., Y. W.; Heeger, A. J.; Shirakawa, H.; Louis, E. J.; Gau, S. C.; MacDiarmid, A. G. Electrical conductivity in doped polyacetylene. Phys. Rev. Lett., 1977, 39, 1098–1101.

18. Thejokalyani, N.; Dhoble, S. Importance of eco-friendly OLED lighting. Defect Diffus. Forum, 2014, 357, 1–27.

19. Karzazi, Y. Organic light emitting diodes: devices and applications. J. Mater. Environ. Sci., 2014, 5, 1–12.

20. Thejo Kalyani, N.; Dhoble, S. Organic light emitting diodes: energy saving lighting technology-a review. Renew Sustain Energy Rev., 2012, 16, 2696–2723.

21. Geffroy, B.; Le Roy, P.; Prat, C. Organic light-emitting diode (OLED) technology: materials, devices and display technologies. Polym. Int., 2006, 55(6), 572–582.

22. Zheludev, N. The life and times of the LED—a 100-year history. Nat. Photon., 2007, 1, 189–192.

23. James, R. B.; Pittman, G. US Patent 3293513, "Semiconductor Radiant Diode", Filed on August 8th, 1962, Issued on December 20th, 1966.

24. Holonyak, N.; Bevacqua, S. F. Coherent (visible) light emission from Ga(As1-xPx) junctions. Appl. Phys. Lett., 1962, 1, 82–83.

25. Nair, G.B.; Dhoble, S. 'Current trends and innovations' in The Fundamentals and Applications of Light-Emitting Diodes. The Revolution in the Lighting Industry Woodhead Publishing Series in Electronic and Optical Materials: Elsevier, 2021, 253–270.

26. Bernanose, A.; Comte, M.; Vouaux, P. Sur UN nouveau mode d'émission lumineuse chez certains composes organiques. J. Chim. Phys., 1953, 50, 64–68.

27. Bernanose, A. Electroluminescence of organic compounds. Br. J. Appl. Phys, 1955, 6, S54–S55.

28. Pope, M.; Kallmann, H. P.; Magnante, P. Electroluminescence in organic crystals. J Chem. Phys., 1963, 38, 2042–2043.

29. Tang, C. W.; VanSlyke, S. A. Organic electroluminescent diodes. Appl. Phys. Lett., 1987, 51, 913–915.

30. Kaji, H.; Suzuki, H.; Fukushima, T.; Shizu, K.; Suzuki, K.; Kubo, S.; Komino, T.; Oiwa, H.; Suzuki, F.; Wakamiya, A.; Murata, Y.; Adachi, C. Purely organic electroluminescent material realizing 100% conversion from electricity to light. Nat. Commun., 2015, 6, 1.

31. Lee, J.G.; Kim, Y.; Jang, S.H.; Soon-Nam Kwon, S.N.; Jeong, K. Mixing effect of chelate complex and metal in organic light-emitting diodes. Phys. Lett.,1998, 72, 1757–1759.

32. Tang, C. W.; VanSlyke, S. A. Organic electroluminescent diodes. Appl. Phys. Lett., 1987, 51, 913–915.

33. Adachi, C.; Baldo, M. A.; Thompson, M. E.; Forrest, S. R. Nearly 100% internal phosphorescence efficiency in an organic light-emitting device. J. Appl. Phys., 2001, 90, 5048–5051.

34. Salehi, A.; Fu, X.; Shin, D.H.; So, F. Recent advances in OLED optical design. Adv. Funct. Mater., 2019, 29, 1808803.

35. Chou, P.-T.; Chi, Y. Phosphorescent dyes for organic light-emitting diodes. Chem. Eur. J., 2007, 13, 380–395.

36. Ho, C.-L.; Li, H.; Wong, W.-Y. Red to near-infrared organometallic phosphorescent dyes for OLED applications. J. Organomet. Chem., 2014, 751, 261–285.

37. Jeong, E. G.; Kwon, J. H.; Kang, K. S.; Jeong, S. Y.; Choi, K. C. A review of highly reliable flexible encapsulation technologies towards rollable and foldable OLEDs. J. Inf. Disp., 2020, 21, 19 – 32.

38. Park, C. I.; Seong, M.; Kim, M. A.; Kim, D.; Jung, H.; Cho, M.; Lee, S. H.; Lee, H.; Min, S.; Kim, J.; Kim, M.; Park, J.-H.; Kwon, S.; Kim, B.; Kim, S. J.; Park, W.; Yang, J.-Y.; Yoon, S.; Kang, I. World's first large size 77-in. transparent flexible OLED display. J. Soc. Inf. Disp., 2018, 26, 287–295.

39. Kalyani N. T.; Dhoble, S. J. Novel materials for fabrication and encapsulation of OLEDs. Renew Sustain Energy Rev., 2015, 44, 319–347.

40. Hong, G.; Gan, X.; Leonhardt, C.; Zhang, Z.; Seibert, J.; Busch, J. M.; Bräse, S. A. Brief history of OLEDs-emitter development and industry milestones. Adv. Mater., 2021, 33, 2005630.

41. Yang, Z.; Mao, Z.; Xie, Z.; Zhang, Y.; Liu, S.; Zhao, J.; Xu, J.; Chi, Z.; Aldred, M. P. Recent advances in organic thermally activated delayed fluorescence materials. Chem. Soc. Rev., 2017, 46, 915–1016.

42. Di, D.; Romanov, A. S.; Yang, L.; Richter, J. M.; Rivett, J. P.; Jones, S.; Thomas, T. H.; Abdi Jalebi, M.; Friend, R. H.; Linnolahti, M.; Bochmann, M.; Credgington, D. High-performance light-emitting diodes based on carbene-metal-amides. Science, 2017, 356, 159–163.

43. Haldar, R.; Heinke, L.; Wöll, C. Advanced photo responsive materials using the metal–organic framework approach. Adv. Mater., 2019, 32(20), 1905227.

44. Bizzarri, C.; Spuling, E.; Knoll, D. M.; Volz, D.; Bräse, S. (2018). Sustainable metal complexes for organic light-emitting diodes (OLEDs). Coord. Chem. Rev., 2018, 373, 49–82.

45. Baldo, M. A.; O'Brien, D. F.; You, Y.; Shoustikov, A.; Sibley, S.; Thompson, M. E.; Forrest, S. R. Highly efficient phosphorescent emission from organic electroluminescent devices. Nature, 1998, 395, 151–154.

46. Gocha, A. USGS mineral commodity summary 2015 highlights. Am. Ceram. Soc. Bull., 2015, 94, 33–35.

47. Baranoff, E.; Yum, J.-H.; Graetzel, M.; Nazeeruddin, M. K. Cyclometallated iridium complexes for conversion of light into electricity and electricity into light. J. Organomet. Chem., 2009, 694, 2661–2670.

48. Omae, I. Application of the five-membered ring blue light emitting iridium products of cyclometalation reactions as OLEDs. Coord. Chem. Rev., 2016, 310, 154–169.

49. Li, T.-Y.; Wu, J.; Wu, Z.-G.; Zheng, Y.-X.; Zuo, J.-L.; Pan, Y. Rational design of phosphorescent iridium(III) complexes for emission color tunability and their applications in OLEDs. Coord. Chem. Rev., 2018, 374, 55–92.

50. Mao, H.-T.; Li, G.-F.; Shan, G.-G.; Wang, X.-L.; Su, Z.-M. Recent progress in phosphorescent Ir(III) complexes for nondoped organic light-emitting diodes. Coord. Chem. Rev., 2020, 413, 213283.

51. Wu, X.; Zhu, M.; Bruce, D. W.; Zhu, W.; Wang, Y. An overview of phosphorescent metallomesogens based on platinum and iridium. J. Mater. Chem. C., 2018, 6, 9848–9860.

52. Che, C. M.; Kwok, C.-C.; Lai, S.-W.; Rausch, A. F.; Finkenzeller, W. J.; Zhu, N.; Yersin, H. Photophysical properties and OLED applications of phosphorescent platinum(II) Schiff base complexes. Chem. Eur. J., 2010, 16, 233–247.

53. Li, K.; Tong, G. S. M.; Wan, Q.; Cheng, G.; Tong, W.-Y.; Ang, W.-H.; Kwong, W.-L.; Che, C.-M. Highly phosphorescent platinum(II) emitters: photophysics, materials and biological applications. Chem. Sci., 2016, 7, 1653–1673.

54. Tang, M.-C.; Chan, A. K.-W.; Chan, M.-Y.; Yam, V. W.-W. Platinum and gold complexes for OLEDs. Top. Curr. Chem., 2016, 374, 46.

55. Fleetham, T.; Li, G.; Li, J. Phosphorescent Pt(II) and Pd(II) complexes for efficient, high-color-quality, and stable OLEDs. Adv. Mater., 2017, 29, 1601861.

56. Anastas, P. T.; Warner, J. C. Green Chemistry: Theory and Practice, Oxford University Press, Oxford, New York, 1998.

57. Chan, K. T.; Tong, G. S. M.; Wan, Q.; Cheng, G.; Yang, C.; Che, C.-M. Strongly luminescent cyclometalatedgold(III) complexes supported by bidentate ligands displaying intermolecular interactions and tunable emission energy. Chem. Asian J., 2017, 12, 2104–2120.

58. Di, D.; Romanov, A. S.; Yang, L.; Richter, J. M.; Rivett, J. P. H.; Jones, S.; Thomas, T. H.; Abdi Jalebi, M.; Friend, R. H.; Linnolahti, M.; Bochmann, M.; Credgington, D. High-performance light-emitting diodes based on carbene-metal-amides. Science, 2017, 356, 159–163.

59. Beucher, H.; Kumar, S.; Kumar, R.; Merino, E.; Hu, W.-H.; Stemmler, G.; Cuesta-Galisteo, S.; González, J. A.; Bezinge, L.; Jagielski, J.; Shih, C.-J.; Nevado, C. Phosphorescent κ3-(N^C^C)-gold(III) complexes: synthesis, photophysics, computational studies and application to solution-processable OLEDs. Chem. Eur. J., 2020, 26, 17604–17612.

60. Lima, J. C.; Rodríguez, L. Highlights on gold TADF complexes. Inorganics, 2019, 7, 124. Parker, R. R.; Liu, D.; Yu, X.; Whitwood, A. C.; Zhu, W.; Williams, J. A. G.; Wang, Y.; Lynam, J. M.; Bruce, D. W. Synthesis, mesomorphism, photophysics and device performance of liquidcrystalline pincer complexes of gold(III). J. Mater. Chem. C., 2021, 9, 1287–1302.

61. Malmberg, R.; von Arx, T.; Hasan, M.; Blacque, O.; Shukla, A.; McGregor, S. K. M.; Lo, S.-C.; Namdas, E. B.; Venkatesan, K. Tunable light-emission properties of solution-processable N-heterocyclic carbene cyclometalatedgold(III) complexes for organic light-emitting diodes. Chem. Eur. J., 2021, 27, 7265–7274.

62. Xu, X.; Liao, Y.; Yu, G.; You, H.; Di, C.; Su, Z.; Ma, D.; Wang, Q.; Li, S.; Wang, S.; Ye, J.; Liu, Y. Charge carrier transporting, photoluminescent, and electroluminescent properties of zinc(II)-2- (2-hydroxyphenyl)benzothiazolate complex. Chem. Mater., 2007, 19, 1740–1748. Dumur, F. Zinc complexes in OLEDs: an overview. Synth. Met., 2014, 195, 241–251.

63. Sakai, Y.; Sagara, Y.; Nomura, H.; Nakamura, N.; Suzuki, Y.; Miyazaki, H.; Adachi, C. Zinc complexes exhibiting highly efficient thermally activated delayed fluorescence and their application to organic light-emitting diodes. Chem. Commun., 2015, 51, 3181–3184.

64. Wai-Pong, T.; Glenna, G.C. So Ming Tong, S.M.; Zhou, D.; Che, C.M. Recent advances in metal-TADF emitters and their application in organic light-emitting diodes. Front. Chem., 2020, 8, 653.

65. Robin, M. B.; Day, P. Mixed valence chemistry-a survey and classification. Adv. Inorg. Radiochem., 1967, 10, 247–422.

66. Kushwaha, A.; Prasad R. L. Synthesis, characterization and *ab initio* calculation of π–conjugated ligand & its coordination polymer, Thesis, 2010, 1–196.

67. Wu, M. K.; Ashburn, J. R.; Torng, C. J.; Hor, P. H.; Meng, R. L.; Gao, L.; Huang, Z. J.; Wang, Y. Q.; Chu, C. W. Superconductivity at 93 K in a new mixed-phase Y-Ba-Cu-O compound system at ambient pressure. Phys. Rev. Lett., 1987, 58, 908–910.

68. Whangbo, M. H.; Torardi, C. C. Hole density dependence of the critical temperature and coupling constant in the cuprate superconductors. Science, 1990, 249, 1143–1146.

69. Shriver, D.F.; Atkins, P.W.; Langford, C.H. Inorganic Chemistry, Oxford University Press, Walton Street, Oxford, 1990, 588.

70. Creutz, C.; Taube, H. Binuclear complexes of ruthenium ammines. J. Am. Chem. Soc., 1973, 95, 1086–1094.

71. Stebler, A.; Ammeter, J.H.; Furholz, U.; Ludi, A. The Creutz-Taube complex revisited: a single-crystal EPR study. Inorg. Chem., 1984, 23, 2764–2767.

72. Ward, M. D. Metal-metal interactions in binuclear complexes exhibiting mixed valency; molecular wires and switches. Chem. Soc. Rev., 1995, 24, 121–134.

73. Zhu, S.S.; Swager, T. Conducting metallorotaxanes: metal ion mediated enhancement in conductivity and charge localization. J. Am. Chem. Soc., 1997, 119, 12568–12577.

74. Yamamoto, T.; Maruyama, T.; Zhou, Z-H.; Ito, T.; Fukuda, T.; Yoneda, Y.; Begum, F.; Ikeda, T.; Sasaki, S.; Takezoe, H.; Fukuda, A.; Kubota, K. π-Conjugated poly(pyridine–2,5–diyl), poly(2,2'–bipyridine–5,5'–diyl), and their alkylderivatives. Preparation, linear structure, function as a ligand to form their transition metal complexes, catalytic reactions, n-type electrically conducting properties, optical properties and alignment on substrates. J. Am. Chem. Soc., 1994, 116, 4832–4845.

75. Dei, A.; Gatteschi, D.; Pardi, L. Dinuclear ruthenium complexes with bridging 1,4,5,8–tetraoxonaphthalene: redox properties and mixed-valence interactions. Inorg. Chem., 1990, 29, 1442–1444.

76. Ward, M. D. A dinuclearruthenium(II) complex with the dianion of 2,5– dihydroxy–1, 4–benzoquinone as bridging ligand. redox, spectroscopic, and mixed – valence properties. Inorg. Chem., 1996, 35, 1712–1714.

77. Lee, G.H.; Ciana, L. D.; Haim, A. Intramolecular electron transfer from pentacyanoferrate(II) to pentaamminecobalt(III) with 3,3'–dimethyl–4,4'– bipyridine,4,4' – bipyridylacetylene, 1,4–bis(4–pyridyl)butadiyne, 2,7-diazapyrene, and 3,8– phenanthroline as bridging ligands: adiabaticity and the role of distance. J. Am. Chem. Soc., 1989, 111, 2535–2541.

78. Bencini, A.; Ciofini, I.; Daul, C.A.; Ferretti, A. Ground and excited state properties and vibronic coupling analysis of the Creutz Taube ion, $[(NH_3)_5Ru - pyrazine - Ru(NH_3)_5]^{5+}$, Using DFT. J. Am. Chem. Soc., 1999, 121, 11418–11424.

79. Coronado, E.; Day, P., Magnetic molecular conductors. Chem. Rev., 2004, 104, 5419–5448.

80. Koshevoy, I. O.; Karttunen, A. J.; Lin, Y. C.; Lin, C. C.; Chou, P. T.; Tunik, S. P.; Haukka, M.; Pakkanen, T. A. Synthesis, photophysical and theoretical studies of luminescent silver(i)–copper(i) alkynyl-diphosphine complexes. Dalton Trans., 2010, 39, 2395–2403.

81. Xu, L.-J.; Zhang, X.; Wang, J. Y.; Chen, Z. N. High-efficiency solution-processed OLEDs based on cationic Ag_6Cu heteroheptanuclear cluster complexes with aromatic acetylides. J. Mater. Chem. C., 2016, 4, 1787–1794.

82. Xu, L.-J.; Wang, J.-Y.; Zhu, X.-F.; Zeng, X.-C.; Chen, Z.-N. Phosphorescent cationic $Au_4 Ag_2$ alkynyl cluster complexes for efficient solution-processed organic light-emitting diodes. Adv. Funct. Mater., 2015, 25, 3033–3042.

83. Volz, D.; Baumann, T.; Flugge, H.; Mydlak, M.; Grab, T.; Bachle, M.; Barner, K. C.; Bräse, auto-catalysed crosslinking for next-generation OLED-design. J. Mater. Chem., 2012, 22, 20786–20790.

84. Volz, D.; Chen, Y.; Wallesch, M.; Liu, R.; Fléchon, C.; Zink, D.M.; Friedrichs, J.; Flügge, H.; Steininger, R.; Göttlicher, J.; Heske, C.; Weinhardt, L.; Bräse, S.; So, F.; Baumann, T. Bridging the efficiency gap: fully bridged Dinuclear cu(i)-complexes for singlet harvesting in high-efficiency OLEDs. Adv. Mat., 2015, 27, 2538–2543.

85. Volz, D.; Wallesch, M.; Grage, S. L.; Göttlicher, J.; Steininger, R.; Batchelor, D.; Vitova, T.; Ulrich, A. S.; Heske, C.; Weinhardt, L.; Baumann, T.; Bräse, S. Labile or stable: can homoleptic and heterolepticpyrphos–copper complexes be processed from solution? Inorg. Chem., 2014, 53, 7837–7847.

86. Wallesch, M. A.; Verma, C.; Fléchon, H.; Flügge, D. M.; Zink, S. M.; Seifermann, J. M.; Navarro, T.; Vitova, J.; Göttlicher, R.; Steininger, L.; Weinhardt, M.; Zimmer, M.; Gerhards, M.; Heske, C.; Bräse, S.; Baumann, T.; Volz, D. Frontispiece: towards printed organic light-emitting devices: a solution-stable, highly soluble Cu^I-NHetPHOS** complex for inkjet processing. Chem. Eur. J., 2016, 22, 46.

87. Ferraris, J.; Cowan, D. O.; Walatka, V. V.; Perlstein, J. H. Electrontransferina new highly conducting donor-acceptor complex. J. Am. Chem. Soc., 1973, 95, 948–949.

88. Jeromc, D.; Mazaud, A.; Ribault, M.; Bcchgaard, K. Superconductivity in a synthetic organic conductor (TMTSF) 2PF$_6$. J. Phys. Lett., 1980, 41, 95–98.

89. Bousscau, M.; Valade, L.; Legros, J. P.; Cassoux, P.; Garbauskas, M.; Interrantc, L. V. Highly conducting charge-transfer compounds of tetrathiafulvalene and transition metal-"dmit" complexes. J. Am. Chem. Soc., 1986, 108, 1908–1916.

90. Brossard, L.; Ribaut, M.; Bousscau, M.; Valade, L.; Cassoux, P. A new type of molecular superconductor: TTF[Ni(DMIT)$_2$]$_2$, C. R. Acad. Sci., Paris, Ser., 1986, 302, 205.

91. Skotheim, T. A.; Elsenbaumer, R. L.; Reynolds, J. R. Handbook of conducting polymers, 2nd ed. Marcel Dekker, Inc., New York, 1997.

92. Chen, C.-T.; Suslick, K. S. One-dimensional coordination polymers: applications to material science. Coord. Chem. Rev., 1993, 128, 293–322.

93. Aumuller, A.; Erk, P.; Klebe, G.; Hunig, S.; Schutz, J. U. V.; Werner, H.-P. A radical anion salt of 2,5-dimethyl-N,N′-dicyanoquinonediimine with extremely high electrical conductivity. Angew. Chem. Int. Ed. Engl., 1986, 25, 740–741.

94. Zhong, J. C.; Misaki, Y.; Munakata, M.; Kuroda-Sowa, T.; Maekawa, M.; Suenaga, Y.; Konaka, H. Silver(I) coordination polymer of 2,5-bis- (4',5'-bis(methylthio)-1',3'-dithiol-2'-ylidene)-1,3,4,6-tetrathiapentalene (TTM-TTP) and its highly conductive iodine derivative. Inorg. Chem., 2001, 40, 7096–7098.

95. Miller, J. S.; Epstein, A. J. Organic and organometallic molecular magnetic materials—designer magnets. Angew. Chem. Int. Ed. Engl., 1994, 33, 385–415.

96. Ferlay, S.; Mallah, T.; Ouahes, R.; Veillet, P.; Verdaguer, M. A room-temperature organometallic magnet based on Prussian blue. Nature, 1995, 378, 701–703.

97. Hirao, T. Conjugated systems composed of transition metals and redox active π-conjugated ligands. Coord. Chem. Revs., 2002, 226, 81–91.

98. Bredas, J. L.; Beljonne, D.; Coropceanu, V.; Cornil, J. Charge-transfer and energy-transfer processes in π-conjugated oligomers and polymers: a molecular picture. Chem. Rev., 2004, 104, 4971–5003 and references therein.

99. Burroughes, J. H.; Bradley, D. D. C.; Brown, A. R.; Marks, R. N.; Friend, R. H.; Burns, P. L.; Holmes, A. B. Light-emitting diodes based on conjugated polymers. Nature, 1990, 347, 539–541.

100. Sirringhaus, H.; Kawase, T.; Friend, R. H.; Shimoda, T.; Inbasekaran, M.; Wu, W.; Woo, E. P. High-resolution inkjet printing of all-polymer transistor circuits. Science, 2000, 290, 2123–2126.

101. Baldo, M. A.; Thompson, M. E.; Forrest, S. R. High–efficiency fluorescent organic light-emitting devices using a phosphorescent sensitizer. Nature, 2000, 403, 750–753.

102. Nattestad, A. J. M.; Fischer, M. K. R.; Cheng, Y.-B.; Mishra, A.; Bäuerle, P.; Bach, U. Highly efficient photocathodes for dye sensitized tandem solar cells. Nat. Mater., 2010, 9, 31–35.

103. Gratzel, M. Solar energy conversion by dye-sensitized solar cells. Inorg. Chem., 2005, 44, 6841–6851.

104. Meyer, G. J. Molecular approach to solar energy conversion with coordination compounds anchored to semiconductor surfaces. Inorg. Chem., 2005, 44, 6852–6864.

105. Chakravorty, S.; Wadas, T. J.; Hester, H.; Schmehl, R.; Eisenberg, R. Platinum chromophore-based system for photoinduced charge separation: a molecular design approach for artificial photosynthesis. Inorg. Chem., 2005, 44, 6865–6878.

106. Horowitz, G. Organicfield-effect transistors. Adv. Mater., 1998, 10, 365–377.

107. Huitema, H. E. A.; Gelinck, G. H.; van der Putten, J. B. P. H.; Kuijk, K. E.; Hart, K. M.; Cantatore, E.; de Leeuw, D. M. Active-matrix displays driven by solution processed polymeric transistors. Adv. Mater., 2002, 14, 1201–1204.

108. Brabec, C. J.; Sariciftic, N. S.; Hummelene, J. C. Plastic solar cells. Adv. Funct. Mater., 2001, 11, 15–26.

109. McQuade, D.T.; Pullen, A.E.; Swager, T. M. Conjugated polymer-based chemical sensors. Chem. Rev., 2000, 100, 2537–2574.
110. McQuade, D. T.; Hegedus, A. H.; Swager, T. M. Signal amplification of a "Turn – On" sensor: harvesting the light captured by a conjugated polymer. J. Am. Chem. Soc., 2000, 122, 12389–12390.
111. Miller, J. S. Conducting polymers-materials of commerce. Adv. Mater., 1993, 5, 671–676.

9 Synthesis of Electroluminescent Polymer for OLEDs

Subhasis Roy and Rituraj Dubey

9.1 INTRODUCTION

Emission of light from an active material by optical absorption and relaxation by radiative decay of an excited state is known as electroluminescence (EL) and in this process light is generated in a nonthermal way from the application of an electric field to a substrate.[1] In an organic light emitting diode (OLED), an organic compound is used as a light-emissive electroluminescent layer, which produces light by the application of an electric current.[2] Electroluminescent small molecules, oligomers and polymers are used in OLED technology throughout the world for making various digital displays in screens, mobile phones, etc. Applications in flat-panel displays are promising for replacing cathode ray tubes (CRTs) or liquid crystal displays (LCDs).[3–6]

Scientists first observed the phenomenon of EL in the organic single crystals of anthracene in 1963,[7] but in this device a high working voltage was required due to having large crystal thickness as well as poor electrical contact quality. Thus, efficiency of the device was significantly lower than the inorganic semiconductor systems. Later, in 1987, Tang and Van Slyke[8] made a breakthrough by showing an efficient and low-voltage OLED using films of vapour-deposited organic materials. After this report, in 1990, Friend et al.[9] reported a highly fluorescent conjugated polymer, poly(*p*-phenylenevinylene) (PPV) as the active material in a single-layer OLED. Discovery of this polymer-based device overcame the limitations of expensive and technologically inconvenient vapour deposition of fluorescent dyes and inorganic semiconductors and started a new era in OLED technology.

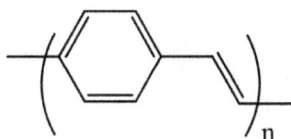

PPV

Different types of polymeric materials are now widely used to fabricate highly efficient optical, electrical and electro-optical materials and to introduce new functions to the devices. Some examples of this kind of polymer (Figure 9.1) include poly(phenyleneethynylene) (PPE), poly(*p*-phenylene) (PPP), poly(*p*-phenylenevinylene)

DOI: 10.1201/9781003260417-9

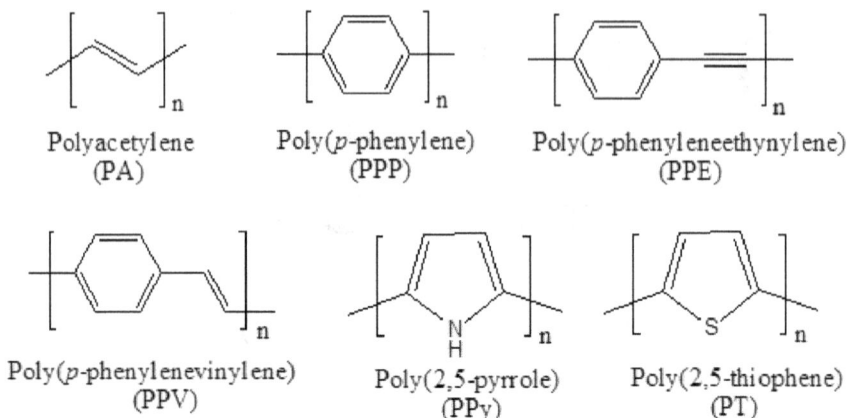

FIGURE 9.1   Main classes of organic conjugated polymers: structures and acronyms.

(PPV) and polythiophene (PT), which has been used in OLEDs,[10] organic field effect transistors[11] and polymer solar cells.[12]

This chapter provides a brief discussion of the progress of OLED technology, types of OLED, material design, and working principle. How these electroluminescent polymeric systems are synthesized to have highly efficient properties and new functions of OLEDs are described in detail.

9.2   ORGANIC LIGHT-EMITTING DIODES (OLEDs)

9.2.1   ADVANCEMENT OF OLEDs

The phenomenon of electroluminescence was observed first in an inorganic system in 1936. Destriau et al.[13] observed a high field electroluminescence by placing ZnS phosphor powder dispersed in an isolator between two electrodes. After that, during the early 1960s, light emitting diodes (LEDs) were discovered in which the inorganic semiconductor material GaAsP was used.[14] Various coloured LEDs were developed by changing the energy gap of the semiconducting material in the active region of the LED.[15] Generally, compounds from group III and group V elements of the periodic table, such as GaAs, GaP, AlGaAs, InGaP, GaAsP, GaAsInP, have been frequently used. Due to the requirement of a large energy gap, the development of blue LEDs was difficult, but the use of SiC, ZnSe or GaN semiconducting systems resolved this difficulty.[15]

Observation of the electroluminescence phenomenon from the organic system introduced a new horizon for this research. In 1963, a crystal of anthracene was first observed to show the electroluminescence behaviour in place of inorganic semiconducting materials.[7] However, the lifetime and efficiency of this system were found to be significantly lower than the inorganic systems. Later, in the late 1980s, two groups, viz. Tang and VanSlyke[8] and Saito et al.[16] developed new generations of light emitting diodes by using organic fluorescent dyes and revived research interest in using organic materials. Simultaneously, in 1990, Friend et al.[9] showed their work related to the discovery of OLEDs. Their work clearly showed the financial and technological limitations of using fluorescent dyes and inorganic semiconductors, and opened up a new horizon by using a highly fluorescent conjugated polymer, poly(p-phenylenevinylene) (PPV), as the active material in a single-layer OLED.

9.2.2 WORKING PRINCIPLE OF OLEDs

Organic electroluminescent materials containing π-conjugated molecules have conductivity levels ranging from insulator to conductor and these are considered organic semiconductors. Basically, an OLED is assembled by positioning a layer of organic material between two electrodes, the anode and cathode, all of which are deposited on a substrate (Figure 9.2). Like the valence band and conduction band in an inorganic semiconductor, organic electroluminescent materials consist of lowest unoccupied molecular orbitals (LUMOs) and highest occupied molecular orbitals (HOMOs).[17]

In the very beginning, polymer-based OLEDs were designed by placing a single organic layer, with PPV used as the organic material. But later, multilayer OLEDs were fabricated to improve the efficiency of devices. Modern OLEDs are designed using a bilayer, where one is a conductive layer and other is an emissive layer. Quantum efficiency has also been increased in OLEDs by using a graded heterojunction.

Anodes are chosen based on their optical transparency, electrical conductivity and chemical stability. A potential difference is created throughout the OLED where the anode is positive in comparison to the cathode. At the cathode, electrons are pushed into the LUMO of the organic layer, and at the anode electrons are taken up from the HOMO; as a result, a current of electrons flows from cathode to anode.[18] The withdrawal of electrons generates a hole in the HOMO. Recombination of electrons and holes by electrostatic force of attraction forms an exciton, that is, a bound-state electron and hole. This process occurs in the conductive layer, closer to the emissive layer. Decay in excited states causes relaxation of the energy level of the electron.

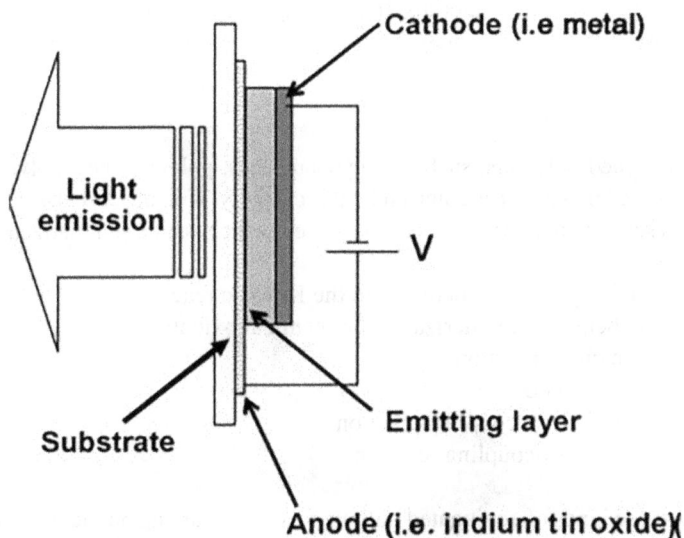

FIGURE 9.2 Basic structure of OLED devices. (Reproduced with permission from Bernard et al.[4])

Thus, the emission radiation generated falls in the visible region. By varying the band gap, that is, changing the distance between the HOMO and LUMO, the frequency of the emission can be changed and different coloured OLEDs are obtained.

Transparency to visible light and high work function make indium tin oxide (ITO) suitable as an anode.[19,20] Group II metals such as barium and calcium are used in the cathode due to their low work function value. As these metals are reactive, a capping layer of aluminium or Mg/Al alloy is used to avoid degradation.[21] Aluminium capping is beneficial as it is robust to electrical contacts as well as the back reflection of emitted light out to the transparent ITO layer. Often a thin LiF layer (~1 nm) is capped with a thicker Al layer for use as the cathode.[22] Many other insulating layers, such as CsF, MgO, Al_2O_3 and NaCl, have been investigated to enhance electron injection.[23,24]

Efficiency, performance and lifetime of the OLEDS are determined mainly by the anode/hole transport layer (HTL) interface topology. Nonemissive dark spots are generated in OLEDs due to imperfections in the surface of the anode, which affects the lifetime of the device. Experiments show that anodes composed of a single-crystal sapphire substrate treated with gold (Au) film gives OLEDs with a longer lifetime and other favourable functions.

9.3 SYNTHESIS OF ELECTROLUMINESCENT POLYMERS

Electroluminescent polymers are mainly conjugated polymers that have attracted huge attention in fabricating OLED devices due to the delocalized π-molecular orbitals along the polymeric chain. By varying the chain length, different colourful OLEDs can be obtained. A large variety of conjugated polymers have been synthesized and are being used in OLED technology. By knowing the basic structure and working principle of the OLEDs targeted properties and high efficiency in the devices can be achieved. Methods for synthesis are set up to influence the stereochemistry and important properties of the materials. In this chapter important synthetic pathways of the main classes of electroluminescent polymers are presented.

9.3.1 POLYARYLENES

Several conjugated polymers, such as polyphenylenes, polyfluorenes, polythiophenes and unsubstituted polyarylenes come under this category. To synthesize these subclasses of polymers several processes have been followed, which can be summarized as:

i)   The soluble precursor method and the Kovacic reaction
ii) Electrochemical polymerization or chemical oxidation
iii) The Yamamoto reaction
iv) The Rieke protocol
v)   The Suzuki cross-coupling reaction
vi) The Stille cross-coupling reaction

Generally, these kinds of conjugated polymeric systems are insoluble in a wide range of organic solvents when they are unsubstituted. Thin films can be made from soluble precursors which, after deposition, may be converted to conjugated polymers.

Ballard et al.[25] showed that a nonconjugated soluble precursor (**2** in Figure 9.3) obtained by the polymerization of (**1**) can produce a conjugated polymer PPP by aromatization via heat). Dehydro-coupling of the aromatic benzene ring (**3**) can generate PPP by catalytic oxidation by Kovacic's route[26] (Figure 9.3).

Electrochemical polymerization or chemical oxidation with FeCl$_3$ in chloroform produces alkyl-substituted or unsubstituted polyfluorenes (PFs) and polythiophenes (Figure 9.4). These processes have some limitations, such as low regioregularity, branched polymer formation, and low average molecular weight.[27,28]

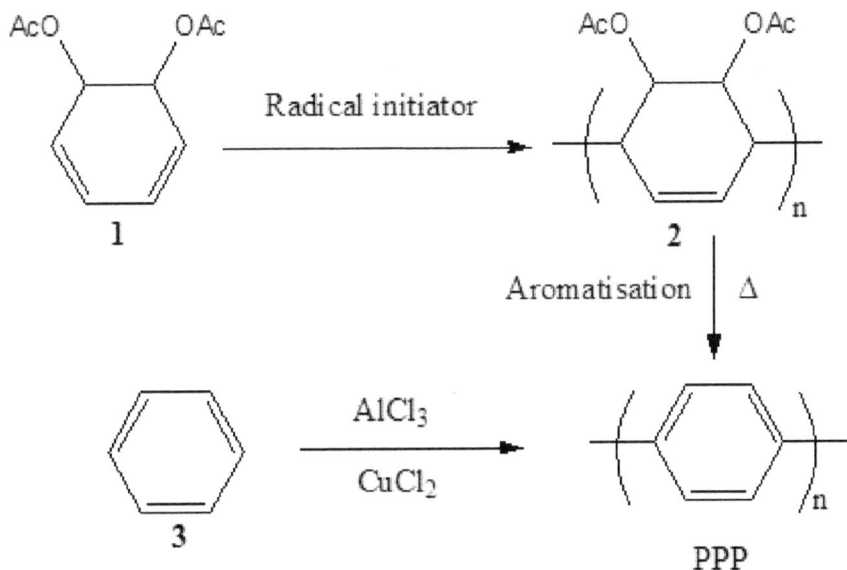

FIGURE 9.3 PPP synthesis using the soluble precursor method and the Kovacic reaction.

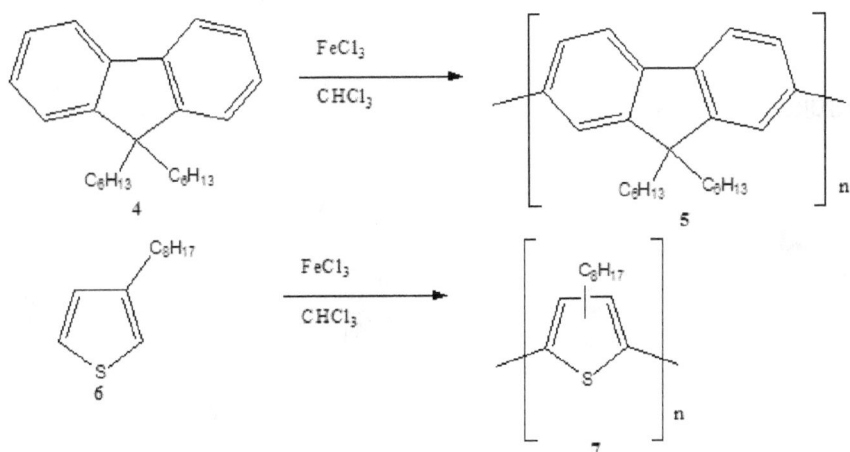

FIGURE 9.4 Electrochemical polymerization or chemical oxidation method for the synthesis of poly(9,9-dialkylfluorene) and poly(3-alkylthiophene).

Coupling reactions catalyzed by transition metal complexes have excellent control on structure over a wide range of substituents. Yamamoto[29] showed that high molecular weight polyarylenes can be obtained by nickel-activated reductive homocoupling of aryl dihalides. The polymeric unit can be obtained from the monomers by using nickel(II) complex [Ni(COD)$_2$] as catalyst and 2,2′-bipyridine as ligand. Figure 9.5 shows the synthetic pathway of a high molecular weight polymer poly(9,9-dialkylfluorene)s (9) used in OLED devices. As the Ni(COD)$_2$ complex was expensive, NiCl$_2$ was introduced later as the catalyst for this type of synthesis. Using this process an emissive polymer (11) has been synthesized and this polymer is used to fabricate a blue OLED device with 3% external quantum efficiency.[30] To get pure blue emission from polyfluorenes is still a concern due to the fluorenone defects caused by oxidation in the fluorine rings of the polymer chain. This can be avoided by careful purification, introducing bulky substituents in position 9, along with capping the ends of the chains with sterically hindering units, leading to an increase in emissive character.[31]

Miteva et al.[32] showed that triarylamino functionalities can improve the mobility of holes during the production of polymer (14) from (13) (Figure 9.6).

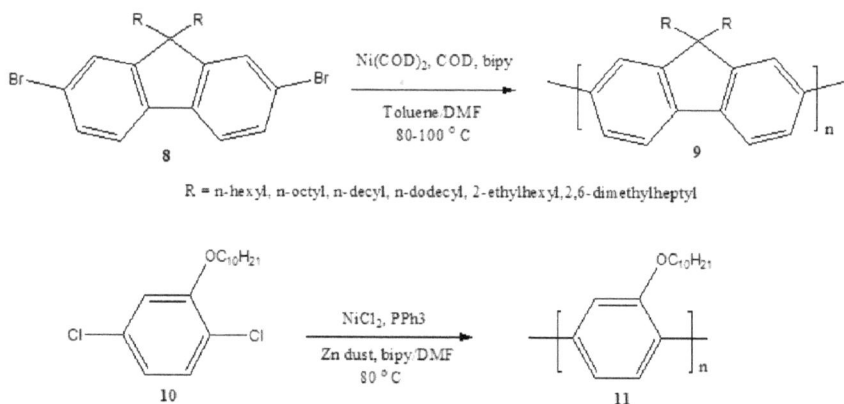

FIGURE 9.5   Route of synthesis of polyarylenes using the Yamamoto reaction.

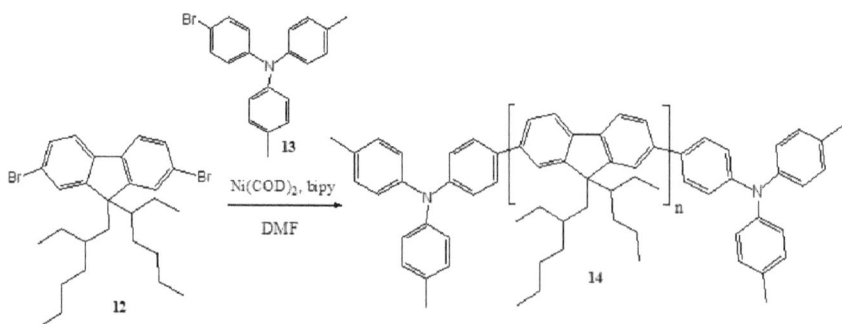

FIGURE 9.6   Synthesis of polyfluorene (14) via the Yamamoto route.

The synthesis of polymers using the Yamamoto process has some limitations related to the removal of metal impurities, which can reduce the emissive nature of the polymers. This difficulty can be overcome by using palladium complexes as catalysts instead of nickel complexes.

Three coupling schemes, head to head (HH), head to tail (HT) and tail to tail (TT), for two 3-substituted thiophene rings are depicted in Figure 9.7. The mode of coupling also affects the emission behaviour of the materials. Coupling can be done via the Yamamoto route but using organometallic complexes leads to more regioregularity. It is seen that when two thiophene rings connect through the HH mode some portion of the ring goes out of the plane and thus there is a loss of polymer conjugation length, which increases the band gap between HOMO and LUMO and results in a blue shift.[33]

It is thus very clear that by introducing bulky substituents and controlling regioregularity the bandgap can be modified and thus the emission colour. Researchers have shown that when alkyl substitution is done in the ring, HH coupling increases the distance, leading to a red shift in the emission maxima.[34] Some examples of polythiophenes are depicted in Figure 9.8.

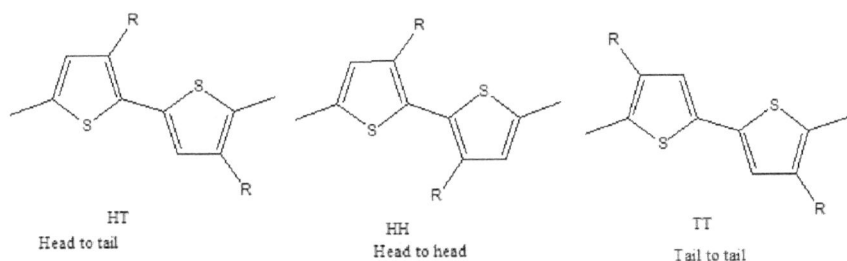

FIGURE 9.7 Different possible modes in poly(3-alkylthiophene)s.

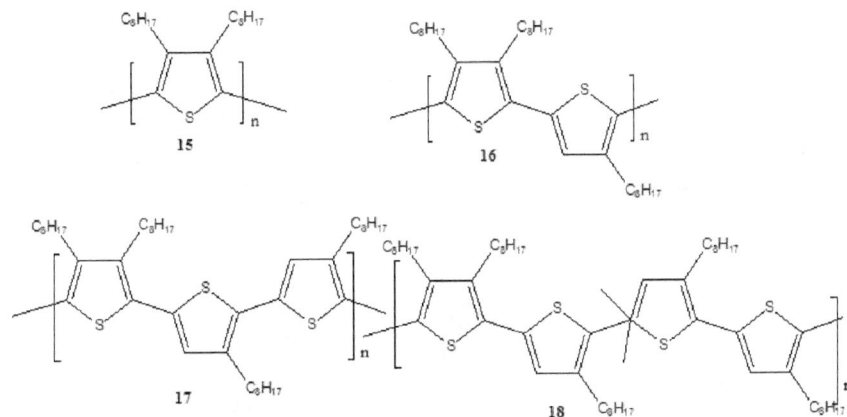

FIGURE 9.8 Structures of some polythiophenes.

The photoluminescence efficiency of poly(3-arylthiophene) (**20** in Figure 9.9) in solution is very similar to the poly(3-alkylthiophene)s (P3ATs) but higher in the solid state, which is due to the bulky aryl groups that cause greater interchain distance.[34] Figure 9.9 represents the chemical oxidation of (**19**) by FeCl₃ to generate the polymer (**20**). This is a regioselective reaction that takes place due to the steric effect of the bulkier aryl substituent.

Chen et al.[35] have demonstrated the preparation of high molecular weight regioregular P3ATs by the Rieke protocol. Grignard metathesis can also be employed for this type of synthesis. Figure 9.9 illustrates that Negishi coupling using organo-zinc reagents is the basis of the Rieke protocol. In this process the starting material is substituted thiophene (**21**).Rieke zinc (Zn*)[35] in activated form adds to the C-Br bond through oxidative addition.

The reaction temperature can control the regioselectivity of the reaction. At −78°C, as high as ~97%–98% regioselectivity can be obtained. In the presence of Ni(dppe) Cl₂ (dppe:1,2-bis(diphenylphosphino)ethane) as catalyst the polymerization is done in a single pot reaction at 0°C.

McCullough et al.[36,37] introduced the Kumada cross-coupling procedure for the synthesis of P3ATs. In this reaction, organomagnesium reagents are used. The procedure is also known as Grignard metathesis. Figure 9.10 shows regioselective metalation of (**21**) with methylmagnesium bromide at position 5. Polymerization is done

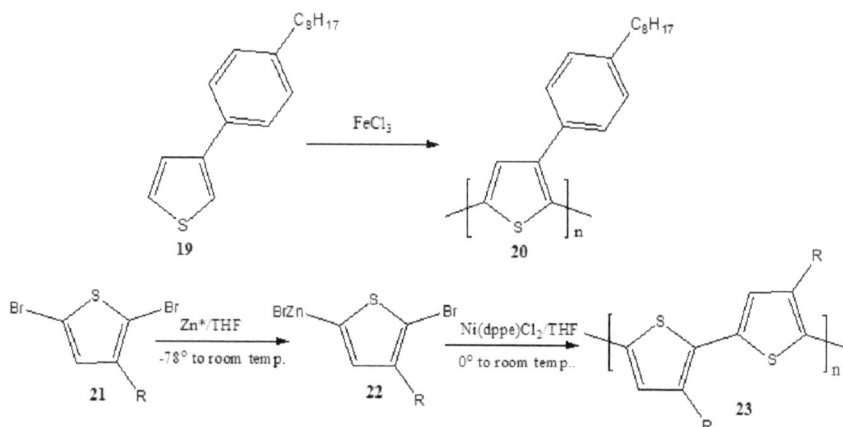

FIGURE 9.9   The chemical oxidation method used for regioselective synthesis of poly (3-arylthiophene) (**20**; top); synthesis of poly(3-alkylthiophene)s by the Rieke protocol (bottom).

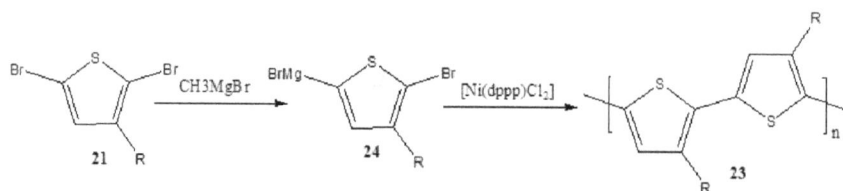

FIGURE 9.10   The Grignard metathesis reaction to synthesize poly(3-alkylthiophene)s.

in the next step with the aid of Ni(dppp)Cl$_2$. A high molecular weight regioregular polymer is obtained by this process, with good yield.

Yamamoto et al. showed the synthesis of poly(phenylene)s by one pot self coupling.[38] Here aryl dihalides are coupled with each other with the use of Mg and mild Ni catalysis.

Synthesis of alternating copolyarylenes has been done by the coupling of bis-organomagnesium derivatives with aryl dihalides. Alternating poly[1,4-(2,5-dioctyloxyphenylene)-2,5-thiophene] (**27**) can be obtained by using Grignard reagent (**25**) and 2,5-dibromothiophene (**26**; Figure 9.11). This polymerization reaction is catalyzed by PdCl$_2$(dppf) (dppf: 1,1′-bis(diphenylphosphino)ferrocene).[39] This polymer has widespread optoelectronic applications.[40,41]

The Stille cross-coupling reaction can be used as an altenative to Suzuki-Miyaura coupling for building polyarylene skeletons.[42] Aryl organostannanes couple with aryl halides in the presence of Pd as catalyst. Blue emitting polymer (**30**; Figure 9.12) has been synthesized by coupling of 2,5-bis(tributylstannyl)thiophene (**28**) with the diiodo derivative (**29**).[43] Here the catalyst Pd(AsPh$_3$)$_4$ is generated in situ. A high molecular weight polymer with high stability and efficiency is obtained.

9.3.2 POLY(ARYLENEETHYNYLENE)S

Poly(aryleneethynylene)s (PAEs), a class of conjugated polymers, show a variety of electronic properties, such as photoluminescence, nonlinear optics and exceptionally high solution-state photoluminescence efficiencies, which are unique compared with other conjugated polymers. Due to the reduced emission efficiencies in the solid state, PAEs are shown to form aggregates. A large potential barrier to the injection and transportation of holes has been observed for the limitation in redox properties in the PAEs. So far in the use of efficient OLED materials, PAEs should be designed in such a way where aggregation will be minimized and the charge-balancing properties will be enhanced. Although PAEs have better luminescence properties in solid as well as in solution state and greater photostability than other classes of luminescent polymers such as poly(arylenevinylene)s (PAVs), their rigid rod structure corresponds to

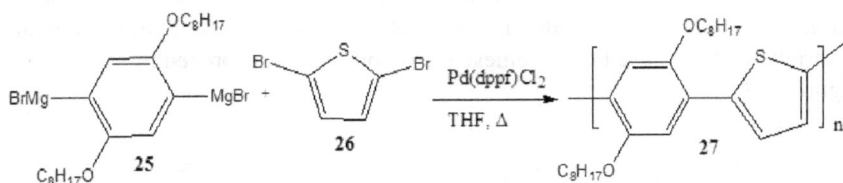

FIGURE 9.11   A synthetic route for copolymer **27**.

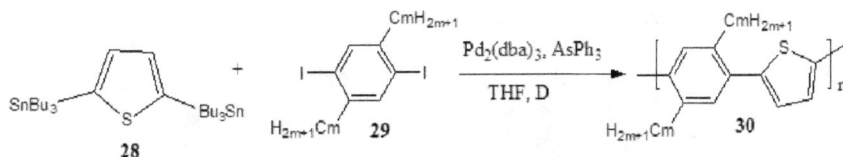

FIGURE 9.12   Synthesis of polymer **30** via the Stille cross-coupling reaction.

more red-shifted emission from solution to the solid state in contrast to PAVs. Due to these issues PAEs have been used less frequently in OLEDs than PPVs.

Some important synthetic routes for the synthesis of PAEs are:

i) The Cassar-Heck-Sonogashira cross-coupling reaction, which is a Pd-catalyzed bond formation reaction, in which a new Csp-Csp^2 bond is generated in the polymeric unit.[44]

ii) The metathesis reaction.[45]

A multifunctional aryleneethynylene backbone polymeric unit is mostly synthesized using Cassar-Heck-Sonogashira cross-coupling.[44] Regular alternating copolymers[46] can be achieved by the base coupling reaction between terminal ethynyl moieties and Csp^2 of aryl or heteroaryl halides. This coupling reaction is catalyzed by palladium and a copper (I) co-catalyst. Synthesis of some PAEs is depicted in Figure 9.13, which follows Cassar-Heck-Sonogashira cross-coupling. Poly(p-aryleneethynylene) (**31**), which shows a blue emission, is prepared using two different monomeric units. This reaction is catalyzed by $Pd(PPh_3)_4$ and CuI, and diisopropylamine is used as a base.[47] Breen et al. have synthesized a similar kind of polymeric system, which has been used to fabricate single-layer blue OLEDs.[48] Negative charge transport properties are developed in the side group of oxadiazole-containing oligoarylene, which enables proficient electron transport. This system provides a luminance efficiency of 0.34 cd A^{-1} with peak efficiency value of 0.29%. In comparison to dialkoxy-substituted polymer (**31**), Breen et al. have synthesized polymers with higher operational efficiency.[48]

Cassar-Heck-Sonogashira cross-coupling-based polymerization has some limitations also. Oxidative homocoupling of ethynyl monomers causes the formation of butadiyne defects in the polymer chain. Kotora and Takahashi showed that this defect is seen due the Glaser coupling activated by copper salts.[49] Difficulties with the synthesis of PAEs by coupling reaction are related to imbalance of the monomers ratio and butadiyne defects, lowering performance in OLEDs. The use of silver oxide as the activator with palladium complex in the coupling of aromatic halides with trimethylsilyl alkynes reduces unwanted homocoupling reaction.[50] A variety of PAEs have been synthesized by following this procedure, as shown in Figure 9.14.

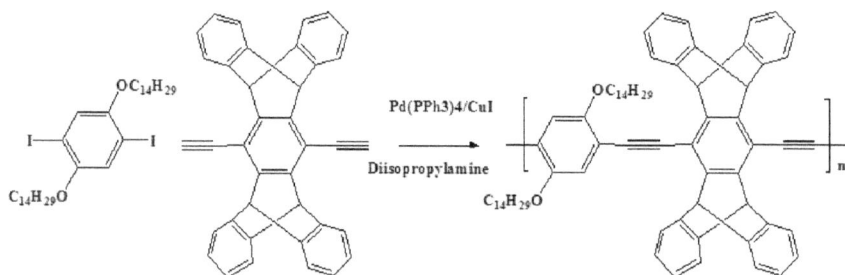

FIGURE 9.13   Cassar-Heck-Sonogashira cross-coupling reaction for the synthesis of PAE (**31**).

Another good method of synthesis of PAEs is the metathesis reaction (ADI-MET = Acyclic DIyne METathesis), which has been widely used.[45] This method can generate polymers without butadiyne defects. In the ADIMET reaction dipropynyl arene is used as the substrate and $(tBuO)_3$ WtC-tBu complex as the catalyst. Figure 9.15 shows synthesis of PAEs by this method. Sometimes the degree of polymerization has reached almost 100 repeating units.[51] Here synthesis of the catalyst is somewhat difficult, and it is sensitive to air and moisture. This difficulty can be overcome by using molybdenum-based complexes.[52] Molybdenum complexes, which are air stable, working in wet solvent can be obtained by mixing molybdenum hexacarbonyl with 4-chlorophenol or 4-trifluoromethylphenol. An example of using this molybdenum-based catalyst to synthesize PPV-PPE alternating copolymer is shown in Figure 9.15.

9.3.3 POLY(ARYLENEVINYLENE)S

The evolution of OLEDs was published by Friend et al. in 1990.[9] They showed the limitations of using inorganic semiconductors and introduced a highly fluorescent conjugated polymer, poly(p-phenylenevinylene) (PPV) in simplest poly(arylenevinylene) as an active material in the single-layer OLED device. As PPV was insoluble, Friend et al. made PPV OLEDs by using a soluble precursor polymer by thermoconversion. This synthesis helped to open a large area in LED technology. Since then, interest in this class of materials has risen steadily.

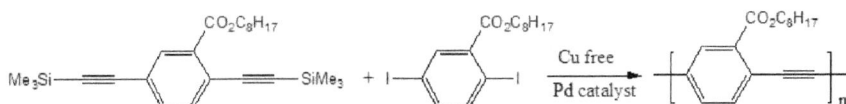

FIGURE 9.14 Synthesis of functionalized PAE.

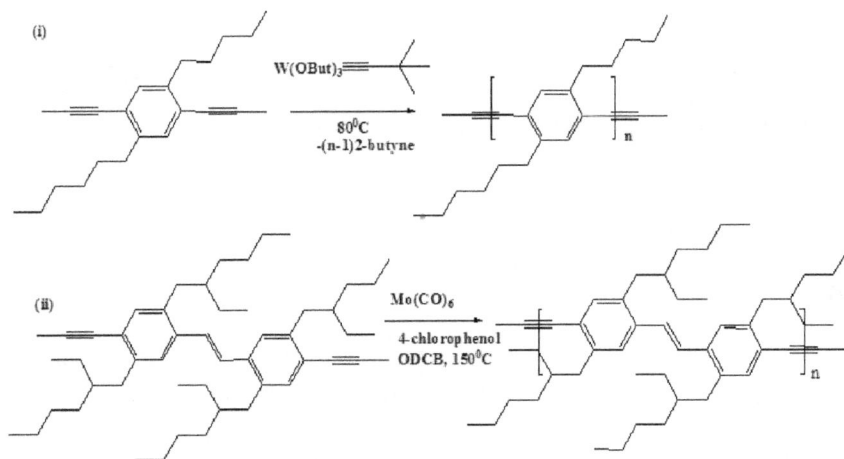

FIGURE 9.15 PAE synthesis using metathesis reactions.

So far PPVs have been synthesized by the following synthetic routes:

   i)   The soluble precursor method
   ii) The Heck coupling reaction
   iii) The Suzuki cross-coupling reaction
   iv) The Stille cross-coupling reaction
   v)   Metathesis reactions

In this section, synthesis of some important PPVs widely used in OLEDs are discussed.

   The soluble precursor method is a process in which a nonconjugated soluble precursor can be converted into the desired product by a thermal or chemical elimination reaction. One important route is the Wessling route. Thermal treatment is used to prepare PPV by using a sulfonium precursor (**32** in Figure 9.16).[53]

   Alternatively, the Gilch route[54] can be used for in situ preparation of p-quinodimethane (**33**) by using benzyl dihalides and in the final step base-promoted condensation produces PPV (Figure 9.16). For large-scale production this method is useful but structural defects arise due to the noncomplete efficiency, regioregularity and stereospecificity of the elimination processes. For the preparation of high molecular weight PPVs this is still a good process.

   A base catalyzed Heck coupling reaction between aryl dihalides and alkenes has been used to synthesize arylenevinylene polymers.[55] The Heck reaction gives better control over the stereochemistry of the vinylene units. Several substituted PPVs have been prepared by this method. Figure 9.17 shows some examples of synthesis

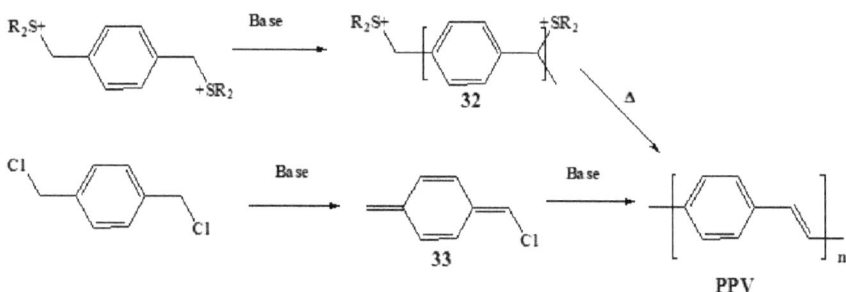

FIGURE 9.16   Soluble precursor methods for the synthesis of PPV.

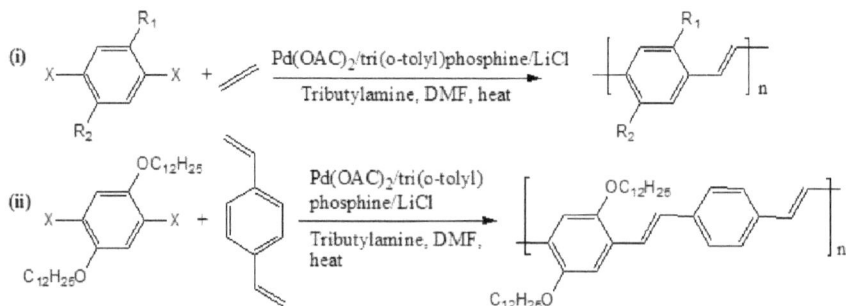

FIGURE 9.17   Heck coupling reactions for the generation of PPVs.

of PPVs prepared by this method. In this process structural defects are fewer, but α,α-vinylidene defects are present. Figure 9.17 represents the route of synthesis of polymers by the Heck coupling method using ethylene, divinylbenzene the *trans* position of the vinylene units has been confirmed with the help of Fourier-transform infrared spectroscopy.

Figure 9.18 shows PPV polymer (**34**) synthesiszed by Sengupta et al. where phosphine-free Pd(OAc)$_2$ is used.[56] PPV polymers are also obtained from the reaction of arenediazonium tetrafluoroborate salts with trialkoxysilylethene or with bis(α-trimethylsilylvinyl) arene. A double aryldesilylation process is applied to get a polymer (**35**) on the α,α'-bis(trimethylsilyl) derivative with arenebisdiazonium tetrafluoroborate.[57,58] Here a phosphine-free palladium catalyst Pd(dba)$_2$ complex is employed for the arylation reaction.

Another important reaction to prepare PPVs is Suzuki-Miyaura cross-coupling.[59,60] *Cis*-poly[(arylenevinylene)-*alt*-(2,5-dioctyloxy-1,4-phenylenevinylene)] (**36** in Figure 9.19) has been synthesized by this method by using a Pd(PPh$_3$)$_4$ catalyst in basic medium with the help of a phase-transfer catalyst.[59]

Arylenevinylene-type polymers[61,62] can be prepared by Stille cross-coupling. Figure 9.20 shows the route of synthesis of polymers by the reaction of 1,2-bis(tributylstannyl)ethene (**37**) with aryl dihalides. Linear, branched or bridged alkoxy and/or polyethereal chains-based poly(*p*-phenylenevinylene)s (**38a–d**) have been synthesized by the coupling of **37** with aryldihalide in the presence of Pd(PPh$_3$)$_4$ as the catalyst.[61,62]

The position of the substituents on PPV is an important fact to control the energy levels of the highest occupied molecular orbital (HOMO) and the lowest unoccupied

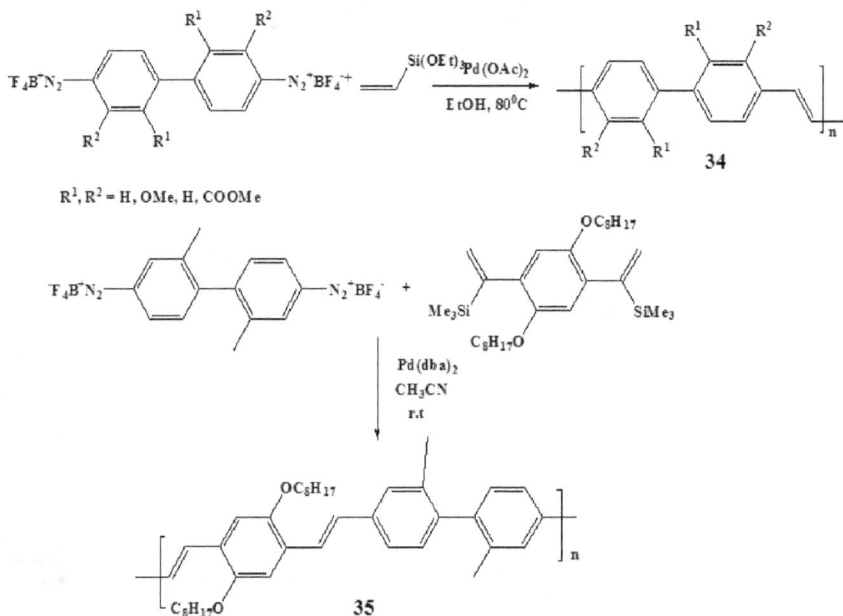

FIGURE 9.18 Synthetic route for the synthesis of PPVs **34** and **35**.

molecular orbital (LUMO) of the polymers. In an electroluminescent device the energy of the LUMO can be lowered by the use of an electron-withdrawing entity, which in turn will be helpful for good electron injection at the cathode.[63] At the cathode C-H bonds can be substituted by C-F bonds, which helps to decrease the HOMO and LUMO levels of PPV polymers. In the soluble precursor route, fluorination has some limitations but Stille polymerization overcomes this difficulty during fluorination.[64,65] Coupling of 1,2-bis(tributylstannyl) ethene (37) with tetrafluoro-1,4-diiodobenzene (39) gives polymer (40) (Figure 9.21). Polymer (40) has been used in OLED materials that show green emission. A blue shifting in green light

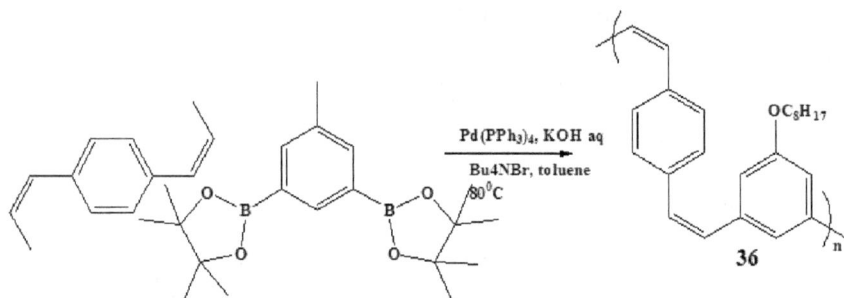

FIGURE 9.19   Suzuki cross-coupling reaction for the synthesis of PPV **36**.

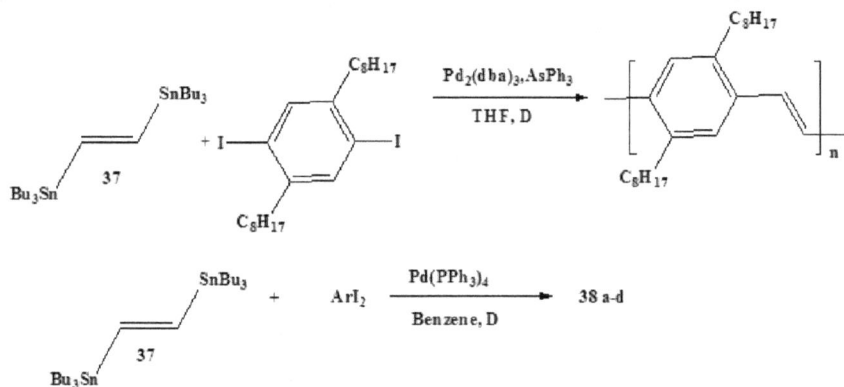

FIGURE 9.20   Stille cross-coupling reaction used for the synthesis of PPVs.

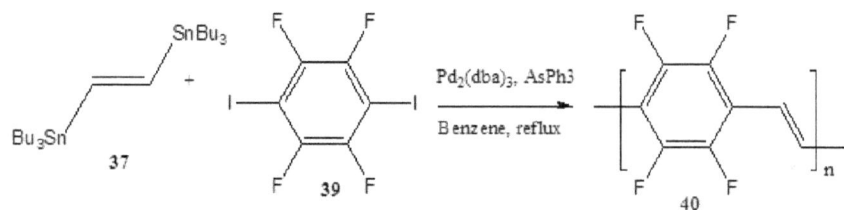

FIGURE 9.21   Stille cross-coupling reaction used for the synthesis of fluorinated PPVs.

emission has been achieved by substitution of the alkoxy group in PPVs with red-or-ange electroluminescence.

Other examples of fluorinated PPV polymers, MEH-PPV (**41a**) and the fluori-nated MEH-PPDFV (**41b**), prepared via Stille cross-coupling,[66,67] are depicted in Figure 9.22. Fluorine atoms on the vinylene units cause a large blue shift of the emis-sion wavelength. Losurdo et al. have shown that when MEH-PPDFV (**41b**) is used as the emitting layer in an OLED, the device is more stable and gives blue emission in comparison to the nonfluorinated material (**41a**).[68]

Another process for the synthesis of PPVs is metathesis reactions. Arylenevinylene polymers is the product synthesized from dienes or alkenes in the presence of molyb-denum alkylidene complex as catalyst. This process enables high molecular weight polymers with less polydispersivity to be produced. Figure 9.23(i) shows that poly-mer (**43**) can be prepared from bicyclo[2.2.2]octadiene (**42**) and thermal treatment helps to convert it into an unsubstituted PPV with a base or acid catalyst.[69] Thermal treatment isomerizes all *cis*-double bonds to their *trans*-configuration. This process has some limitations to prepare thin films for optoelectronics. ADMET (acyclic diene

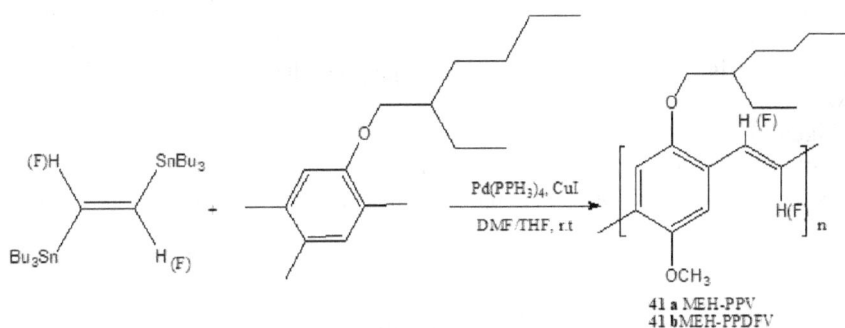

FIGURE 9.22 Stille cross-coupling reaction used for the synthesis of fluorinated PPVs.

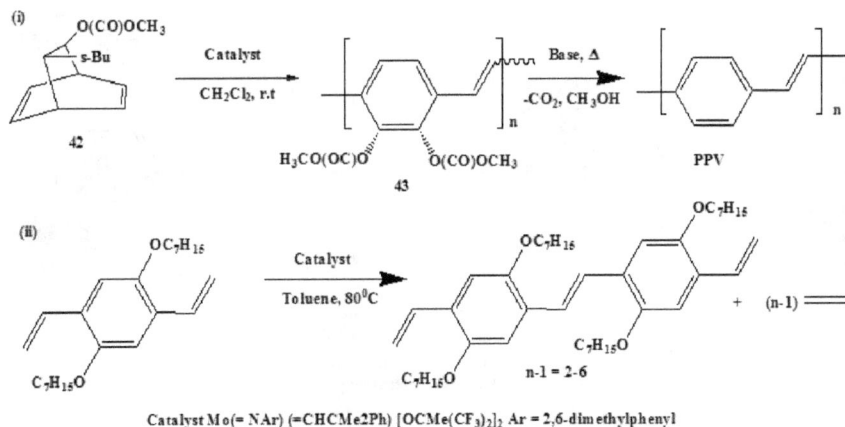

FIGURE 9.23 Synthesis of PPVs via metathesis reactions.

metathesis) has also been used to polymerize divinylbenzenes. Figure 9.23(ii) shows that oligomers with two to six repeating units can be formed.[70]

9.3.4 CONJUGATED COPOLYMERS

In the preparation of conjugated copolymers, different types of monomeric units are linked in the same polymeric chain to generate a copolymer. Delocalized π-molecular orbitals along the polymeric chain in conjugated polymers act as organic semiconductors. Tailoring the properties along with charge transport and light emission colour can be done by using their structural variety. Many properties can be optimized in single polymeric structures to perform several functions simultaneously. Fewer layers in the OLEDs are desirable from the viewpoint of value and simplicity of performance.

In this section the synthesis of some copolymers that have been widely used in OLED technology is discussed. Synthesis of PPE and PPV has already been discussed; by combining these two, one can obtain PPE-PPVs. Alternative repetition of PPE and PPV units is seen in the polymeric chain. Copolymers of PPE and PPV are better than the individual units. They show bathochromic shifted absorption and emission spectra coupled with higher fluorescence quantum yield in the solid state than PPEs. Higher electron affinity of this copolymer than PPVs leads to LEDs with low turn-on voltages.[71]

Figure 9.24 shows the route of synthesis of some π-bond bridged alkoxy-substituted copolymers (**44a-d**). These can be produced by polycondensation or Cassar-Heck-Sonogashira cross-coupling. It has been observed that for the synthesis

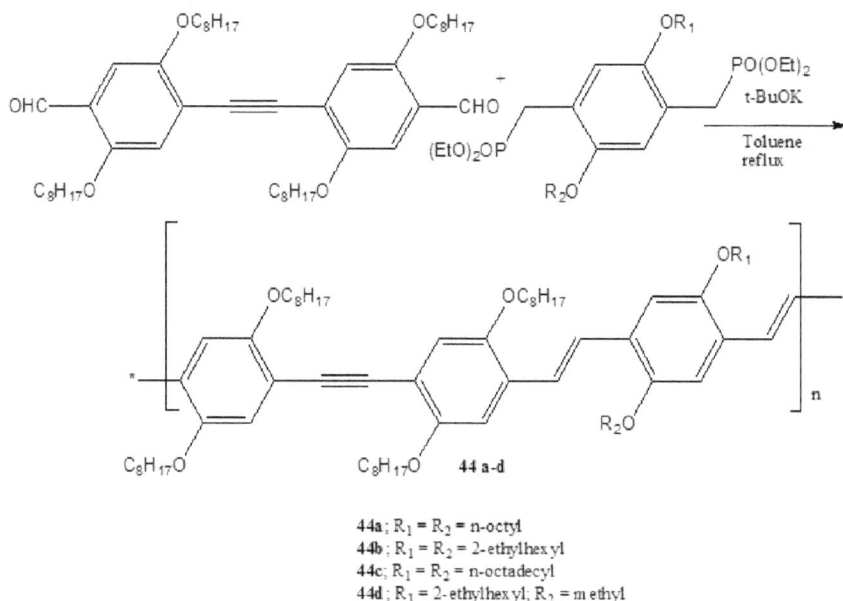

44a; $R_1 = R_2$ = n-octyl
44b; $R_1 = R_2$ = 2-ethylhexyl
44c; $R_1 = R_2$ = n-octadecyl
44d; R_1 = 2-ethylhexyl; R_2 = methyl

FIGURE 9.24 Horner-Wadsworth-Emmons reaction for the generation of copolymers **44a-d**.

of this kind of polymer Horner-Wadsworth-Emmons polycondensation is better as it is devoid of butadiyne defects and yields higher molecular weight polymers.[72] Improved OLED performances have been achieved from the copolymers, which consist of a long linear octadecyloxy or branched 2-ethylhexyloxy side groups.

Copolymers made from mixing of PPE and PPV units have higher thermal stability. The position and number of alkoxy side groups, size, geometry, etc., can affect the solid-state photophysical properties.

Polymer (**44c**) has been synthesized by Egbe et al. and is used in devices for green emission. The construction of the device is like the glass substrate/ITO/PEDOT:PSS/polymer/Ca/Al configuration [PEDOT:PSS = poly(3,4-ethylenedioxythiophene) poly(styrenesulfonate)].[73] It expresses a turn-on voltage of 5 V along with a quantum efficiency of 2.15%.

Synthesis of copolymers (**45**) and (**46**) are depicted in Figure 9.25; synthesis is accomplished by the Stille cross-coupling reaction.[74] These thiophene-based polymers have shown green and red emission and have better efficiency than classic polythiophenes. The drawback of using homopolymer polythiophenes is the strong interaction between the polymer chains in the solid state, which reduces emission efficiency. The heavy atom effect of the sulphur atom leads to intersystem crossing.[75]

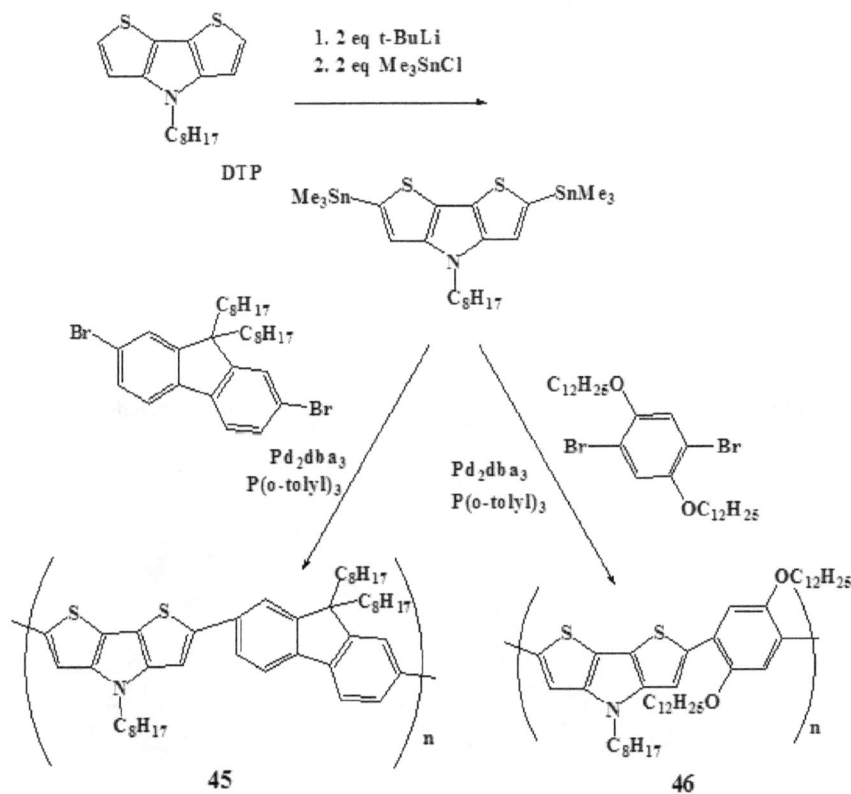

FIGURE 9.25 Copolymers **45** and **46** synthesized by the Stille cross-coupling reaction.

A Suzuki cross-coupling reaction is used for the preparation of copolymer (**47**) from diboronic ester and 2,7-dibromo-9,9-dihexylfluorene (Figure 9.26).[76]

The advantage of DTP over typical thiophene-based systems is seen in the device having configuration ITO/PEDOT:PSS/**45**/Alq 3/LiF/Al, which shows higher luminance (~1.6 times) compared to the devices developed by the analogous dithienylfluorene copolymers.[77]

Alternating copoly(arylenevinylene)s (**53–57**)[78] can be formed by the Heck coupling between aromatic dibromides (**48–52**) and 9,9-dihexyl-2,7-divinylfluorene (Figure 9.27).

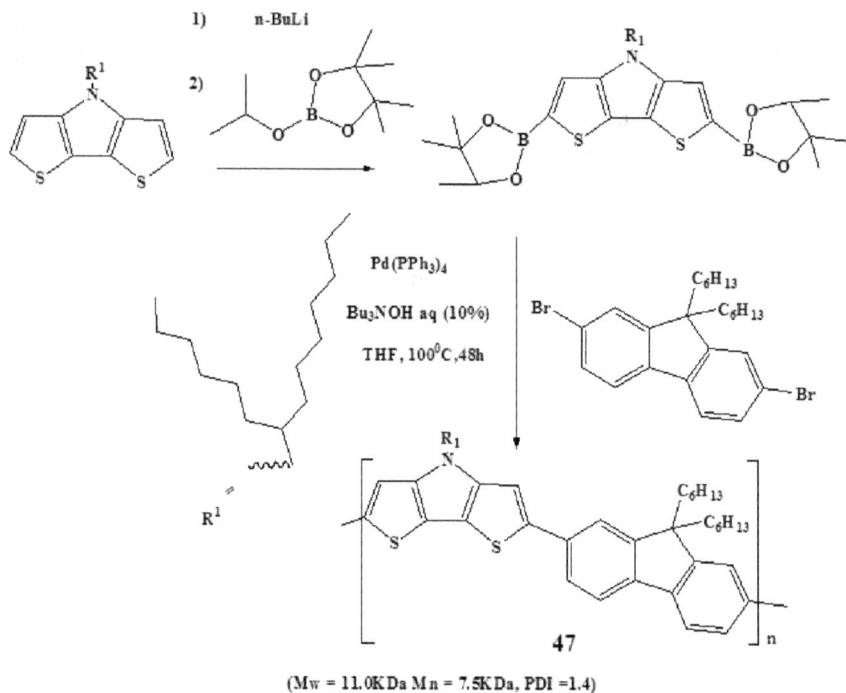

FIGURE 9.26   Synthesis of copolymer **47** via the Suzuki cross-coupling reaction.

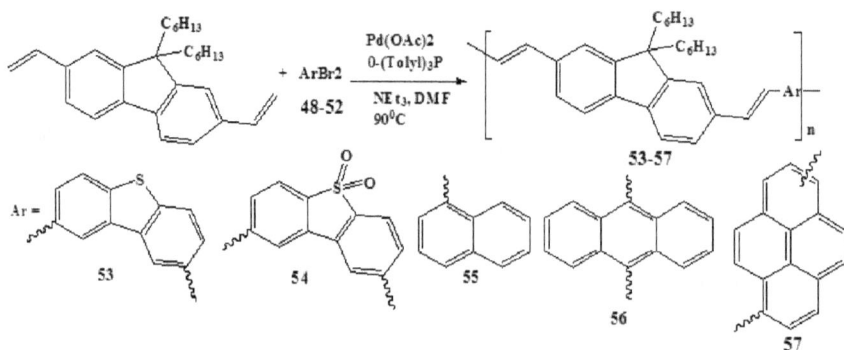

FIGURE 9.27   Heck coupling reaction used for the synthesis of copolymers **53–57**.

Copolymers (**53–57**), when designed like the ITO/PEDOT:PSS/copolymer/Mg:Ag/Ag, showed electroluminescence ranging from green to blue with low quantum efficiencies. This low value may be due to the solid state π–π interactions between polynuclear aromatic rings. Electroluminescent copolymers are more popular in OLED technology because of their white light emission.[79] When the proper ratio and arrangement of conjugated units are used the copolymer obtained from this path displays white electroluminescence. For the white light production control over the ratio, connection between two units is necessary which help in the energy transfer process.

9.3.5 COORDINATION POLYMERS

From the viewpoint of inorganic chemists, coordination polymers can be thought as one-, two- or three-dimensional structures in the solid state and metal coordination interactions help to hold together the molecular building blocks.[80] Macromolecules with stable polymeric structures in solid as well as in solution state can be called coordination polymers or metallo-supramolecular conjugated polymers.[81] The main chain of these polymer consists of organometallic complexes (Figure 9.28a) or sometimes organometallic complexes are attached as side groups (Figure 9.28b).

This class of polymers can be combined with soluble conjugated polymers with desired properties with potential applications.

There are primarily two routes for the synthesis of coordination polymers:

i) In the first process, metal ion complexation occurs by two terminal coordinating units containing organic building blocks. "Main-chain" coordination polymers (Figure 9.29a) have been synthesized in this way where at the time of the polymerization organometallic complexes are formed.

ii) Two (or more) monomers can form this type of polymer by cross-coupling or other condensation method. The original monomeric unit may already

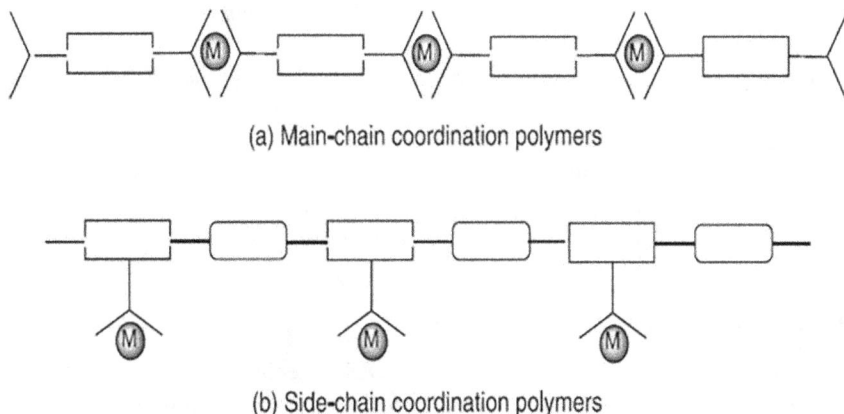

(a) Main-chain coordination polymers

(b) Side-chain coordination polymers

FIGURE 9.28 Structures of coordination polymers. (Reproduced with permission from Ragni et al.[6])

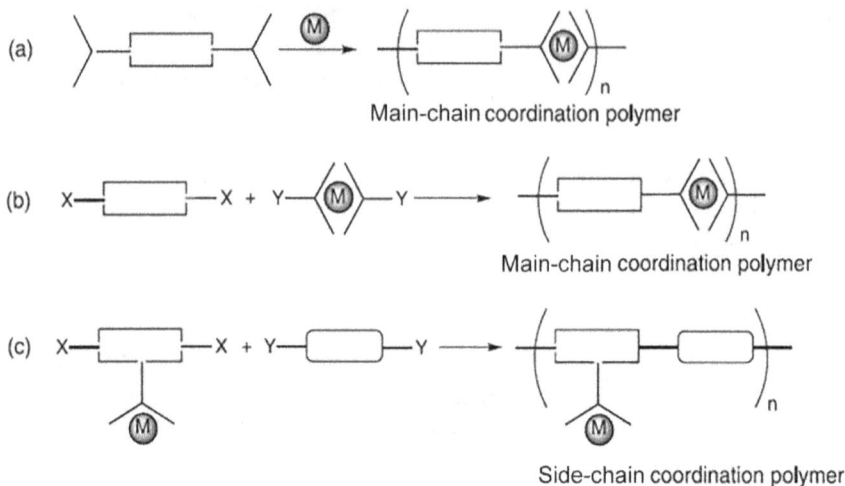

(a)

Main-chain coordination polymer

(b)

Main-chain coordination polymer

(c)

Side-chain coordination polymer

FIGURE 9.29 Main approaches to the synthesis of coordination polymers. (Reproduced with permission from Ragni et al.[6])

have an organometallic complex (Figure 9.29b) or it may be grafted covalently as a pendant group (Figure 9.29c). Synthesis of main-chain as well as side-chain coordination polymers can be accomplished using this method.

To date, very few OLED devices have been fabricated using coordination polymers as light emitting materials, yet researchers are engaged to develop new coordination polymers suitable for OLED devices.[82,83] The routes of synthesis of some coordination polymers have been discussed in this chapter.

Polymer (**58**) is obtained using 3,5-bis(8-hydroxyquinoline Schiff base)-benzoic acid in the presence of AlCl$_3$ (Figure 9.30). This polymer-based configuration of ITO/PEDOT/MEH-PPV/**122**/LiF/Al can give higher performance in comparison to the simple AlQ$_3$ complex.[84]

Another main-chain coordination polymer (**59**),[85] is synthesized via ligand exchange in the compound *tris*(acetylacetonate)aluminium by oligofluorenes (Figure 9.31).

A microwave-assisted Yamamoto polycondensation reaction yields polyfluorenes, main-chain platinum-salen chromophores (**60**)[86] (Figure 9.32). When this polymer was used to design OLED devices it showed efficiencies as high as 6 cd·A^{-1}.

9.4 CONCLUSION

This chapter describes the advancement of OLED technology from the beginning, classification of OLEDs, their working principles and how an electroluminescent polymer can be used as an active material in OLEDs. From these basic ideas the types of polymers suitable for use in OLEDs have been determined and a variety of routes have been used to synthesize different types of

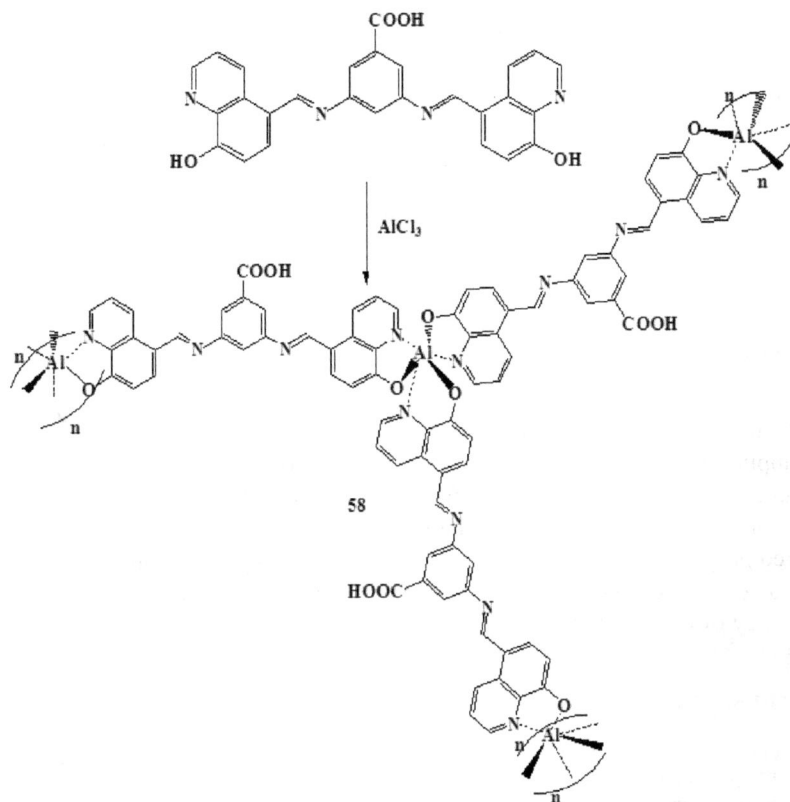

FIGURE 9.30 Strategy for the synthesis of polymer **58**.

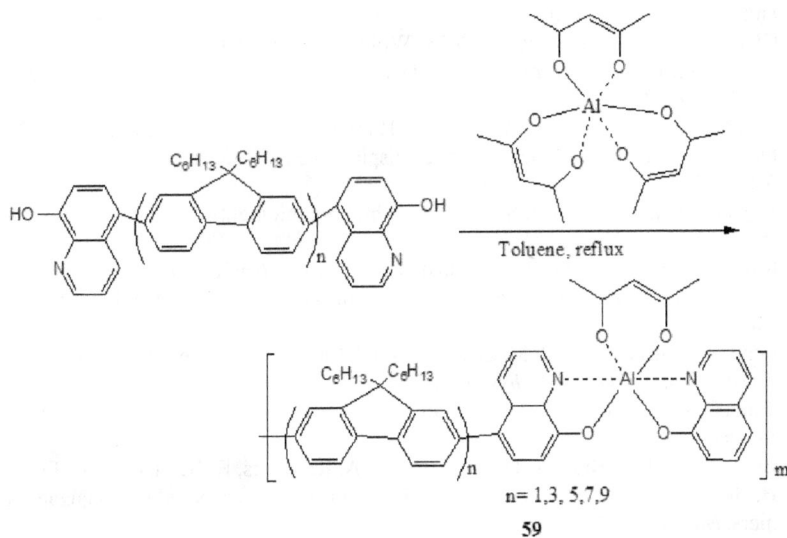

FIGURE 9.31 Synthesis of coordination polymer **59**.

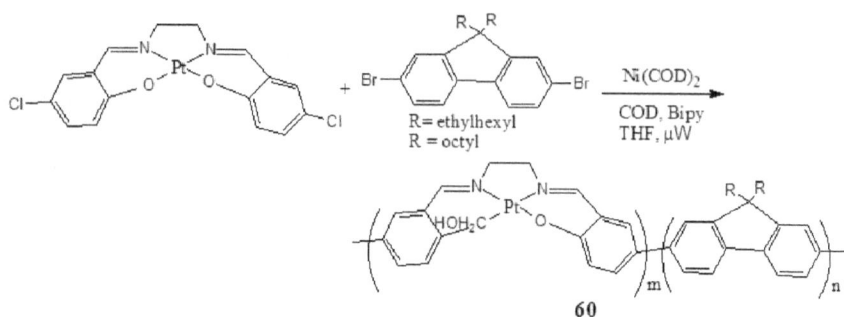

FIGURE 9.32 Synthesis of coordination polymer **60**.

electroluminescent polymers with suitable desired properties, and their potential applications explored. Several transition-metal-catalyzed cross-coupling and metathesis reactions are shown. How stereo- and regioselectivity of the reaction can be controlled to prepare the specific structure of the polymer to get the desired property has been a focus. Special attention is given to precise functions, such as diverse emission colour, charge transport and high electroluminescence efficiency of these polymers.

REFERENCES

1. Ono, Y. A., Trigg, G. L., ed. 1993. *Electroluminescence in Encyclopedia of Applied Physics.* VCH.5: 295.
2. Organic Electroluminescence-R&D. 2019. In-house development including OLED material design and fabrication, and panel prototyping. Aiming for pioneering high-performance OLED displays Semiconductor Energy Laboratory. https://www.sel.co.jp/en/technology/oled.html
3. Chang, Y. L., Zheng-Hong, L. 2013. White organic light-emitting diodes for solid-state lighting. *Journal of Display Technology* 99. doi:10.1109/JDT.2013.2248698. S2CID19503009.
4. Bernard, G., Philippe, le R., Christophe, P. 2006. Organic light-emitting diode (OLED) technology: materials, devices and display technologies. *Polymer International* 55: 57–582.
5. D'Andrade, B. W., Forrest, S. R. 2004. White organic light-emitting devices for solid-state lighting. *Advanced Materials* 18: 1585–1595.
6. Ragni, R., Operamolla, A., Farinola, G. M. 2013. *Synthesis of Electroluminescent Conjugated Polymers for OLEDs.* University of Bari Aldo Moro; Woodhead Publishing Limited.
7. Pope, M. Kallmann, H. P., Magnante, P. 1963. Electroluminescence in organic crystals. *The Journal of Chemical Physics* 38: 2042.
8. Tang, C. W., VanSlyke, S. A. 1987. Organic electroluminescent diodes. *Applied Physics Letters* 51: 913.
9. Burroughes, J. H. Bradley, D. D. C. Brown, A. R. Marks, R. N. Mackay, K. Friend, R. H. Burns, P. L., Holmes, A. B. 1990. Light-emitting diodes based on conjugated polymers. *Nature* 347: 539.
10. Kraft, A. Grimsdale, A. C., Holmes, A. B. 1998. Electroluminescent conjugated polymers-seeing polymers in a new light. *Angewandte Chemie* 37: 402–428.

11. Operamolla, A., Farinola, G. M. 2011. Molecular and supramolecular architectures of organic semiconductors for field-effect transistor devices and sensors: a synthetic chemical perspective. *European Journal of Organic Chemistry* 423–450.
12. Günes, S., Neugebauer, H., Sariciftc, I. N. S. 2007. Conjugated polymer-based organic solar cells. *Chemical Reviews* 107: 1324–1338.
13. Destriau, G., Chim, J. 1936. *Phys.* 33: 587.
14. Holonyak, N. Jr, Bevacqua, S. F. 1962. *Applied Physics Letters* 1: 82.
15. Craford, M. G., Steranka, F. M. 1994. *Light-Emitting Diodes in Encyclopedia of Applied Physics.* ed. G. L. Trigg. VCH, 485.
16. Adachi, C., Tsutsui, T., Saito, S. 1990. Blue light-emitting organic electroluminescent devices. *Applied Physics Letters* 56: 799.
17. Kho, M-J., Javed, T., Mark, R., Maier, E., David, C. 2008. *Final Report: OLED Solid State Lighting-Kodak European Research.* MOTI (Management of Technology and Innovation) Project, Judge Business School of the University of Cambridge and Kodak European Research, Final Report presented on 4 March 2008, at Kodak European Research at Cambridge Science Park, Cambridge, 1–12.
18. Lin, K. P., Ramadas, K., Burden, A., Soo-Jin, C. 2006. Indium-tin-oxide-free organic light-emitting device. *IEEE Transactions on Electron Devices* 536: 1483–1486.
19. Wu, C. C., Wu.C. I., Sturm, J. C., Kahn, A. 1997. Surface modification of indium tin oxide by plasma treatment: An effective method to improve the efficiency, brightness, and reliability of organic light emitting devices. *Applied Physics Letters* 70: 1348.
20. Kim, J. S., Cacialli, F., Cola, A., Gigli, G., Cingolani, R. 1999. Increase of charge carriers density and reduction of Hall mobilities in oxygen-plasma treated indium–tin–oxide anodes. *Applied Physics Letters* 75: 19.
21. Tang, C. W. V., Slyke, S. A., Chen, C. H. 1989. Electroluminescence of doped organic thin films. *Journal of Applied Physics* 65: 3610.
22. Hung, L. S., Tang, C. W., Mason, M. G. 1997. Recent progress in organic light-emitting diodes. *Applied Physics Letters* 70: 152.
23. Jabbour, G. E., Kippelen, B., Armstrong, N. R., Peyghambarian, N. 1998. Aluminum based cathode structure for enhanced electron injection in electroluminescent organic devices. *Applied Physics Letters* 73: 1185.
24. Kang, S. J., Park, D. S., Kim, S. Y., Whang, C. N., Jeong, K., Im, S. 2002. Enhancing the electroluminescent properties of organic light-emitting devices using a thin NaCl layer. *Applied Physics Letters* 81: 2581.
25. Ballard, D. G. H., Courtis, A., Shirley, I. M., Taylor, S. C. 1988. Synthesis of polyphenylene from a cis-dihydrocatechol biologically produced monomer. *Macromolecules* 21: 294–304.
26. Kovacic, P., Jones, M. B. 1987. Dehydro coupling of aromatic nuclei by catalyst-oxidant systems: poly(p-phenylene). *Chemical Reviews* 87:357–379.
27. Österholm, J. E., Laakso, J., Nyholm, P., Isotalo, H., Stubb, H. 1989. Melt and solution processable poly(3-alkylthiophenes) and their blends. *Synthetic Metals* 28: 435–444.
28. Fukuda, M., Sawada, K., Yoshino, K. 1993. Synthesis of fusible and soluble conducting polyfl uorene derivatives and their characteristics. *Journal of Polymer Science Part A: Polymer Chemistry* 31: 2465–2471.
29. Yamamoto, T. 1992. Electrically conducting and thermally stable π-conjugated poly(arylene) s prepared by organometallic processes. *Progress in Polymer Science* 17: 1153–1205.
30. Yang, Y., Pei, Q., Heeger, A. J. 1996. Efficient blue polymer light-emitting diodes from a series of soluble poly(paraphenylene)s. *Journal of Applied Physics* 79: 934–939.
31. Scherf, U., List, E. J. W. 2002. Semiconducting polyfluorenes-towards reliable structure property relationships. *Advanced Materials* 14: 477–487.
32. Miteva, T., Meisel, A., Knoll, W., Nothofer, H. G., Scherf, U. 2001. Improving the performance of polyfluorene-based organic light-emitting diodes via end-capping. *Advanced Materials* 13: 565–570.

33. Gill, R. E., Malliaras, G. E., Wildeman, J., Hadziioannou, G. 1994. Tuning of photo-and electroluminescence in alkylated polythiophenes with well-defined regio-regularity. *Advanced Materials* 6: 132–135.
34. Andersson, M. R., Thomas, O., Mammo, W., Svensson, M., Theander, M., Inganäs, O. 1999. Substituted polythiophenes designed for optoelectronic devices and conductors. *Journal of Materials Chemistry* 9: 1933–1940.
35. Chen, T. A., Wu, X., Rieke, R. D. 1995. Regiocontrolled synthesis of poly(3-alkylthiophenes) mediated by Rieke zinc: their characterization and solid-state properties. *Journal of the American Chemical Society* 117: 233–244.
36. McCullough, R. D., Lowe, R. D. 1992. Enhanced electrical conductivity in regioselectively synthesized poly(3-alkylthiophenes). *Journal of the Chemical Society, Chemical Communications* 1: 70–72.
37. Loewe, R. S., Khersonsky, S. M., McCullough, R. D. 1999. A simple method to prepare head-to-tail coupled, regioregular poly(3-alkylthiophenes) using Grignard metathesis. *Advanced Materials* 11: 250–253.
38. Yamamoto, T., Hayashi, Y., Yamamoto, A. 1978. A novel type of polycondensation utilizing transition metal-catalyzed C-C coupling. I. preparation of thermostable polyphenylene type polymers. *Bulletin of the Chemical Society of Japan* 51: 2091–2097.
39. Babudri, F., Colangiuli, D., Farinola, G. M., Naso, F. 2002. A general strategy for the synthesis of conjugated polymers based upon the palladium-catalysed cross-coupling of Grignard reagents with unsaturated halides. *European Journal of Organic Chemistry* 2785–2791.
40. Tanese, M. C., Farinola, G. M., Pignataro, B., Valli, L., Giotta, L. 2006. Poly(alkoxyphenylene-thienylene) Langmuir-Schäfer thin films for advanced performance transistors. *Chemistry of Materials* 18: 778–784.
41. Giancane, G., Ruland, A., Sgobba, V., Manno, D., Serra, A. 2010. Aligning single-walled carbon nanotubes by means of Langmuir-Blodgett film deposition: optical, morphological, and photo-electrochemical studies. *Advanced Functional Materials* 20: 2481–2488.
42. Miyaura, N. 2004. Metal-catalyzed cross-coupling reactions of organoboron compounds with organic halides. In deMeijere, A., Diederich, F. eds. *Metal-Catalyzed Cross-Coupling Reactions*, 2nd ed. Wiley-VCH, 41–123.
43. Bao, Z., Chan, W. K., Yu, L. 1995. Exploration of the Stille coupling reaction for the synthesis of functional polymers. *Journal of the American Chemical Society* 117: 12426–12435.
44. Marsden, J. A., Haley, M. M. 2004. Cross-coupling reactions to sp carbon atoms. In Diederich, F., Stang P. J. eds. *Metal-Catalyzed Cross-Coupling Reactions*. Wiley-VCH, 317–394.
45. Bunz, U. H. F. 2001. Poly(p-phenyleneethynylene)s by alkyne metathesis. *Accounts of Chemical Research* 34: 998–1010.
46. Bunz, U. H. F. 2000. Poly(aryleneethynylene)s: syntheses, properties, structures and applications. *Chemical Reviews* 100: 1605–1644.
47. Yang, J. S., Swager, T. M. 1998. Porous shape persistent fluorescent polymer films: an approach to TNT sensory materials. *Journal of the American Chemical Society* 120: 5321–5322.
48. Breen, C. A., Rifai, S., Bulovic, V., Swager, T. M. 2005. Blue electroluminescence from oxadiazole grafted poly(phenylene-ethynylene)s. *Nano Letters* 5: 1597–1601.
49. Kotora, M., Takahashi, T. 2002. Palladium-catalyzed homocoupling of organic electrophiles or organometals. In Negishi, E., ed. *Handbook of Organopalladium Chemistry for Organic Synthesis*. John Wiley & Sons, 317–394.
50. Mori, A., Kondo, T., Kato, T., Nishihara, Y. 2001. Palladium-catalyzed cross-coupling polycondensation of bisalkynes with dihaloarenes activated by tetrabutylammonium hydroxide or silver(I) oxide. *Chemistry Letters* 30: 286–287.

51. Weiss, K., Michel, A., Auth, E., Bunz, U. H. F., Mangel, T., and Müllen, K. 1997. Acyclic diyne metathesis (ADIMET), an efficient route to poly(phenylene)ethynylenes (PPEs) and nonconjugated polyalkynylenes of high molecular weight. *Angewandte Chemie* 36: 506–509.

52. Brizius, G., Pschirer, N. G., Steffen, W., Stitzer, K., zur Loye, H.-C., Bunz, U. H. F. 2000. Alkyne metathesis with simple catalyst systems: efficient synthesis of conjugated polymers containing vinyl groups in main or side chain. *Journal of the American Chemical Society* 122: 12435–12440.

53. Wessling, R. A. 1985. The polymerization of xylylene bisdialkyl sulfonium salts. *Journal of Polymer Science: Polymer Symposia* 72: 55–66.

54. Gilch, H. G., Wheelwright, W. L. 1966. Polymerization of α -halogenated p-xylenes with base. *Journal of Polymer Science Part A: Polymer Chemistry* 4: 1337–1349.

55. Bräse, S., de Meijere, A. 1998. Palladium-catalyzed coupling of organyl halides to alkenes-the Heck reaction. In Diederich, F., Stang, P. J. eds. *Metal-Catalyzed Cross-Coupling Reactions*. Wiley-VCH, 98–166.

56. Sengupta, S., Battacharyya, S., Sadhukhan, S. K. 1998. Synthesis of trans-4,4′- diiodostilbene and other symmetrical trans-stilbenes by Heck reaction of arene-diazonium salts with vinyltriethoxysilane. *Journal of the Chemical Society, Perkin Transactions* 1: 275–278.

57. Ikenaga, K. Kikukawa, K., Matsuda, T. 1986. Reaction of diazonium salts with transition metals. Part 11. Palladium-catalyzed aryldesilylation of alkenylsilanes by arenediazonium salts. *Journal of the Chemical Society, Perkin Transactions* 1: 1959–1964.

58. Sengupta, S., Bhattacharyya, S. 1993. Heck reaction of arenediazonium salts: a palladium- catalysed reaction in an aqueous medium. *Journal of the Chemical Society, Perkin Transactions* 1: 1943–1944.

59. Wakioka, M., Ikegami, M., Ozawa, F. 2010. Stereocontrolled synthesis and photoisomerization behavior of all-Cis and all-Trans poly(m-phenylenevinylene)s. *Macromolecules* 43: 6980–6985.

60. Babudri, F., Cardone, A., Farinola, G. M., Martinelli, C., Mendichi, R., Naso, F., Striccoli, M. 2008. Synthesis of poly(arylenevinylene)s with fluorinated vinylene units. *European Journal of Organic Chemistry* 1977–1982.

61. Babudri, F., Cicco, S. R., Chiavarone, L., Farinola, G. M., Lopez, L. C., Naso, F., Scamarcio, G. 2000. Synthesis and optical investigations of low molecular weight alkoxy-substituted poly(p-phenylenevinylene)s. *Journal of Materials Chemistry* 10: 1573–1579.

62. Bolognesi, A., Botta, C., Babudri, F., Farinola, G. M., Hassan Omar, O., Naso, F. 1999. Silicon-substitued PPV for LED preparation. *Synthetic Metals* 102: 919.

63. Bredas, J. L., Heeger, A. J. 1994. Influence of donor and acceptor substituents on the electronic characteristics of poly(paraphenylene vinylene) and poly(paraphenylene). *Chemical Physics Letters* 217: 507–512.

64. Babudri, F., Cardone, A., Chiavarone, L., Ciccarella, G., Farinola, G. M., Naso, F., Scamarcio, G. 2001. Synthesis and characterization of poly(2,3,5,6-tetrafluoro-1,4-phenylenevinylene). *Chemical Communications* 1940–1941.

65. Babudri, F., Cardone, A., Farinola, G. M., Naso, F., Cassano, T., Chiavarone, L., Tommasi, R. 2003. Synthesis and optical properties of a copolymer of tetrafluoro-and dialkoxy-substituted poly(p-phenylenevinylene) with a high percentage of fluorinated units. *Macromolecular Chemistry and Physics* 204: 1621–1627.

66. Babudri, F., Cardone, A., Cassano, T., Farinola, G. M., Naso, F., Tommasi, R. 2008. Synthesis and optical properties of a poly(2′,5′-dioctyloxy-4,4′,4″- terphenylenevinylene) with high content of (Z) vinylene units. *Journal of Organometallic Chemistry* 693: 2631–2636.

67. Cardone, A., Martinelli, C., Pinto, V., Babudri, F., Losurdo, M., Bruno, G., Cosma, P., Naso, F., Farinola, G. M. 2010. Synthesis and characterization of perfluorinated arylenevinylene polymers. *Journal of Polymer Science, Part A: Polymer Chemistry* 48: 285–291.

68. Losurdo, M., Giangregorio, M. M., Capezzuto, P., Cardone, A., Martinelli, C., Farinola, G. M., Babudri, F., Naso, F., Büchel, M., Bruno, G. 2009. Blue-gap poly(p-phenylene vinylene)s with fluorinated double bonds: interplay between supramolecular organization and optical properties in thin films. *Advanced Materials* 21: 1115–1120.

69. Conticello, V. P., Gin, D. L., Grubbs, R. H. 1992. Ring-opening metathesis polymerization of substituted bicyclo[2.2.2]octadienes: a new precursor route to poly(1,4-phenylenevinylene). *Journal of the American Chemical Society* 114: 9708–9710.

70. Thorn-Csany, E., Kraxner, P., Strachota, A. 1998. Synthesis of soluble all-trans oligomers of 2,5-diheptyloxy-p-phenylenevinylene via olefin metathesis. *Macromolecular Rapid Communications* 19: 223–228.

71. Egbe, D. A. M., Neugebauer, H., Sariciftci, N. S. 2011. Alkoxy-substituted poly(aryleneethynylene)-alt-poly(arylene-vinylene)s: synthesis, electroluminescence and photovoltaic applications. *Journal of Materials Chemistry* 21: 1338–1349.

72. Egbe, D. A. M., Tillmann, H., Birckner, E., Klemm, E. 2001. Synthesis and properties of novel well-defined alternating PPE/PPV copolymers. *Macromolecular Chemistry and Physics* 202: 2712–2726.

73. Egbe, D. A. M., Carbonnier, D., Ding, L., Mühlbacher, D., Birckner, E., et al. 2004. Supramolecular ordering, thermal behavior, and photophysical, electrochemical, and electroluminescent properties of alkoxy-substituted yne-containing poly(phenylenevinylene)s. *Macromolecules* 37: 7451–7463.

74. Evenson, S. J., Mumm, M. J., Pokhodnya, K. I., Rasmussen, S. C. 2011. Highly fluorescent dithieno[3,2- b:2′,3′- d] pyrrole-based materials: synthesis, characterization, and OLED device applications. *Macromolecules* 44: 835–841.

75. Perepichka, I. F., Perepichka, D. F., Meng, H. 2009. Thiophene-based materials for electroluminescent applications. In Perepichka, I. F., Perepichka, D. F. eds. *Handbook of Thiophene-Based Materials: Applications in Organic Electronics and Photonics.* John Wiley & Sons.

76. Zhang, W., Li, J., Zhang, B., Qin, J. 2008. Highly fluorescent conjugated copolymers containing dithieno[3,2- b:2′,3′- d] pyrrole. *Macromolecular Rapid Communications* 29: 1603–1608.

77. Donat-Bouillud, A., Lévesque, I., Tao, Y., D'Iorio, M., Beaupré, S., Blondin, P., Ranger, M., Bouchard, J., Leclerc, M. 2000. Light-emitting diodes from fluorene-based π-conjugated polymers. *Chemistry of Materials* 12: 1931–1936.

78. Mikroyannidis, J. A., Fenenko, L., Yahiro, M., Adachi, C. 2007. Alternating copolyfluorenevinyles with polynuclear aromatic moieties: synthesis, photophysics, and electroluminescence. *Journal of Polymer Science Part A: Polymer Chemistry* 45: 4661–4670.

79. Farinola, G. M., Ragni, R. 2011. Electroluminescent materials for white organic light emitting diodes. *Chemical Society Reviews* 40: 3467–3482.

80. Kitagawa, S., Kitaura, R., Noro, S-I. 2004. Functional porous coordination polymers. *Angewandte Chemie* 43: 2334–2375.

81. Weck, M. 2007. Side-chain functionalized supramolecular polymers. *Polymer International* 56: 453–460.

82. Burnworth, M., Knapton, D., Rowan, S. J., Weder, C. 2007. Metallo-supramolecular polymerization: a route to easy-to-process organic/inorganic hybrid materials. *Journal of Inorganic and Organometallic Polymers and Materials* 17: 91–103.

83. Knapton, D., Rowan, S. J., Weder, C. 2006. Synthesis and properties of metallosupramolecular poly(p-phenyleneethynylene)s. *Macromolecules* 39: 651–657.

84. Jiang, P., Zhu, W., Gan, Z., Huang, W., Li, J., et al. 2009. Electron transport properties of an ethanol-soluble AlQ_3-based coordination polymer and its applications in OLED devices. *Journal of Materials Chemistry* 19: 4551–4556.
85. Montes, V. A., Zyryanov, G. V., Danilov, E., Agarwal, N., Palacios, M. A., Anzenbacher, P. Jr 2009. Ultrafast energy transfer in oligofluorene-aluminum bis(8-hydroxyquinoline) acetylacetone coordination polymers. *Journal of the American Chemical Society* 131: 1787–1795.
86. Galbrecht, F., Yang, X. H., Neels, B. S., Neher, D., Farrell, T., Scherf, U. 2005. Semiconducting polyfluorenes with electrophosphorescent on-chain platinumsalen chromophores. *Chemical Communications* 2378–2380.

10 Improvement in the Efficiency of Organic Semiconductors via Molecular Doping for OLEDs Applications

Sunil Kumar, Pankaj Kumar Chaurasia, and Sandeep K. S. Patel

10.1 INTRODUCTION

Organic light emitting diodes (OLEDs) are twofold charge injection optoelectronic devices, comprising integrated supply of both electrons and holes into electroluminescent organic semiconductors. In common, the organic semiconductor compounds possess ambipolar transport characteristics, which opens the possibility of realizing highly advanced, large surface area and low-cost organic optoelectronic devices.[1] In the field of lighting, exclusive external quantum efficiency is required at low operating voltages and high brightness to upgrade energy utilization and consumption[2,3] The unwanted diffusion kinetics of dopant molecules causes hampering of charge transport, OLED lifetime and efficiency.[4] Therefore, high-performance OLEDs can be developed through the immobilization of dopants via covalent functionalization to the organic semiconductor or host structure.[5,6] Strong donor groups are added to achieve a shift in conductivity and Fermi level.[7] Thus, the molecular features permits better control of the diffusion kinetics, for example, covalent functionalization reorients the molecule's morphology (size or shape).[8] A large quantity of emitted photons does not lower efficiency because of total internal reflection and the photons are integrated into the organic structure.[9] Conducting dopants facilitate transport layers to minimize device operating voltages.[10,11] Charge-carrier conduction is a result of transmission of chain of redox (electrochemical) processes between the native molecules and the corresponding radical ions.[12] Electron-donor and electron-acceptor organic groups have potential capacity to conduct holes and electrons, respectively.[13] Common electron-withdrawing groups, that is, carbonyls, cyano, perfluoroalkyl, imide, amide and their analogues are tagged to functionalize organic semiconductors.[14] Bipolar molecules improve performances and operational stability due to the synergistic influence of the emitting layer. Organic functional groups resolve more light-emitting capacity and, hence, emission efficiency through

DOI: 10.1201/9781003260417-10

extended π-conjugation.[15] However, organic groups bring large bathochromic shifts. Therefore, one of the major issues is to prevent the red-shift emission through incorporation of bipolar blue emitters with D-A structures to increase the photoluminescence quantum yield.[16] Lin et al. developed carbazole and dimesitylboraneinto a π-conjugated organic semiconductor. Other amino donors and carbazoles are incorporated to produce short wavelength absorption or emission because of weaker π donors than amine-containing compounds. A broad color display generates blue, green, and red emissions in equal proportions; the color purity and efficiency can be selected into white OLEDs. Carbazole and thiophene are heterocyclic organic groups that prevent transport/injection of holes because of the electron-rich environment.[13] Hole injection and transport features are produced through the introduction of organic electron-withdrawing dopants into a suitable hole-conducting layer. Selective grafting tailors the energy structure for full-color OLEDs. The blue emitter decreases the power consumption to produce light of other colors by an energy cascade of fluorescent or phosphorescent dopants.[17] Organic semiconductors are exclusively versatile compounds that allow selection of low-cost alternatives to the inorganic compounds in light emitting diodes for displays and lighting.[18,19] The extensive p-type (hole-transporting) dopants are strong oxidants that intensify holes through withdrawing electrons from the organic semiconductor.[17] In contrast, n-type (electron-transporting) chemical dopants can be achieved via the incorporation of electron donor species (reducing agents). The dopant charge carrier saturates deep trap states. Doping introduces charge-donating or -accepting atoms or molecules, generating sufficient charge carriers to control the charge conduction and negate the influence of unwanted impurities.[20] Conductivity dopants are used in OLEDs to reduce the operating voltage and consequently improve the power efficiency.[10] High external quantum efficiency at low operating voltages and high brightness are required for improved energy utilization. The purpose of doping is to induce electron transfer between host and dopant, which introduces additional charge carrier concentration and increases the conductivity of the host materials. Carrier type and density can be controlled by doping. Increased free carrier density by doping also modifies the contact properties at the junctions with an electrode or other semiconductors. Increasing the number of mobile charge carriers in transparent layers at the contacts helps in reducing injection and extraction barriers and facilitates charge transport from absorber to emission layers. Bipolar materials stabilize excitons formation and broaden the formation zone in the emitting layer, and consequently enhance the performances and operational stability in OLEDs. Both the electron-donating (D) and electron-accepting (A) moieties in a single molecule can effectively stabilize exciton formation and balance the hole and electron charge in the emitting layer. A multilayered structure retains continuous charge transport because its emission characteristics achieve more organized charge recombination than single-layer devices. The smooth injection of electrons and holes intensifies the charge recombination for better OLED function. The efficiency of OLEDs requires controlled engineering of the carrier transport into organic semiconductors. The interfacial contacts frequently hold an energetic barrier that delays the flow of charge carriers.[21] Interfacial charge-injection barriers bring about high driving voltages and low power efficiencies in organic optoelectronic devices. OLED emission efficiency or color is expressed as a function of the band gap of the organic semiconductor. A wide energy gap is required for deep-blue emission into OLEDs from the organic host.[22] The anode/OSc/cathode requires tuning of the interface barrier for efficient hole and electron injection into the organic semiconductor. After injection of charges from the two electrodes into the organic semiconductor, the π covalent system of the organic semiconductor is subjected to disturbance, resulting in the generation of regular deformation, which traps the excitons (polarons). Holes and electrons are simultaneously transported via the organic hole-transporting layer (HTL) and an electron-transporting layer (ETL) into an emissive layer (EML), causing a recombination reaction.[23] The strong electron-withdrawing features of triazine compared with other heterocyclic compounds, that is, aza-aromatic compounds, pyridine, pyradine, and pyrimidine, prefer emission peaks of longer wavelength, which ultimately creates difficulty in the design of deep-blue emittive OLEDs. The electron-deficient features of pyrimidine compared with benzene and pyridine permits bipolar host compounds owing to the presence of greater electronegative character, which offers suitability for better electron transport and injection capacity. The deeply stabilized ionization energy (IE) difficult to accommodate energy level with better alignment between anodes (indium tin oxide (ITO) = 4.70 eV or metals, e.g., silver = 4.73 eV) with emitting materials.[24] The interface barrier causes degradation of hole injection with deep low-lying highest occupied molecular orbital (HOMO) of the emissive layer.[25] Therefore, assembly requires a dopant exhibiting high electron affinity. Thus, the degree of charge transfer depends strongly on the electron affinity EA and the ionization energy IE of the dopant and host molecule, respectively. A complete charge transfer from the dopant to the host is noticeable ($Z = 1$).[24]

10.2 OLEDs AND THEIR FUNCTIONAL COMPONENTS

An OLED is an optoelectronic device that transmits selective light when an external voltage is applied. OLED display systems contain organic electroluminescent material that emits wavelength of selective light under stimulation of external electricity (Figure 10.1a).[17] An OLED is a solid-state semiconductor device composed of conducting and emissive layers 100 to 500 nm in thickness, with an integrated design. The conducting layer is fabricated of organic plastic molecules to facilitate hole from the anodic electrode. The film-shaped emissive layer is crafted of organic materials to conduct electrons from the cathode with rapid light emission as a response to an electrical current. The conducting organic layer functions by π-electron delocalization over the surface of the conjugated structure.[26] The cathode, having shallow lowest unoccupied molecular orbital (LUMO) levels, can inject electrons effectively into the emissive layer. The anode, having deep HOMO levels, can inject holes to the emissive layer effectively.

10.3 ELECTROLUMINESCENCE MECHANISM AND FUNCTION OF OLEDs

Multilayered interfaces are developed to facilitate charge injection in a channelwise fashion and rapid carrier transport into the light emitting film.[27] The emissive layer of a functional OLED involves electronic and excitonic transitions, that is, charge

(a)   (b)

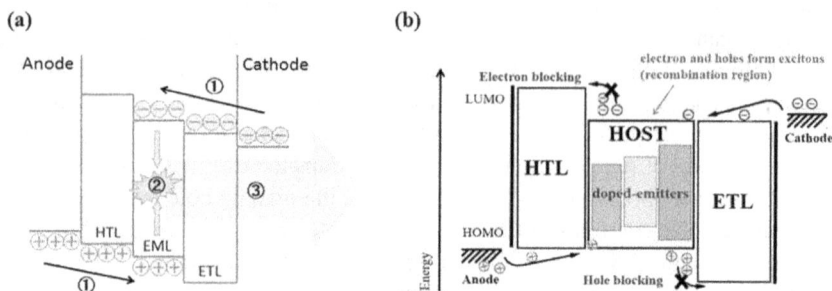

FIGURE 10.1 (a) Systematic function and mechanism of electroluminescence in an OLED: (1) Electrons and holes are injected from the electrodes into the organic emissive layer. (2) Generation of electron-hole pair (exciton). (3) The radiative recombination of excitons causes photon (light) emission.[1] (b) Schematic of OLED and its charge transport mechanism for efficient emission. EML = emissive layer; ETL = electron-transporting layer; HOMO = highest occupied molecular orbital; HTL = hole-transporting layer; LUMO = lowest unoccupied molecular orbital. (Reprinted with permission from Karzazi (a) and Lu et al. (b).[28]

transport, exciton formation and excitonic energy transfer (Figure 10.1b). Two electrodes inject charges into the organic layers leading to disorganization of the π-covalent bond within organic semiconductors, resulting in the generation of confined distortion to ambush the charges. The physical mechanism relies heavily on the energetics of the organic host and the dopant environment. The energetic interface between the two layers causes recombination of the injected hole–electron within a confined dimension to produce electroluminescence. Thus, OLEDs require synchronous amalgamation of both electrons and holes to the electroluminescent compound sandwiched between the anode and the cathode. In two-layer OLEDs, first, electrons are injected from the cathode into the conduction band (LUMO) of the organic compound at the electrode-organic interface, and, second, holes shoot up from the anode into the valence band (HOMO) of the organic compound to produce the recombination process. The extent of excitons is diluted through a photoemissive mechanism and thereby, undergoes degradation radiatively to the ground state by unsolicited emission. Fundamentally, OLED efficiency is decided by charge saturation, radiative degradation of excitons, and light extraction. In OLEDs, the hole current is regulated through injection kinetics, while the electron current is strongly altered through the presence of a trap state of the interface of metal–organic compound. The energy barrier at the interface of the metal electrode and organic layer suppresses effective carrier injection into the active layer and results in unbalanced electron-hole currents flowing through the OLED.

10.4   ORGANIC SEMICONDUCTOR AS A LIGHT EMITTER MEDIUM

High-efficiency OLEDs require strong and continuous emission through the organic semiconductor.[15] The control of emission region into emissive layer is due to effective energetics, interactions and reasonable charge injection. The color variation of the light depends on the type of organic structure, chromophore into emissive layer.[29]

The organic emissive layer maintains higher quantum efficiency, exciton injection and conducting capacity. The organic semiconductor behaves as a light emitter because of its ambipolar nature, with sufficient exciton transfer capacity. Deep-blue-colored OLEDs require an organic host with a broad energy gap for selective recombination.[22] Using a doped organic semiconductor improves the mobility of the charge carrier with respect to the hole-injecting layer and the electron-injecting layer. However, various π-conjugated rich organic compounds, such as anthracene, pyrene, fluorine and di (styryl)arylene functionalized have been selected as blue emitters for higher efficiency and stability in OLEDs. The structural fluorophores extend higher emission efficiency (nearly unity) in a dilute solution. However, concentration quenching further lowers the emission efficiency in the solid state owing to molecular conglomeration. The concentration quenching is regulated through conjugated molecules as dopants to enhance the emission capacity of the organic semiconductor host. Liu et al.[30] developed a highly conducting organic semiconductor (2, 6-diphenylanthracene) as a strong light emitter for OLED. The developed single-crystal organic molecule exhibited higher mobilities ($34\,cm^2 V^{-(1} s^{-1})$ and photoluminescence quantum yield (41.2%) under functional state of OLED. The resultant OLED emits pure blue emission with brightness capacity of $6{,}627\,cd\,m^{-2}$ under turn-on voltage at 2.8 V. Thanks to the structural diversity into organic molecule which inspires to develop wavelength selective light emitter into OLEDs science and technology.

10.5 ROLE OF CHROMOPHORES, CONJUGATION AND CHARGE TRAP IN ORGANIC SEMICONDUCTORS

The chromophores stabilize the color quality of organic emitters. The emission colors can be simply tailored by functionalizing the organic structures.[31] The amalgamation of weak donor/acceptor groups produces blue emission owing to poor charge-transfer (CT) interaction, while the combination of strong donor/acceptor groups causes red emission because of strong CT interaction.[32] Suitable molecular design and choice of chemical structures provides different optoelectronic characteristics. The introduction of charge-conducting groups and chromophores intensifies emission efficiency because of functional interactions within the same molecules (Figure 10.2 and Table 10.1). Electron-withdrawing and electron-donating groups facilitate the carrier conductivity and saturate them to recombine impressively. The increase of effective

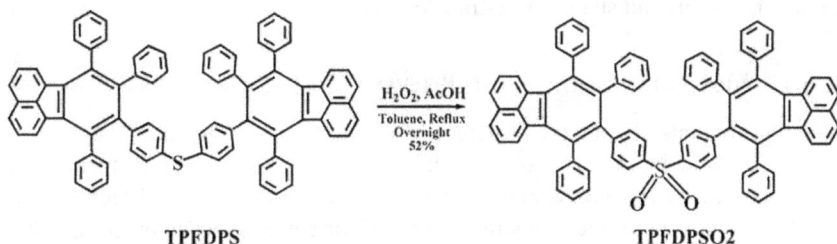

TPFDPS   TPFDPSO2

FIGURE 10.2 Functionalization of an organic semiconductor to create an efficient chromophore. (Reproduced with permission from Kumar and Patil.[31])

TABLE 10.1

Electroluminescence Properties of a Functionalized Organic Emissive Layer[31]

Compound	Electroluminescence (nm)	Brightness I_{max} (cd/m²)	External Quantum Efficiency (%)
TPFDPS	495	4,427	0.12
TPFDPSO2	487	18,140	0.50

conjugation length causes red shifts in absorption/emission spectra, which enhances carrier conduction and lowers the material energy gap. Carbazole and thiophene are electron-donating groups which lead to enhancement of the hole-transport process. The electron-deficient nitrogenic group, that is, oxadiazole, diazole, imidazole, triazine and pyridine used to functionalize organic semiconductor for strengthening interfacial electron transport in OLEDs. Structural and chemical imperfections always create a trap state for charge carriers in organic semiconductors. The traps depress the density of electrons in the transport or conducting level. Functionalization enhances the functional capacity of organic emitters for advancement of OLEDs science.

The charge carriers produced by dopants saturate deep trap states and improve the mobility of excitons.[18] Phenanthroimidazole compounds contain electronegative nitrogen atoms, which makes organic emitters containing these compounds highly polar because of their electron-withdrawing nature. This group effectively transfers electrons to the emissive layer. Electron traps are present within the HOMO-LUMO gap, which can reduce the efficiency of radiative recombination. Trap states depress charge carriers due to the Coulomb force of attraction between ionized dopants and free excitons. Typically, electrostatic attraction lowers the dielectric constant for such an organic semiconductor.[33] Molecular doping merges the trap level either on the HOMO or LUMO level, leading to improvement of luminescence efficiency. Trap deactivation has a marked influence on OLED performance.

10.6 MOLECULAR DOPING

Molecular doping has emerged as a viable alternative where impressive large organic molecules (electron acceptors or donors) ignited as p or n dopants for organic emitter. The interaction mechanisms of dopant and OSc matrix leads to the generation of an ion pair and ground state charge-transfer complex.

10.6.1 Molecular Dopants and Relevant Doping Strategies

Molecular dopants change the structure of organic semiconductors (Figure 10.3). High doping stoichiometry retards the diffusion of dopants, which further destabilizes the organic semiconductor and organic optoelectronic system.[13] Electron-donor organic groups densify electrons to the LUMO or electron-acceptor groups dilute electrons from the HOMO of the organic semiconductor. The organic structure of these species permits regular control of diffusion by covalent functionalization.[34] The exclusive organic dopants, that is, 2,3,5,6-tetrafluro-7,7,8,8-tetracyanoquinodimethane

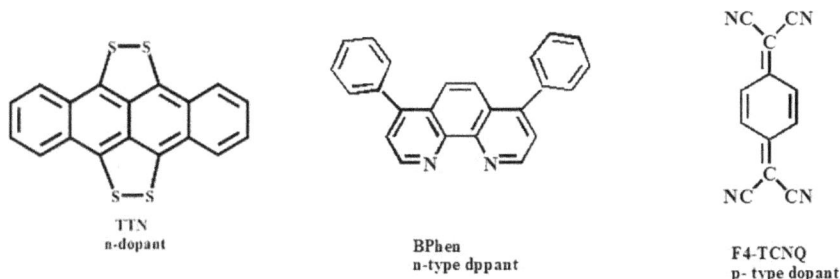

TTN
n-dopant

BPhen
n-type dppant

F4-TCNQ
p- type dopant

FIGURE 10.3 Chemical structure of n- and p-type dopants.[13,24]

(F4-TCNQ), 2,2-(perfluoronaphthalene-2,6,-diylidene)-dimalononitrile (F6-TCNNQ), 1,4,5,8,9,11-hexaazatriphenylenhexacarbonitrile (HATCN) and $C_{60}F_{36}$, have been already emphasized to impressively enhance efficiency activity of OLEDs. Organic dopants enable a strategic tool for reducing ohmic degradation in the charge-transport layers and injection barriers at the interfaces. Polar side groups covalently attached on an organic semiconductor significantly increase its dopant affinity. Polar chains greatly improve the thermal stabilization of organic dopants through elevating the melting temperature. Larger organic dopants provide selective morphological dimensions to stabilize thermally activated diffusion of dopants in the organic host. Charged organic dopants do not allow rapid diffusion owing to electrostatic interactions with the surface structure. Neutral dopants are preferred to increase the efficiency of charge transport. Functional covalent bonds between the dopant and the host molecule support immobilization of both neutral and charged dopants.[5] The asset p-type organic dopant generates deep LUMO sufficient to withdraw an electron from the HOMO level of an organic semiconductor. In a similar fashion, n-type organic dopant is a molecule with shallow HOMO level adequate to donate an electron into the LUMO of an organic semiconductor. The intensive dopants having higher electron affinity (EA) compared to the ionization energy (IE) of the host were selected to be beneficial for p-type dopants.[20]

10.6.2 ROLE OF INTERFACIAL DOPING

Either breaking or reforming of chemical bonds in organic compounds are catalyzed through chemically reactive or strongly interacting interfaces. Interfacial dopants are prone to localize chemical bonds via strong hybridization between electrode surface, electronic states and molecular orbitals.[35] The eminence of the charge-injection barrier is connected with the relative energetics of donor and acceptor levels on either side of the interface or organic emitter.[35] Havare et al. developed 4-[(3-methylphenyl) (phenyl)amino]benzoic acid (MPPBA) to facilitate hole injection into OLEDs (Figure 10.4). The OLED exhibited brightness of 52.9 cd/m² and turn-on voltage of 4.5V, while a bare ITO-assembled OLED device produces brightness of 14.8 cd/m² and turn-on voltage 5.5V. Multilayered interfaces facilitate more charge-transport, recombination and emission than single-layered OLEDs.[36] A multilayer configuration

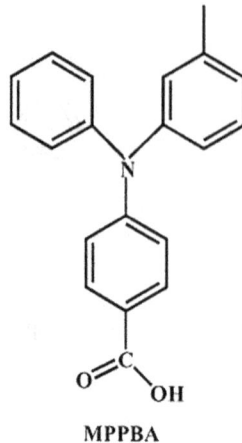

MPPBA

FIGURE 10.4 Chemical structure of 4-[(3-methylphenyl)(phenyl)amino]benzoic acid (MPPBA) as surface dopant. (Reproduced with permission from Havare et al.[36])

facilitates rapid injection of holes and electrons into the emitting region. Multiple interfaces reduce the energy barrier to facilitate charge transport and overall OLED efficiency.[36] The efficiency of integrated OLEDs can be advanced via doping emitters into an organic or hosting layer.[36] The higher the degree of work function of the OLED layers, the greater the chances of charge injection.[37] Thus, OLEDs or organic semiconductors having high hole or electron mobility are responsible for impressive charge transport.

10.6.3 INFLUENCE OF CONDUCTING POLYMER-DOPED ORGANIC SEMICONDUCTORS

The energy structure between organic hosts and dopants governs physical mechanisms such as charge transport and exciton formation for OLEDs science.[38] A conjugated conducting polymer-doped emissive layer (the organic semiconductor) improves the emitted color quality and efficiency of OLED.[39] A conducting polymer enhances the mobility of polarons for efficient recombination. The extension of color variation depends on conjugated dimensions and integrated interactions.

10.6.4 IMPACT OF COMPOSITE DOPED METAL OXIDE AND ORGANIC METAL COMPLEX ON ORGANIC SEMICONDUCTORS

Transition-metal-dependent metal oxides (TMOs), such as WO_3, MoO_3, V_2O_5, and ReO_3, have been broadly investigated as p-type inorganic dopants because of their deep-lying electronic energy. It has been widely explored that TMO dopants improve injection and conducting efficiency of organic semiconductors, which in turn decrease the operating voltage as a result of highly efficient tuned OLEDs.[40] The TMO dopant shifts the energetics of the Fermi level toward the HOMO level of the organic semiconductor.[41] White et al.[42] reported that an extended band gap shaped organic compound, that is, 4,4'-N,N'-dicarbazole-biphenyl (CBP), with deep ionization energy of 6.0 eV, is treated as a selective species that forms charge transfer complexes with higher electron affinity TMOs, that is, MoO_3 (EA = 6.9eV).Rapid diffusion causes enhancement of the efficiency of OLEDs. However, in contrast to the metal oxide dopants, the organic dopant homogeneously concentrated into an organic host leads to high charge generation efficiency and recombination efficiency due to similar physicochemical structure.[42] Tris [1,2-bistrifluoromethylethane-1,2-dithiolene] $[Mo(tfd)_3]$ organic p-dopant, which has a low LUMO level of 5.59 eV demonstrates beneficial combination with organic semiconductors in OLED devices.

10.6.5 INTEGRATION OF HIGH CHARGE DENSITY AND LUMINESCENCE IN ORGANIC SEMICONDUCTORS

Organic semiconductors require adequate mobility and strong fluorescence emission to achieve high-efficiency OLEDs.[43] Liu et al. developed OLEDs based on 2,6-diphenylanthracene (DPA) with pure blue emission and high brightness (6,627 cdm^{-2}) and turn-on voltage of 2.8V. The impressive result is associated with high charge carrier mobility (34 cm^2 V s^{-1}) on the single-crystal level. The organic molecule DPA contains strong fluorescence emission from anthracene tagged with phenyl groups at the 2-, 6- positions to extend the p conjugation and molecular packing. The emissive layer exhibited florescence quantum yield of 41.2%.

10.6.6 CHARGE GENERATION EFFICIENCY OF p/n-DOPANTS IN ORGANIC SEMICONDUCTORS

Efficient charge generation can be achieved through tuning of IE or EA for host or dopant, respectively.[44] Doping or chemical substitution makes it easy to tailor the ionization potential of the host organic semiconductor.[45] The diffusion of organic dopants into the light-emitting layer causes decreased lifetime via a phenomenon called exciton quenching. The exciton quenching process is caused by the lower band gap of the radical anion of the dopant compared to the emitter. The valence electrons (HOMO electrons) for p-doped semiconductors are excited into acceptor states (LUMO) of the p-dopant than into the conduction band (LUMO) of the organic semiconductor due to poor diffusion tendency.[46] The organic host (e.g., N,N'-bis(3-methylphenyl)-N,N'-diphenylbenzidine, TPD) with higher ionization energy improves charge transfer efficiency (0.64) for the combination TPD/F4-TCNQ. The charge transfer efficiency of functionalized TPD can be enhanced by lowering the ionization energy, which is possible via the electron-pushing effect of methoxy groups at the outer benzene rings, as in MeO-TPD.[47] The charge transfer ratio (0.74) is quantized for the functionalized MeO-TPD/F4-TCNQ species. The molecule tetrathianaphthacene (TTN, HOMO = 4.7 eV) can be used as n-dopant for hexadecafluoro-zinc-phthalocyanine (F16ZnPc) but not for Alq3. The energy levels of F16ZnPc (LUMO = 4.5 eV) and Alq3 (LUMO = 2.5 eV). It is not surprising that TTN works more efficiently as electron donor for the F16ZnPc.

10.7   ROLE OF BIPOLAR/BIFUNCTIONAL ORGANIC EMITTERS

One advanced technique to realize bipolar transport in organic molecules favors the embodiment of both electron-donating and electron-withdrawing blocks in the same organic molecule (Figure 10.5a).[15,48] Current density can be changed by variation of the structure and applied voltage (Figure 10.5b). Trap-free bipolar charge conduction happens in the range of a potential window of EI (6 eV) to EA (3.6 eV). Bipolar transport is simply associated with the position of energy levels with reference to the trap-free window.[23] Bipolar emitters stabilize exciton generation into the emitting layer. The emission color is a function of the energy gap between the HOMO (donor) and the LUMO (acceptor). Chen et al.[49] developed fashionating bipolar organic host 11-(3-(4,6-diphenyl-1,3, 5-triazin-2-yl)phenyl)-12,12-dimethyl-11,12-dihydroindeno[2,1-a] carbazole (DPDDC) (Figure 10.5c, d). Hole transport favors denocarbazole and electron-withdrawing 2,4-diphenyl-1,3,5-triazine units into the nonconjugated benzene ring. The OLED composed of DPDDC as organic host produces an outstanding emissive performance, with high power efficiency (92.3 lm/W) and high external quantum efficiency (EQE) of 23.6%.

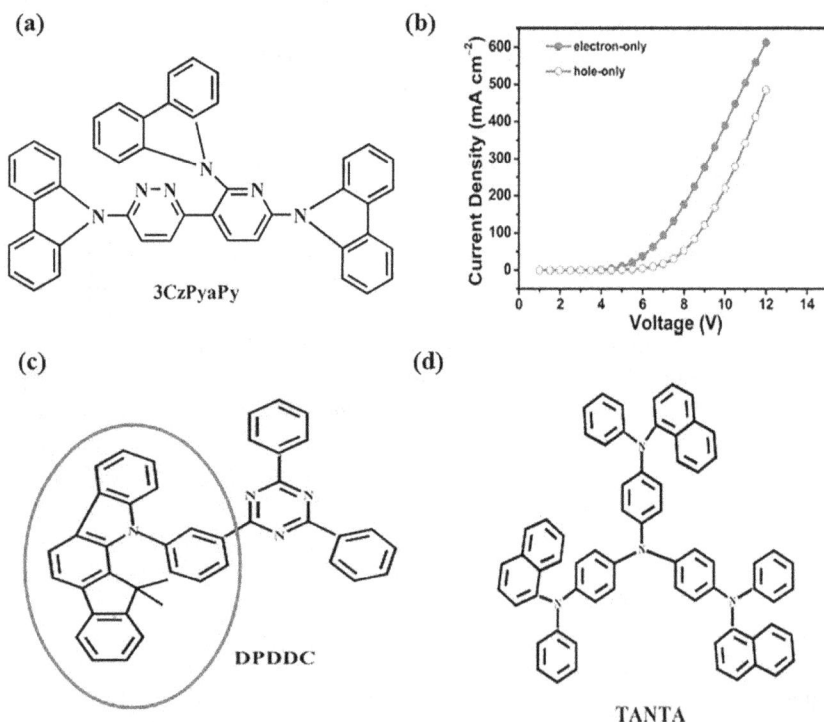

FIGURE 10.5   (a) Chemical structure of D-A type organic semiconductor; (b) current density vs. voltage curve for hole-only and electron-only device for 3CzPyaPy;[16] (c) structure of bipolar host.[50] (d) TANTA.[51]

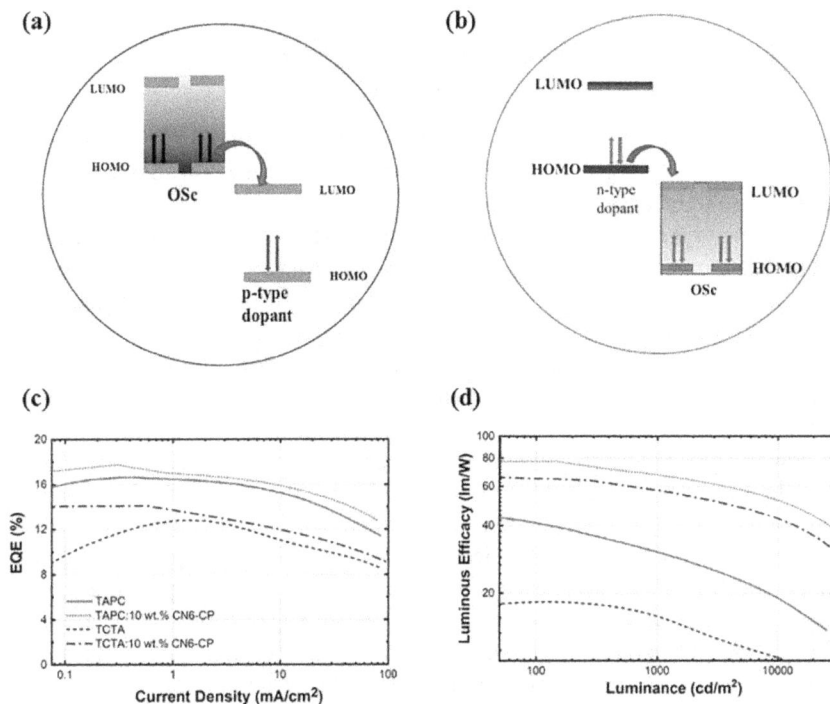

FIGURE 10.6 (a)-(b) Doping of the organic semiconductor (OSc) and its charge transport for effective generation of charges (holes or electrons).[38] (c) External quantum efficiency (EQE) along with current density. (d) Electroluminescence efficiency of the OLED assumes different hole injection and transport layers.[30]

10.7.1 CHARGE TRANSFER IN P- AND N-TYPE DOPED ORGANIC SEMICONDUCTORS

Doping converts the crystalline phase into an amorphous phase, which exhibits high mobility.[52] Interfacial charge transport occurs through interfacial band bending and alignment, which shifts the Fermi level or energy gap (positive or negative) of the organic semiconductor. The OLED is composed of HTL/EML/ETL layers in which the hole transport material enhances the hole density in the EML because the electron-donating group intensifies the HOMO, causing more attraction of holes injected from the anode (Figure 10.6a, b).[53] The electron transport material, having an electron-withdrawing group, accelerates the electron deficiency on the HOMO, which attracts the injected electron intensely from the cathode.[54] In this way the HTL and the ETL create a depletion region in the EML for effective recombination. For *p*-doping, the HOMO electron of the organic semiconductor is subjected to hop into the LUMO (electron acceptor) and vice versa for *n*-doping, the HOMO electron of the electron donor is expected to inject into the LUMO of the organic semiconductor.

10.8   STRUCTURAL AND FUNCTIONAL STABILIZATION EFFICIENCY IN ORGANIC SEMICONDUCTORS VIA MOLECULAR ENGINEERING

The controlled molecular structure intensifies the OLED properties. Functional design that emphasizes hole- or electron-conducting moieties into the emitting organic structure would increase the efficiency of the organic semiconductor. The functional energetics of organic semiconductors are the HOMO and the LUMO centered on the organic group, which determines the emission wavelength. The hole transport phenomenon is necessary for reversible operation of the device.[55] The doping efficiency is elucidated as the ratio of the number of free charge carriers to the number of dopants. The doping efficiency always results in below unity, while all dopant molecules transfer charges into a pristine molecule. The technological principle of doping into the organic semiconductor is to chemicalize strong electron-donating or electron-withdrawing groups, which either transfer an electron to the LUMO of a host organic molecule to generate a free electron density (n-type doping) or remove an electron from the HOMO of a host molecule to upgrade a free hole (p-type doping). The injection kinetics of both carriers (electrons and holes) must be equivalent for higher OLED efficiency. Otherwise, excess electrons or holes will not recombine, leading to low functional efficiency. In the emissive organic layer, the charge conduction is positively dependent on the energetics of these frontier orbitals between the host and the dopant group. The hole (or electron) is trapped naturally inside the dopant when the energy of the HOMO or LUMO of the dopant is higher (or lower) than that of the organic host.[56] However, charge trapping could ease the direct exciton evolution inside the dopant molecule. Doping controls the Fermi-level location, which is associated with the number of free charge carriers produced in the organic semiconductor. To reach coherent doping, the chargetransfer process must be energetically favorable, which means that the electron affinity (EA) of the dopant is equal to or higher than the ionization energy (IE) of the host molecule. High driving voltage causes degradation of efficiency in OLEDs. Liue et al.[30] developed novel p-dopant (CN6-CP) structured as hexacyano-trimethylene-cyclopropane (EA = 5.87 eV) for OLED applications. The CN6-CP molecule works not only as a dopant in the conventional hole transport structure 4,4'-cyclohexylidenebis[N,N-bis(4-methylphenyl)benzamine] (TAPC, IE = 5.50 eV) but also, importantly, dopes the organic host compound tris(4-carbazoyl-9-ylphenyl)amine (TCTA, IE = 5.85 eV), producing a conductivity of 1.86×10^{-4} S/cm at a molar ratio of 0.25 (Figure 10.6 c, d). Hole injection can be promoted by an electron-sufficient chemical structure, while electron injection can be facilitated through an electron-deficient chemical and functional structure. Thus, saturated excitons undergo recombination to emit bright color. Hole transport compounds exhibit a shallow HOMO level and electron transport materials have deeper HOMO levels organic emitting layers. The EQE of an OLED can be expressed as:

$$EQE = \gamma \eta_{oc} \eta_{exc} \Phi_{PL} \qquad (1.1)$$

where the symbol Υ is the charge recombination factor; η_{oc} defines optical out-coupling efficiency (0.2 to 0.3 for traditional devices). η_{exc} indicates exciton utilization efficiency. Frequently, Φ_{PL} exerts a large influence on electron-to-photon conversion efficiency. The electrical features (charge-injection capacity and mobility) of an organic emitter importantly influence the Φ_{PL} factor. Φ_{PL} depends on the chemical structure of the organic emitter. A bipolar emitter has a very high Φ_{PL} value.

The distance between the transport states and the Fermi-level E_F can be measured by the Seebeck effect. The Seebeck coefficient S depends on the distance between the transport states $E\,\mu$ and the Fermi-level E_F.[57]

$$S(T) = \frac{k_b}{e}\left[\left(\frac{E_F(T)-E_\mu}{k_BT}\right)+A\right] \qquad (1.2)$$

where A refers to a numerical factor that is insignificant to the organic structure. k_B denotes the Boltzmann constant, T the temperature, and e corresponds to elementary charge. Seebeck measurements justify the difference between the Fermi state and transport level experiences depressions with doping concentrations.

10.8.1 IMPROVEMENT OF INJECTION PROPERTIES OF DOPED HOLE/ELECTRON TRANSPORT MATERIALS

The p-doped hole transport material populates the Fermi states of dopant towards the hole conducting level of the emissive organic layer.[54] Hole injection and transport efficiency can be intensified by merging of an organic electron-withdrawing group into the hole-conducting layer. The n-doped electron transport material populates the Fermi level (dopant) towards the electron transport level of the emissive layer.[44] Electron injection and transport efficiency can be magnified by the functionalization of a suitable electron transport layer with electron-donating functional dopant (Figure 10.7).[58] Doping science increases the number of free conducting

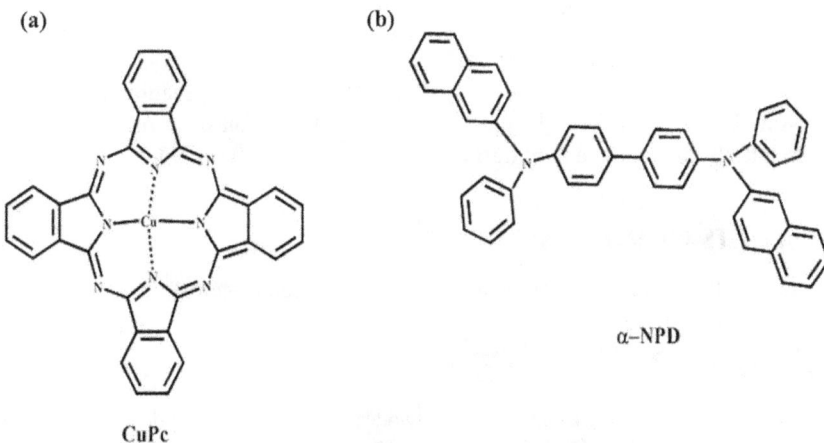

FIGURE 10.7   Chemical structure of different electron transport layers: (a) CuPc and (b) α -NPD used in OLEDs.[26,51,57]

charges in organic layers resulting in impressive enhancement in electrical conductivity.[41] In this way electron-hole pairs are intensified in the organic semiconductor for effective emission.[59]

The multilayered structures facilitate charge injection from interfacial contacts of functional OLEDs and other energy devices.[60, 61] The emissive efficiency emerges due to sufficient carrier conduction into the functionalized emitting organic layer. The pyrimidine possesses a high electron-deficient character compared with benzene and pyridine. Such an organic molecular structure creates an attractive electron-injecting functional moiety for renovating OLEDs science and technology.

10.9   CONCLUSION

Molecular doping improves the conductivity of organic semiconductors by incorporating excess electrons or hole. The structural and chemical imperfections produce trap states for recombination, which lowers the efficiency of emitting charge carriers in the active organic molecule. Polarons are stabilized for effective recombination leading to efficient light emission. Interfacial barrier heights are minimized and as a result charge injection efficiency is increased. The chemical functionalization of organic semiconductors improves the density of states the Fermi level and occupancy factor, which in turn determines the doping efficiency. This technology presents the key challenge of OLED science and diversity. Doped species shift the energy gap, which broadens the emission wavelength. Thus, the doped or controlled molecular structure can notably increase the efficiency of charge recombination contacts in OLED technology. The chemical technology has concluded that adopting n- and p-type doped regions can produce efficient recombination in functionalized organic molecule. Thus, molecular doping is a functional technique to improve the efficiency of OLEDs and other advanced optoelectronic devices.

ACKNOWLEDGMENTS

The authors (Dr. Sunil Kumar and Dr. Pankaj Kumar Chaurasia) gratefully acknowledge B.R.A. Bihar University, Muzaffarpur, for their kind cooperation in publishing this work. Dr. Sandeep K. S. Patel recognizes B.H.U. Varanasi for making available research tools and management during the preparation of the chapter.

CONFLICTS OF INTEREST

The authors have declared there is no conflict of financial interest.

REFERENCES

1. Karzazi, Y. Organic Light Emitting Diodes: Devices and Applications. *J. Mater. Environ. Sci.* **2014**, *5* (1), 1–12.
2. Tagare, J.; Vaidyanathan, S. Recent Development of Phenanthroimidazole-Based Fluorophores for Blue Organic Light-Emitting Diodes (OLEDs): An Overview. *J. Mater. Chem. C.* **2018**, *6* (38), 10138–10173. https://doi.org/10.1039/C8TC03689F.

3. Prakash, R.; Kumar, S.; Maiti, P. Carbon Nanotube Based Nanomaterials for Solar Energy Storage Devices. In *Current and Future Developments in Nanomaterials and Carbon Nanotubes: Applications of Nanomterials in Energy Storage and Electronics*; Manik, G., Sahoo, S. K., Eds.; Bentham Science, 2022; pp 1–18. https://doi.org/10.2174 /9789815050714122030004.

4. Kumar, S.; Yadav, P. K.; Prakash, R.; Santra, A.; Maiti, P. Multifunctional Graphene Oxide Implanted Polyurethane Ionomer Gel Electrolyte for Quantum Dots Sensitized Solar Cell. *J. Alloys Compd.* **2022**, *922*, 166121. https://doi.org/10.1016/j. jallcom.2022.166121.

5. Reiser, P.; Benneckendorf, F. S.; Barf, M. M.; Müller, L.; Bäuerle, R.; Hillebrandt, S.; Beck, S.; Lovrincic, R.; Mankel, E.; Freudenberg, J.; Jänsch, D.; Kowalsky, W.; Pucci, A.; Jaegermann, W.; Bunz, U. H. F.; Müllen, K. N-Type Doping of Organic Semiconductors: Immobilization via Covalent Anchoring. *Chem. Mater.* **2019**, *31* (11), 4213–4221. https://doi.org/10.1021/acs.chemmater.9b01150.

6. Kumar, S.; Prakash, R.; Maiti, P. Redox Mediation Through Integrating Chain Extenders in Active Ionomer Polyurethane Hard Segments in CdS Quantum Dot Sensitized Solar Cell. *Sol. Energy.* **2022**, *231*, 985–1001. https://doi.org/10.1016/j.solener.2021.12.043.

7. Nollau, A.; Pfeiffer, M.; Fritz, T.; Leo, K. Controlled N-Type Doping of a Molecular Organic Semiconductor: Naphthalenetetracarboxylic Dianhydride (NTCDA) Doped with Bis(Ethylenedithio)-Tetrathiafulvalene (BEDT-TTF). *J. Appl. Phys.* **2000**, *87* (9), 4340–4343. https://doi.org/10.1063/1.373413.

8. Jacobs, I. E.; Moulé, A. J. Controlling Molecular Doping in Organic Semiconductors. *Adv. Mater.* **2017**, *29* (42), 1–39. https://doi.org/10.1002/adma.201703063.

9. Frischeisen, J.; Yokoyama, D.; Endo, A.; Adachi, C.; Brütting, W. Increased Light Outcoupling Efficiency in Dye-Doped Small Molecule Organic Light-Emitting Diodes with Horizontally Oriented Emitters. *Org. Electron.* **2011**, *12* (5), 809–817. https://doi. org/10.1016/j.orgel.2011.02.005.

10. Koech, P. K.; Padmaperuma, A. B.; Wang, L.; Swensen, J. S.; Polikarpov, E.; Darsell, J. T.; Rainbolt, J. E.; Gaspar, D. J. Synthesis and Application of 1,3,4,5,7,8-Hexafluorotetracyanonaphthoquinodimethane (F6-TNAP): A Conductivity Dopant for Organic Light-Emitting Devices. *Chem. Mater.* **2010**, *22* (13), 3926–3932. https://doi.org/10.1021/cm1002737.

11. Kumar, S.; Prakash, R.; Maiti, P. Advanced Batteries and Charge Storage Devices Based on Nanowires. In *Current and Future Developments in Nanomaterials and Carbon Nanotubes: Applications of Nanomterials in Energy Storage and Electronics*; Manik, G., Sahoo, S. K., Eds.; Bentham Science, 2022; pp 159–175. https://doi.org/10. 2174/9789815050714122030012.

12. Yadav, P. K.; Kumar, S.; Maiti, P. Conjugated Polymers for Solar Cell Applications. In *Conjugated Polymers for Next-Generation Applications, Volume 2*; Kumar, V., Sharma, K., Sehgal, R., Kalia, S., Eds.; Woodhead Publisher (An Imprint of Elsevier), 2022; pp 367–401. https://doi.org/10.1016/B978-0-12-824094-6.00004-2.

13. Bruder, I.; Watanabe, S.; Qu, J.; Müller, I. B.; Kopecek, R.; Hwang, J.; Weis, J.; Langer, N. A Novel P-Dopant with Low Diffusion Tendency and Its Application to Organic Light-Emitting Diodes. *Org. Electron.* **2010**, *11* (4), 589–593. https://doi.org/10.1016/j. orgel.2009.12.019.

14. Komatsu, R.; Sasabe, H.; Kido, J. Recent Progress of Pyrimidine Derivatives for High-Performance Organic Light-Emitting Devices. *J. Photonics Energy.* **2018**, *8* (3), 1. https://doi.org/10.1117/1.jpe.8.032108.

15. Kim, S. J.; Zhang, Y.; Zuniga, C.; Barlow, S.; Marder, S. R.; Kippelen, B. Efficient Green OLED Devices with an Emissive Layer Comprised of Phosphor-Doped Carbazole/Bis-Oxadiazole Side-Chain Polymer Blends. *Org. Electron.* **2011**, *12* (3), 492–496. https:// doi.org/10.1016/j.orgel.2010.12.006.

16. Liu, S.; Zhang, X.; Ou, C.; Wang, S.; Yang, X.; Zhou, X.; Mi, B.; Cao, D.; Gao, Z. Structure-Property Study on Two New D-A Type Materials Comprising Pyridazine Moiety and the OLED Application as Host. *ACS Appl. Mater. Interfaces* **2017**, *9* (31), 26242–26251. https://doi.org/10.1021/acsami.7b04859.

17. Kovac, J.; Peternai, L.; Lengyel, O. Advanced Light Emitting Diodes Structures for Optoelectronic Applications. *Thin Solid Films* **2003**, *433* (1–2 SPEC.), 22–26. https://doi.org/10.1016/S0040-6090(03)00314-6.

18. Abate, A.; Staff, D. R.; Hollman, D. J.; Snaith, H. J.; Walker, A. B. Influence of Ionizing Dopants on Charge Transport in Organic Semiconductors. *Phys. Chem. Chem. Phys.* **2014**, *16* (3), 1132–1138. https://doi.org/10.1039/c3cp53834f.

19. Kumar, S.; Yadav, P. K.; Maiti, P. Renewable Cathode Materials Dependent on Conjugated Polymer Composite Systems. In *Conjugated Polymers for Next-Generation Applications, Volume 2*; Kumar, V., Sharma, K., Sehgal, R., Kalia, S., Eds.; Woodhead Publisher (An Imprint of Elsevier), 2022; pp 55–90. https://doi.org/10.1016/B978-0-12-824094-6.00002-9.

20. Nell, B.; Ortstein, K.; Boltalina, O. V.; Vandewal, K. Influence of Dopant-Host Energy Level Offset on Thermoelectric Properties of Doped Organic Semiconductors. *J. Phys. Chem. C.* **2018**, *122* (22), 11730–11735. https://doi.org/10.1021/acs.jpcc.8b03804.

21. Kumar, S.; Srivastva, A. N. Application of Carbon Nanomaterials Decorated Electrochemical Sensor for Analysis of Environmental Pollutants. In *Analytical Chemistry—Advancement, Perspectives and Applications*; Srivastva, A. N., Ed.; IntechOpen, 2021. https://doi.org/10.5772/intechopen.96538.

22. Kobayashi, N.; Kuwae, H.; Oshima, J.; Ishimatsu, R.; Tashiro, S.; Imato, T.; Adachi, C.; Shoji, S.; Mizuno, J. A Wide-Energy-Gap Naphthalene-Based Liquid Organic Semiconductor Host for Liquid Deep-Blue Organic Light-Emitting Diodes. *J. Lumin.* **2018**, *200*, 19–23. https://doi.org/10.1016/j.jlumin.2018.03.072.

23. Kotadiya, N. B.; Mondal, A.; Blom, P. W. M.; Andrienko, D.; Wetzelaer, G. J. A. H. A Window to Trap-Free Charge Transport in Organic Semiconducting Thin Films. *Nat. Mater.* **2019**, *18* (11), 1182–1186. https://doi.org/10.1038/s41563-019-0473-6.

24. Pfeiffer, M.; Leo, K.; Zhou, X.; Huang, J. S.; Hofmann, M.; Werner, A.; Blochwitz-Nimoth, J. Doped Organic Semiconductors: Physics and Application in Light Emitting Diodes. *Org. Electron.* **2003**, *4* (2–3), 89–103. https://doi.org/10.1016/j.orgel.2003.08.004.

25. Kumar, S.; Chaurasia, P. K. Functionalized Nanomaterials Based Efficient Photocatalyst for Renewable Energy and Sustainable Environment. In *Versatile Solicitations of Materials Science in Diverse Fields*; Tripathi, M., Srivastava, A., Awasthi, K., Eds.; Nova Science, 2021; pp 173–192.

26. Hohnholz, D.; Steinbrecher, S.; Hanack, M. Applications of Phthalocyanines in Organic Light Emitting Devices. *J. Mol. Struct.* **2000**, *521* (1–3), 231–237. https://doi.org/10.1016/S0022-2860(99)00438-X.

27. Noh, S.; Suman, C. K.; Hong, Y.; Lee, C. Carrier Conduction Mechanism for Phosphorescent Material Doped Organic Semiconductor. *J. Appl. Phys.* **2009**, *105* (3), 1–6. https://doi.org/10.1063/1.3072693.

28. Lu, S. Y.; Mukhopadhyay, S.; Froese, R.; Zimmerman, P. M. Virtual Screening of Hole Transport, Electron Transport, and Host Layers for Effective OLED Design. *J. Chem. Inf. Model.* **2018**, *58* (12), 2440–2449. https://doi.org/10.1021/acs.jcim.8b00044.

29. Liu, J.; Zhang, H.; Dong, H.; Meng, L.; Jiang, L.; Jiang, L.; Wang, Y.; Yu, J.; Sun, Y.; Hu, W.; Heeger, A. J. High Mobility Emissive Organic Semiconductor. *Nat. Commun.* **2015**, *6* (May), 0–7. https://doi.org/10.1038/ncomms10032.

30. Liu, Y.; Nell, B.; Ortstein, K.; Wu, Z.; Karpov, Y.; Beryozkina, T.; Lenk, S.; Kiriy, A.; Leo, K.; Reineke, S. High Electron Affinity Molecular Dopant CN6-CP for Efficient Organic Light-Emitting Diodes. *ACS Appl. Mater. Interfaces.* **2019**, *11* (12), 11660–11666. https://doi.org/10.1021/acsami.8b21865.

31. Kumar, S.; Patil, S. Fluoranthene-Based Molecules as Electron Transport and Blue Fluorescent Materials for Organic Light-Emitting Diodes. *J. Phys. Chem. C.* **2015**, *119* (33), 19297–19304. https://doi.org/10.1021/acs.jpcc.5b03717.

32. Tagare, J.; Ulla, H.; Satyanarayan, M. N.; Vaidyanathan, S. Efficient Non-Doped Bluish-Green Organic Light Emitting Devices Based on N1 Functionalized Star-Shaped Phenanthroimidazole Fluorophores. *J. Photochem. Photobiol. A Chem.* **2018**, *353*, 53–64. https://doi.org/10.1016/j.jphotochem.2017.11.001.

33. Salzmann, I.; Heimel, G.; Oehzelt, M.; Winkler, S.; Koch, N. Molecular Electrical Doping of Organic Semiconductors: Fundamental Mechanisms and Emerging Dopant Design Rules. *Acc. Chem. Res.* **2016**, *49* (3), 370–378. https://doi.org/10.1021/acs.accounts.5b00438.

34. Duan, L.; Qiao, J.; Sun, Y.; Qiu, Y. Strategies to Design Bipolar Small Molecules for OLEDs: Donor-Acceptor Structure and Non-Donor-Acceptor Structure. *Adv. Mater.* **2011**, *23* (9), 1137–1144. https://doi.org/10.1002/adma.201003816.

35. Greiner, M. T.; Lu, Z. H. Thin-Film Metal Oxides in Organic Semiconductor Devices: Their Electronic Structures, Work Functions and Interfaces. *NPG Asia Mater.* **2013**, *5* (7), 1–16. https://doi.org/10.1038/am.2013.29.

36. Havare, A. K.; Can, M.; Demic, S.; Okur, S.; Kus, M.; Aydin, H.; Yagmurcukardes, N.; Tari, S. Modification of ITO Surface Using Aromatic Small Molecules with Carboxylic Acid Groups for OLED Applications. *Synth. Met.* **2011**, *161* (21–22), 2397–2404. https://doi.org/10.1016/j.synthmet.2011.09.007.

37. Lee, J. H.; Kim, J. J. Interfacial Doping for Efficient Charge Injection in Organic Semiconductors. *Phys. Status Solidi Appl. Mater. Sci.* **2012**, *209* (8), 1399–1413. https://doi.org/10.1002/pssa.201228199.

38. Li, P.; Lu, Z. H. Energy Levels of Molecular Dopants in Organic Semiconductors. *Adv. Mater. Interfaces.* **2020**, *7* (17), 1–9. https://doi.org/10.1002/admi.202000720.

39. AlSalhi, M. S.; Alam, J.; Dass, L. A.; Raja, M. Recent Advances in Conjugated Polymers for Light Emitting Devices. *Int. J. Mol. Sci.* **2011**, *12* (3), 2036–2054. https://doi.org/10.3390/ijms12032036.

40. Meyer, J.; Kidambi, P. R.; Bayer, B. C.; Weijtens, C.; Kuhn, A.; Centeno, A.; Pesquera, A.; Zurutuza, A.; Robertson, J.; Hofmann, S. Metal Oxide Induced Charge Transfer Doping and Band Alignment of Graphene Electrodes for Efficient Organic Light Emitting Diodes. *Sci. Rep.* **2014**, *4*, 1–7. https://doi.org/10.1038/srep05380.

41. Li, H.; Duan, L.; Qiu, Y. Mechanisms of Charge Transport in Transition Metal Oxide Doped Organic Semiconductors. *J. Phys. Chem. C.* **2014**, *118* (51), 29636–29642. https://doi.org/10.1021/jp510575q.

42. White, R. T.; Thibau, E. S.; Lu, Z. H. Interface Structure of MoO_3 on Organic Semiconductors. *Sci. Rep.* **2016**, *6* (February), 1–9. https://doi.org/10.1038/srep21109.

43. Liu, J.; Zhang, H.; Dong, H.; Meng, L.; Jiang, L.; Jiang, L.; Wang, Y.; Yu, J.; Sun, Y.; Hu, W.; Heeger, A. J. High Mobility Emissive Organic Semiconductor. *Nat. Commun.* **2015**, *6* (May), 0–7. https://doi.org/10.1038/ncomms10032.

44. Lou, Y.; Okawa, Y.; Wang, Z.; Naka, S.; Okada, H. Efficient Electron Transport in 4,4′-Bis[N-(1-Napthyl)-N-Phenyl-Amino] Biphenyl and the Applications in White Organic Light Emitting Devices. *Org. Electron.* **2013**, *14* (3), 1015–1020. https://doi.org/10.1016/j.orgel.2013.01.029.

45. Ramar, M.; Tyagi, P.; Suman, C. K.; Srivastava, R. Enhanced Carrier Transport in Tris(8-Hydroxyquinolinate) Aluminum by Titanyl Phthalocyanine Doping. *RSC Adv.* **2014**, *4* (93), 51256–51261. https://doi.org/10.1039/c4ra09116g.

46. Meerheim, R.; Walzer, K.; He, G.; Pfeiffer, M.; Leo, K. Highly Efficient Organic Light Emitting Diodes (OLED) for Diplays and Lighting. *Org. Optoelectron. Photonics II.* **2006**, *6192*, 61920P. https://doi.org/10.1117/12.666905.

47. Tian, J.; Cao, G. Semiconductor Quantum Dot-Sensitized Solar Cells. *Nano Rev.* **2015**, (17), 1–10. https://doi.org/10.3402/nano.v4i0.22578.

48. Lin, S. L.; Chan, L. H.; Lee, R. H.; Yen, M. Y.; Kuo, W. J.; Chen, C. T.; Jeng, R. J. Highly Efficient Carbazole-π-Dimesitylborane Bipolar Fluorophores for Nondoped Blue Organic Light-Emitting Diodes. *Adv. Mater.* **2008**, *20* (20), 3947–3952. https://doi.org/10.1002/adma.200801023.

49. Chen, W. C.; Zhu, Z. L.; Lee, C. S. Organic Light-Emitting Diodes Based on Imidazole Semiconductors. *Adv. Opt. Mater.* **2018**, *6* (18), 1–43. https://doi.org/10.1002/adom.201800258.

50. Chen, M.; Zhao, Y.; Tang, Z.; Zhang, B.; Wei, B. Multifunctional Organic Emitters for High-Performance and Low-Cost Organic Light-Emitting Didoes. *Chem. Rec.* **2019**, *19* (8), 1768–1778. https://doi.org/10.1002/tcr.201900005.

51. Salzmann, I.; Heimel, G. Toward a Comprehensive Understanding of Molecular Doping Organic Semiconductors (Review). *J. Electron Spectros. Relat. Phenomena.* **2015**, *204*, 208–222. https://doi.org/10.1016/j.elspec.2015.05.001.

52. Lüssem, B.; Riede, M.; Leo, K. Doping of Organic Semiconductors. *Phys. Status Solidi A.* **2013**, *210*. https://doi.org/10.1002/pssa.201228310.

53. Tietze, M. L.; Burtone, L.; Riede, M.; Lüssem, B.; Leo, K. Fermi Level Shift and Doping Efficiency in P-Doped Small Molecule Organic Semiconductors: A Photoelectron Spectroscopy and Theoretical Study. *Phys. Rev. B Condens. Matter Mater. Phys.* **2012**, *86* (3), 1–12. https://doi.org/10.1103/PhysRevB.86.035320.

54. Zheng, X.; Wu, Y.; Sun, R.; Zhu, W.; Jiang, X.; Zhang, Z.; Xu, S. Efficiency Improvement of Organic Light-Emitting Diodes Using 8-Hydroxy-Quinolinato Lithium as an Electron Injection Layer. *Thin Solid Films.* **2005**, *478* (1–2), 252–255. https://doi.org/10.1016/j.tsf.2004.08.020.

55. Kumar, S.; Maurya, I. C.; Prakash, O.; Srivastava, P.; Das, S.; Maiti, P. Functionalized Thermoplastic Polyurethane as Hole Conductor for Quantum Dot-Sensitized Solar Cell. *ACS Appl. Energy Mater.* **2018**, *1* (9), 4641–4650. https://doi.org/10.1021/acsaem.8b00783.

56. Huseynova, G.; Lee, S. H.; Joo, C. W.; Lee, Y. S.; Lim, Y. J.; Park, J.; Yoo, J. M.; Cho, N. S.; Kim, Y. H.; Lee, J.; Lee, J. H. Dye-Doped Poly(3,4-Ethylenedioxythiophene)-Poly(Styrenesulfonate) Electrodes for the Application in Organic Light-Emitting Diodes. *Thin Solid Films.* **2020**, *707*, 138078. https://doi.org/10.1016/j.tsf.2020.138078.

57. Lüssem, B.; Riede, M.; Leo, K. Doping of Organic Semiconductors. *Phys. Status Solidi Appl. Mater. Sci.* **2013**, *210* (1), 9–43. https://doi.org/10.1002/pssa.201228310.

58. Gao, Z. Q.; Xia, P. F.; Lo, P. K.; Mi, B. X.; Tam, H. L.; Wong, M. S.; Cheah, K. W.; Chen, C. H. P-Doped p-Phenylenediamine-Substituted Fluorenes for Organic Electroluminescent Devices. *Org. Electron.* **2009**, *10* (4), 666–673. https://doi.org/10.1016/j.orgel.2009.02.025.

59. Fathollahi, M.; Ameri, M.; Mohajerani, E.; Mehrparvar, E.; Babaei, M. Organic/Organic Heterointerface Engineering to Boost Carrier Injection in OLEDs. *Sci. Rep.* **2017**, *7*, 1–11. https://doi.org/10.1038/srep42787.

60. Saraswat, A.; Kumar, S. A Topical Study of Electrochemical Response of Functionalized Conducting Polyaniline: An Overview. *Eur. Polym. J.* **2023**, *182* (3 January 2023), 111714. https://doi.org/10.1016/j.eurpolymj.2022.111714.

61. Kumar, S.; Kumar, S.; Rai, R. N.; Lee, Y.; Hong Chuong Nguyen, T.; Young Kim, S.; Van Le, Q.; Singh, L. Recent Development in Two-Dimensional Material-Based Advanced Photoanodes for High-Performance Dye-Sensitized Solar Cells. *Sol. Energy* **2023**, *249*, 606–623. https://doi.org/10.1016/j.solener.2022.12.013.

11 Recent Development of Blue Fluorescent Organic Materials for OLEDs

Pawan Kumar Sada, Pranshu K. Gupta, Tanu Gupta, Abhishek Rai, and Alok Kumar Singh

11.1 INTRODUCTION

An organic light emitting diode (OLED), or organic electroluminescent diode (OELD), belongs to a class of light emitting diodes (LEDs) where the emissive electroluminescent sheet is formed of a thin sheet of an organic compound having carbocyclic and heterocyclic moieties which lie between two conducting electrodes, that is, an anode and a cathode. On passing an electric current between these two electrodes, light is emitted. OLEDs are used for digital displays in the screens of gadgets such as televisions, computers and smartphones.

The basic OLED cell structure in which these organic compounds are stacked between the electrodes and layers is arranged as described in the following.

The common OLED structure is:

- Plastic, glass, or metal foil is the substrate, which forms the basis of the OLED.
- Anode and cathode (depending on the type of OLED these may or may not be transparent). Due to the absence of electrons, holes are created that are injected into the organic layer that is the basic structure of the OLED device.
- Next is the hole injection layer (HIL), which is placed above the anode, and its function is to accept holes from the anode and insert these holes deeper into the OLED layers known as the hole transport layer (HTL) and the electron transport layer (ETL), providing support for the transport of holes and electrons across it and giving the proper path for the emissive layer.
- Next is the emissive layer, which is of the utmost importance because the color-defining emitter is doped into the host and light is created due to the direct transformation of electrical energy into light.
- To improve OLED technology by limiting electrons to the emissive layer, a blocking layer (BL) is used.

There are two main classes of OLEDs: the first belongs to the class of small molecules and the second is polymers. Two types of scheme control and drive OLED display.

1. Passive-matrix OLED (PMOLED)
2. Active-matrix OLED (AMOLED)

DOI: 10.1201/9781003260417-11

The role of PMOLED is to control all row (and line) in the display, while AMOLED has thin-film transistor backplane. Their function is to provide greater resolution via precise contact and can switch all specific pixels "ON" or "OFF"and also allow larger display sizes.

There is a fundamental difference between LEDs and OLEDs. LEDs have a p-n diode structure that was doped and this doping process causes alteration in the conductivity of the semiconductor in LEDs. In case of OLEDs, not having a p-n structure, the radiative efficiency is also increased by doping via direct change of quantum mechanical recombination frequency. Additionally, the doping is also involved in determining the photon emission wavelength.

As an OLED display that is slimmer and lighter perform deprived of backlight and releases visible light can also present profound dark levels. Compared with liquid crystal displays (LCDs), which adopt fluorescent lamps (cold cathode) or LED backlight, OLED screens have a greater "contrast ratio" with respect to LCD, regardless of whether the OLED displays are manufactured following the same pattern as LCDs, but afterward TFT, addressable grid or ITO segment formation.

11.2 HISTORY

In the early 1950s, André Bernanosem et al. at the Nancy-Université in France first reported that on applying high alternating voltages, various organic compounds such as acridine orange placed on or soluble in cellulose or thin films of cellophane in air showed electroluminescence property either by excitation of dye compounds or electrons.[1]

Pope et al. in 1960, at New York University, developed organic crystals having the inserting electrode[2] and the essential active necessities (work functions) for hole and electron inserting electrode associates, which forms the base of charging in recent OLED devices. The research group of Pope[3] in 1963 recognized direct current (DC) electroluminescence in a pure single crystal of tetracene-doped anthracene via a tiny area silver electrode at 400 V under vacuum. The suggested path is that molecular fluorescence arises due to excitation of electrons on applying the appropriate electric field and causes recombination of holes and electrons. The anthracene-conducting layer has greater energy compared to exciton energy[4] due to the nonappearance of an outside electric field.

By employing hole- and electron-injecting electrodes, twofold injection recombination electroluminescence was produced for the first time by Wolfgang Helfrich and W. G. Schneider, from the National Research Council in Canada in 1965 in an anthracene single crystal.[5] The process of making electroluminescent cells having a slim layer of anthracene along with graphite and tetracene powder using high-voltage, AC-driven, was patented by Dow Chemical researchers.[6]

Roger Partridge created the first polymer LED (PLED) in which he used a thin sheet of poly(N-vinylcarbazole) placed between two electrodes, producing light that was clearly observable in all lighting situations through the polymer. The diode had two limitations, lower conductivity and issues with the injection of electrons. Later conjugated polymers would enable removal of these limitations.[7]

In 1987, scientists at Eastman Kodak, C. W. Tang and S. Van Slyke, synthesized the first practical OLED device using a two-layer structure in which the hole- and electron-carrying layers were separated and recombination and emission took place in the middle organic layer, causing reduction in the operating voltage and also refining the effectiveness of the OLED.[8]

Burroughes et al., in 1990, developed a device based on a green-light-emitting polymer with high efficiency via the use of poly(p-phenylenevinylene)[9] thick films (100 nm). Shifting from molecular to macromolecular constituents not only upgraded loyalty of the organic films in terms of life span and allowed great resolution picture to be made effortlessly.[10]

In 1999, Kodak and Sanyo made the 2.4-inch active-matrix, full-color OLED display.[11] Pioneer Corporation, TDK (2001) and Samsung NEC Mobile Display (SNMD) began manufacturing small-molecule OLEDs.[12]

The first OLED television[13] and inkjet-printed OLED panels[14] were released by Sony XEL-1 and the successor of Panasonic's printable OLED business unit, respectively.

11.3 WORKING PRINCIPLE

As we already know, an OLED consists of a layer of organic compounds, electrically conductive due to conjugation of π-electrons on the entire molecule situated between the anode and cathode. Their conduction stages range from conductors to insulators; they are known as organic semiconductors, in which the highest occupied molecular orbitals (HOMOs) and lowest unoccupied molecular orbitals (LUMOs) are similar to the valence and conduction bands of inorganic semiconductors.[15]

J. H. Burroughes et al. developed the first light-emitting device having a single sheet of poly (p-phenylenevinylene). Multilayer OLEDs having two or more layers improve the efficiency of the device. By using a graded heterojunction[16] in 2011 quantum efficiency (up to 19%) was improved by continuously varying the ratio of hole- and electron-carrier constituents in the emissive layer along with a dopant emitter.[17]

On applying the voltage over the OLED, the anode gets a positive charge against the cathode during operation. Optical transparency, chemical stability and electrical conductivity are the key features in deciding the anode.[18] When electrons are inserted at the cathode where the LUMO of the organic sheet is placed and removed at the anode where the HOMO is placed, the direction of electron flows is from cathode to anode. Electrons and holes are brought toward each other by electrostatic forces, and this bound state is known as exciton. This occurs near to the electron-transport sheet, that is, the fragment of emissive sheet as in organic semiconductors. Holes commonly show more mobility than electrons. The movement of electrons to the lower energy levels occurs owing to the deterioration of the excited state, resulting in emission of radiation, which falls in the frequency of the visible region that is decided by the band gap of the material that corresponds to energy alteration among the HOMO and LUMO.

Depending on spins of the electron and hole, the exciton can exist in a singlet or a triplet state. At the cost of each singlet exciton, three triplet excitons can be generated. Transition from triplet states (phosphorescence) is spin forbidden and takes more time and causes restriction for the internal efficiency of fluorescence. Spin-orbit interactions are used by phosphorescent OLEDs to enable intersystem crossing, causing improvement in the internal efficiency.

Generally, indium tin oxide (ITO), being translucent to visible light and having high level of work function, is used as the anode material. ITO helps in insertion of holes into the HOMO of the organic sheet consisting of PEDOT:PSS,[19] and is added as a second conductive (injection) layer since the HOMO of the compounds normally occur among the work function of ITO and the HOMO of additional frequently used polymers and causes reduction in the energy requirement for hole injection. Metals such as Ba^{2+} and Ca^{2+} with aluminium plating to avoid degradation of the metal which is more reactive and also provide robustness to electrical contacts and having low work functions are commonly used as the cathode as they possess and allow insertion of electrons into the LUMO.

11.4 DEVELOPMENT OF BLUE FLUORESCENT ORGANIC MATERIALS FOR OLEDs

Efficiency, cost and stability are the key indicators to be considered for the fabrication of OLEDs and their character in display and lighting fields, particularly, the blue materials that impact the foundation of OLED claims. In 1987, Tang and Van Slyke[20] reported an organic device having the property of blue-emitting materials as a first multicolor display in which fluorescent styrylamine acted as dopant and distyrylarylene (DSA) acted as host.[21] The device was arranged in the form of a sandwiched structure having properties of high efficiency and longer lifetime. The presence of styrylamine- or amine-doped systems having strong blue fluorescence property helps in enhancing the quality of hole insertion and efficient emission.[22,23] Gradually, improvement in the external quantum efficiency (EQE) of the blue devices was achieved by using triplet-triplet fusion (TTF)[24] and also by using molecular alignment of blue dopants.[25] Because of their lower EQE value in comparison to phosphorescent materials with heavy-metal complexes, intense and bright color and good durability, deep blue fluorescent compounds are very useful and have wide application for both industrial and academic purposes. Pure blue emission was required for full-color displays. Pure blue emission can be achieved by shortening the conjugation in the dopant structure but it is very difficult to extend the lifetime of these blue emitters.[26]

Two technologies have been described for increasing the efficiency of the fluorescent blue OLEDs. The first, the TTF mechanism, was introduced to enhance efficiency. The second was introducing the advanced alignment feature in the emitting sheet of new dopant materials.

11.4.1 TRIPLET-TRIPLET FUSION

Collision among triplet excitons (3A*) causes the generation of singlet excitons (1A*). The reaction is shown in the equation.

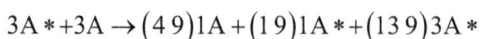

$$3A*+3A \rightarrow (4\ 9)1A+(1\ 9)1A*+(13\ 9)3A*$$

One singlet exciton was generated by five triplet excitons. Additionally, 15% additional singlet excitons (75% × 1/5) would be produced apart from the 25% of singlet excitons which shows that 40% singlet exciton construction would be accomplished via triplet-triplet annihilation (TTA). Upon the impact of two triplets, the generation but not annihilation of additional singlet excitons is called triplet-triplet fusion (TTF). The number of singlet excitons in the sheet produced through TTF is directly dependent on the square of triplet exciton density. Thus, for getting good efficiency, it is necessary to restrain triplet excitons with the emitting layer (EML). Triplet-triplet annihilation processes in fluorescent OLEDs have been reported.[27] Through implanting an additional layer comprising a green dopant and blue host material among the emitting and the electron-transporting sheet, we can get the high EQEs for red and yellow OLEDs, respectively, but it is very difficult to add this layer to the system. For solving this issue, the authors have given their attention to using electron-transporting materials.

11.4.2 MOLECULAR ORIENTATION

Molecular alignment has attracted considerable attention in efforts to achieve high efficiency in OLEDs. Specialized molecular assemblies having horizontal alignment in the substrate have been explored.[28] Additionally, unique dopant and host materials display appropriate molecular assemblies in blue fluorescence materials having high EQEs in doped and undoped emitting organizations.[29] Furthermore, on evaluating current voltage luminance (IVL) features, impedance spectroscopy and transient EL,[30] it was found that improvement in the EQE value turn to enhancement of out-coupling effectiveness by BD alignment.

11.5 RECENT DEVELOPMENT OF BLUE FLUORESCENT ORGANIC MATERIALS FOR OLEDs

A number of modifications has been done in organic materials for getting blue OLED. Some are discussed here.

Jun Yeob Lee et al. have reported the synthesis of a novel rigid chromophore that can be used in deep-blue fluorescent emitters. It was produced through a reaction of 9,9-dimethyl-9H fluorene and indolocarbazole. In addition, 10,10-dimethyl-10H-indeno[2,1-b] indolo[3,2,1- jk]carbazole (1 in Figure 11.1) has also been used as the chromophore for two deep-blue fluorescent emitters which produced a diphenylamine (DPA) auxhochromophore that is mono- or disubstituted by introducing the electron-donating group.

The external quantum efficiency (EQE), longer lifetime, high efficiency and extreme luminescence (L_{max}) of compounds 1, 2 and 3 in Figure 11.1 were enhanced by enhancing the number of DPA units that were studied via step-by-step observation of the electroluminescence properties of these materials. Among them, it was seen that emitter 3 having finest property by extreme EQE of 5.6%, L_{max} of above 47,000 cd/m^2, FWHM of 35 nm through small EQE roll-off at large luminescence and having better lifetime (LT80: above 140 h) than 2 (LT80: 50 h). Therefore, the 1 chromophore has been assigned as a good chromophore for developing deep-blue fluorescent emitters and excellent performance.[31]

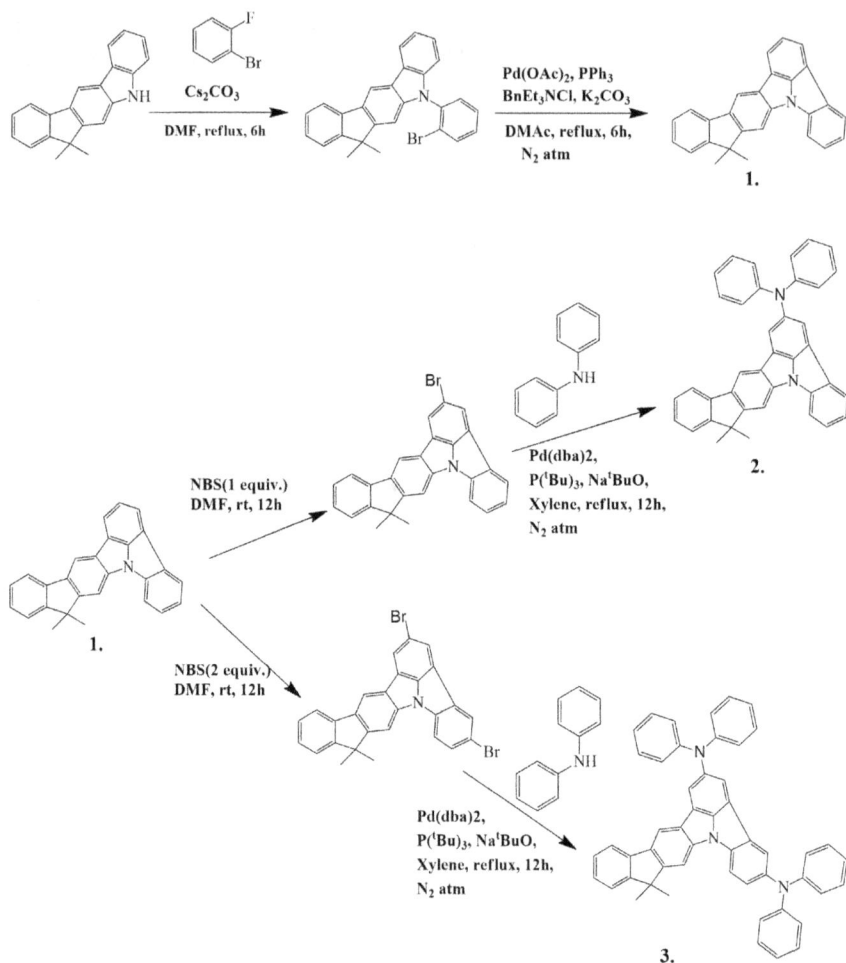

FIGURE 11.1   Synthetic pathway for target molecules.[31]

For designing of thermally activated delayed fluorescence (TADF) molecules **4**, **5** and **6** (Figure 11.2) having carbazole derivative 2,6-diphenylpyrimidine as acceptor was connected to three types of sterically deprived electron-contributing groups that provide outstanding thermal constancy along with large photoluminescence quantum yield (PLQY) around 0.72–0.74 due to its rigid structure.

Large EQE of 8.2% is attained by deep-blue doped material, such as molecule **4**. Similarly, **6** has extreme EQE of 4.9% by a $CIE_{x,y}$(0.15, 0.06). In addition, combination of **6** with TADF sensitizer and fluorescence material TBRb provides an EQE of 19.1%, and having bipolar moving capability and harvest of triplet excitons collected provide extremely competent yellow device which is mainly beneficial in the development of deep-blue TADF resources through stiff assemblies that have multifunctional uses as versatile materials in fluorescence devices (Figure 11.3).[32]

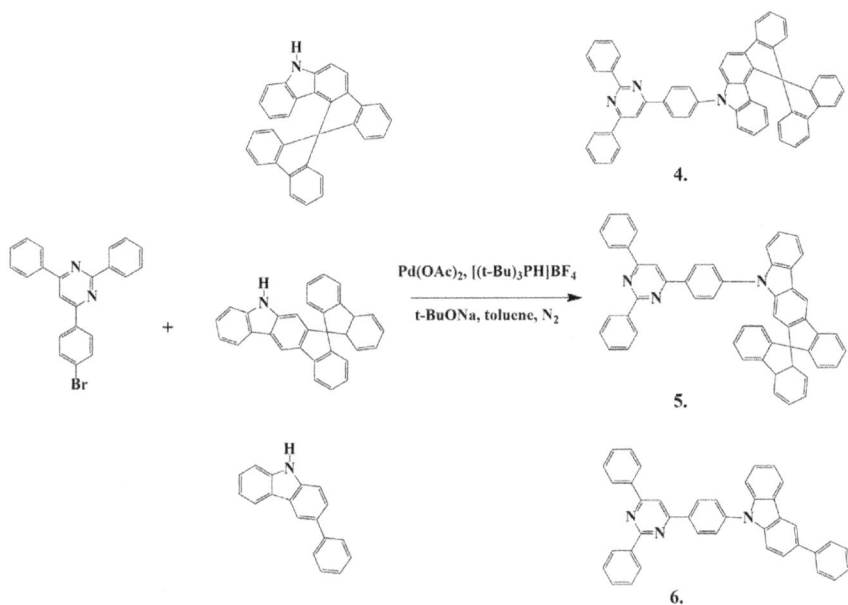

FIGURE 11.2   Synthetic route for molecules.[32]

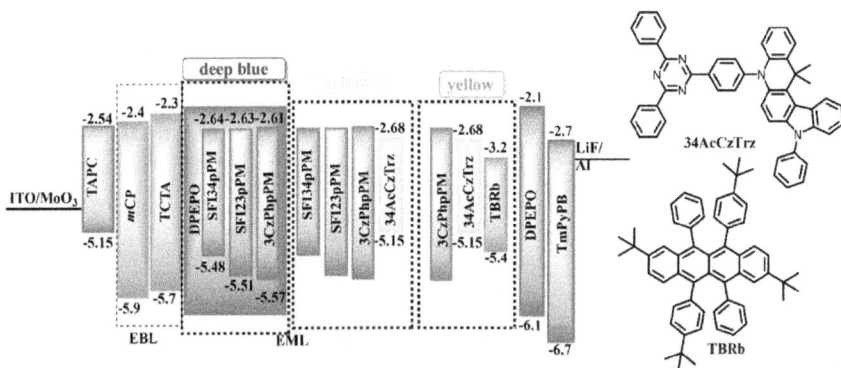

FIGURE 11.3   The detailed device configuration and the structure of 34AcCz-Trz and TBRb. (Lv et al.[32] Copyright 2020 Royal Society of Chemistry.)

On the basis of a core benzo[1,2-*d*:4,5-*d'*]bisoxazole fraction (Figure 11.4), some novel cross-conjugated compounds **7–10** having self-sufficiently tunable HOMO and LUMO stages have been produced by Malika Jeffries-EL et al. The experimental and theoretical analysis of these compounds shows that these are thermally steady with 5% weight loss and exhibiting blue emission in solution over the range 433–450 nm. The host compound, such as 4,4'-bis(9-carbazolyl)-biphenyl (CBP) with various

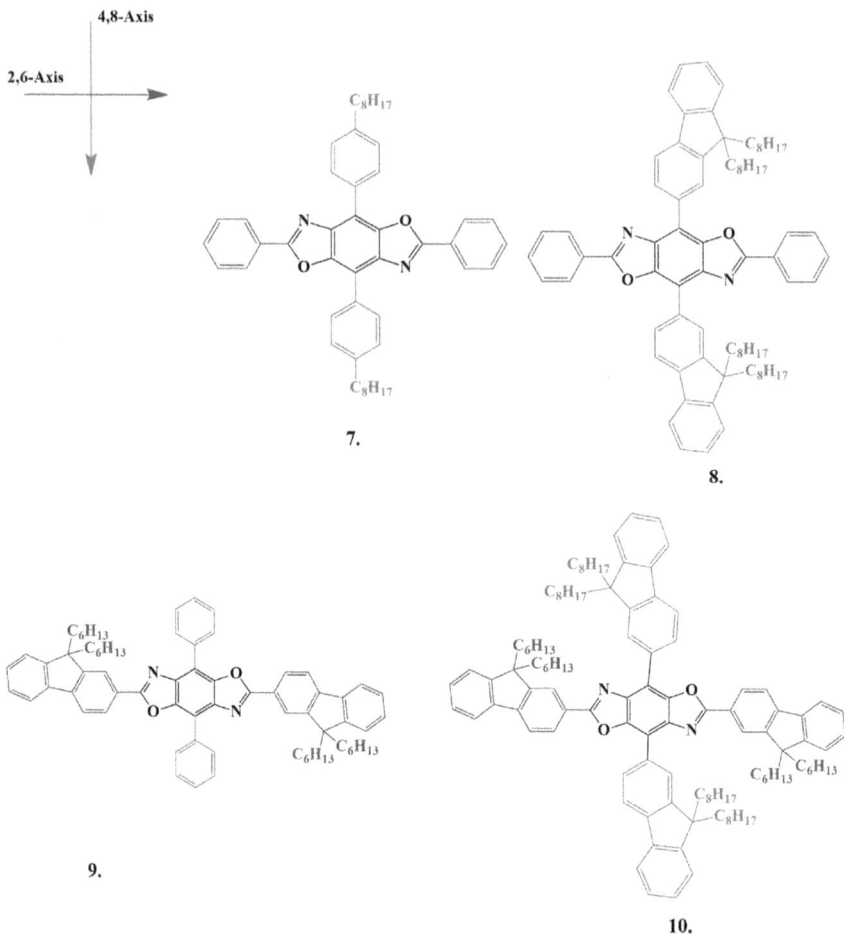

FIGURE 11.4 Cross-conjugated molecules.[33]

concentrations, host-guest OLEDs have been fabricated that exhibiting deep-blue emission with CIE coordinate values ($0.15 \leq x \leq 0.17$, $0.05 \leq y \leq 0.11$), extreme luminance competences ~2 cd/A and demonstrating the efficiency of benzobisoxazole forms as emitters for producing highly valuable deep-blue OLEDs.[33]

Yi Zhao et al. reported the synthesis of some bipolar emitters such as deep-blue **11** and **12** (Figure 11.5) upon addition of the hole via D–π–A architecture in 9,9-dimethylacridan/9,9-diphenylacridan and phenanthroimidazole systems, which have electron-transporting property, and good photoluminescence quantum yields (0.46). **11** and **12** have been employed as emitters in nondoped OLEDs, having CIE values (0.15, 0.07) and (0.15, 0.05), and full width half maximum (FWHM) values of 62 and 59 nm, respectively. These data are very close to the National Television Standards Committee (NTSC) standard blue (0.14, 0.08), possessing a satisfactory good effectiveness of 3.68% for EQE and slight effectiveness roll-off

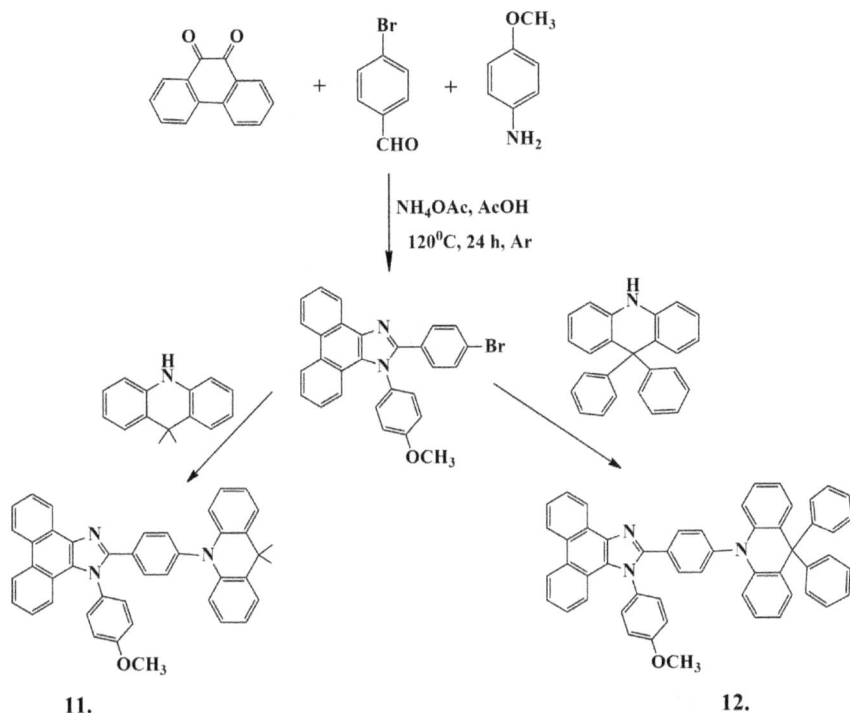

FIGURE 11.5   Synthetic route for molecules **11** and **12**.

by maintaining 3.01% of that even when the luminescence reaches 10,000 cd/m^2. In addition, having excellent thermal stability and charge-transfer performance makes **11** and **12** suitable as alternative host materials in OLED applications and employed as deep-blue emitters.[34]

Dibenzofuran or dibenzothiophene and electron-transporting phenanthroimidazole moieties **13** and **14** (Figure 11.6) having D–π–A assembly have been prepared and characterized, showing reduction of molecular aggregations due to their nonplanar twisting nature, which provides excellent thermal abilities, film-making skills, along with large quantum yields in the solid state and dichloromethane. The compounds exhibit promising performance as emitters in nondoped OLEDs showing the profound blue emission by the Commission International de l'Eclairage (CIE) coordinates (0.15, 0.11) for **13** and (0.15, 0.10) for **14**. A device using **14** as the emitting layer (EML) with favorable energy levels and increasing ability to transport the electron, possesses suitable values of 1.34 cd/m^2 for CE, 0.82 lm/W for PE and 1.63% for EQE used as the blue emitter and the host for a yellow emitter (PO-01) to fabricate white organic light emitting diodes (WOLEDs), giving forward-viewing maximum CE of 8.12 cd/m^2 and CIE coordinates of (0.339, 0.330).[35]

Pyrazolone-based azomethine-zinc complexes have been synthesized and characterized as organic electroluminescent heterostructures and explored computationally

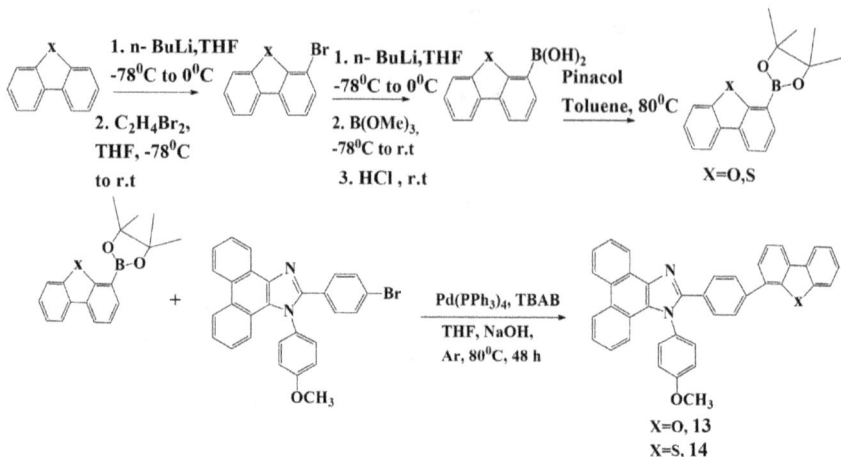

FIGURE 11.6 Synthetic pathways for molecules **13** and **14**.

by various physicochemical techniques. They have found use as a functional sheet in light emitting diodes (LEDs). They have excellent properties such as great thermal constancy and notable photoluminescence in liquid as well as in solid state. Quantum chemical calculations have shown that intraligand charge transfer within the Schiff bases causes luminescence in target complexes. The assembled LEDs expose illumination of 4,300–11,600 Cd/m², low input by 3.2–4.0 V, and EQE of 3.2%, and shows appropriate data for a purely fluorescent OLED system. This complex is used extensively in the formation of deep-blue OLED materials.[36]

Bin Zhang et al. reported the synthesis of various bis(benzothiophene-*S*,*S*-dioxide) fused blue releasing compounds **15–17**, having symmetrical structure designed via naphthalene, spirobifluorene, and 9,9,90,90-tetraoctyl-9H,90H-2,20-difluoren, respectively (Figure 11.7). successfully for creating the new fluorescent cores which is used for development of high-performance blue OLED materials having high thermal stability due to presence of polar sulfonyl groups and polycyclic aromatic conformation in the **FBTO** moiety. For controlling the conjugation *Bis*(benzothiophene-S,S-dioxide) provides better thermal effect, photophysicaland electrochemical possessions, number of electron-donating units and achieved solution-processible with excellent-performance.[37]

Vinich Promarak et al. have reported the preparation of some novel molecules **18–20** having a pyrene backbone, in which the pyrene core is functionalized and categorized like a smooth nondoped solution that is efficiently used as blue emitters in OLEDs. The compounds are *N*-dodecyl-3,6-di(pyren-1-yl)carbazole (**18**), *N*-dodecyl-1,3,6-tri(pyren-1-yl)carbazole (**19**) and *N*-dodecyl-1,3,6,8-tetra(pyren-1-yl) carbazole (**20**), which is generally prepared by the reaction of multibromo-9-dodecylcarbazole and pyren-1-broronic acid via Suzuki coupling (Figure 11.8). These compounds have good blue emissive property in solution as well as in the solid state

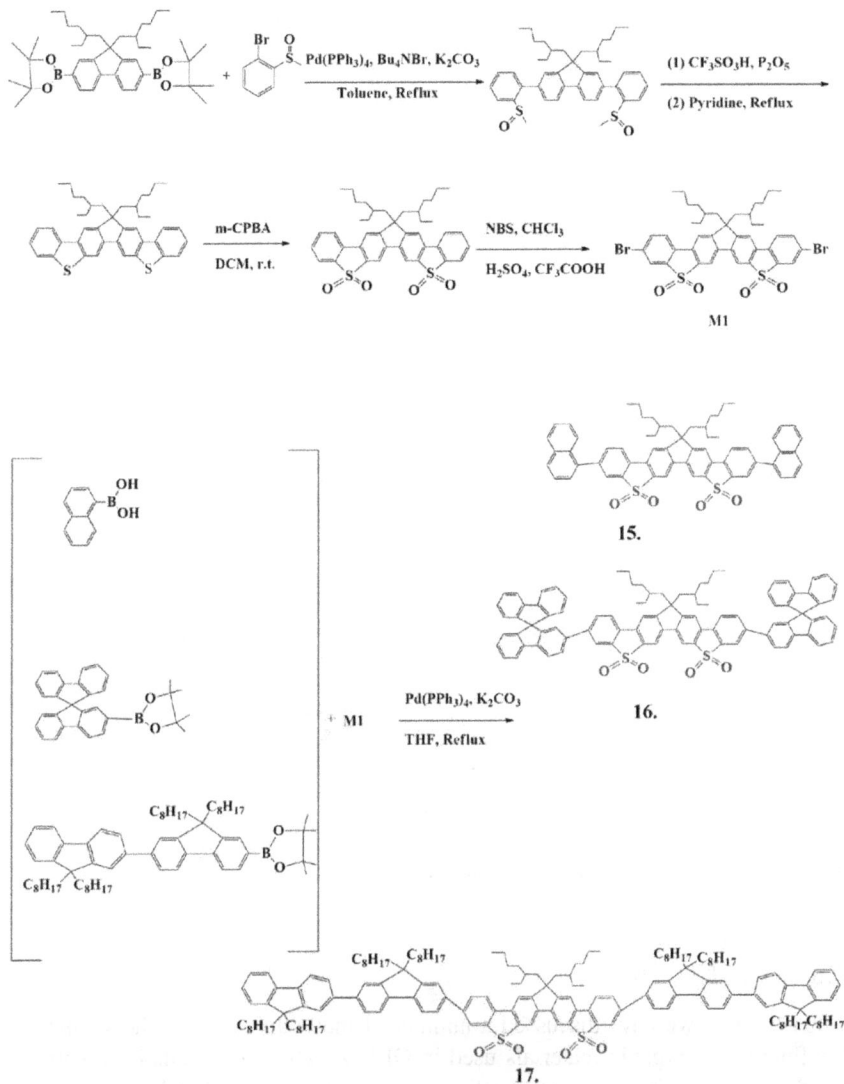

FIGURE 11.7   Synthetic pathway for target molecules.[37]

that was attained by doing a number of substitutions on the pyrene ring that was attached to the carbazole ring. properties of these compounds are amorphous stability and solubility that provides good fluorescence quantum efficiencies. It is also formed as an amorphous thin sheet that is morphologically very stable. Among these compounds, **18** shows a very high rate of blue emission color and very low turn-on voltage. **19** and **20** also have very low turn-on voltage and very good luminance efficiencies with outstanding performance.[38]

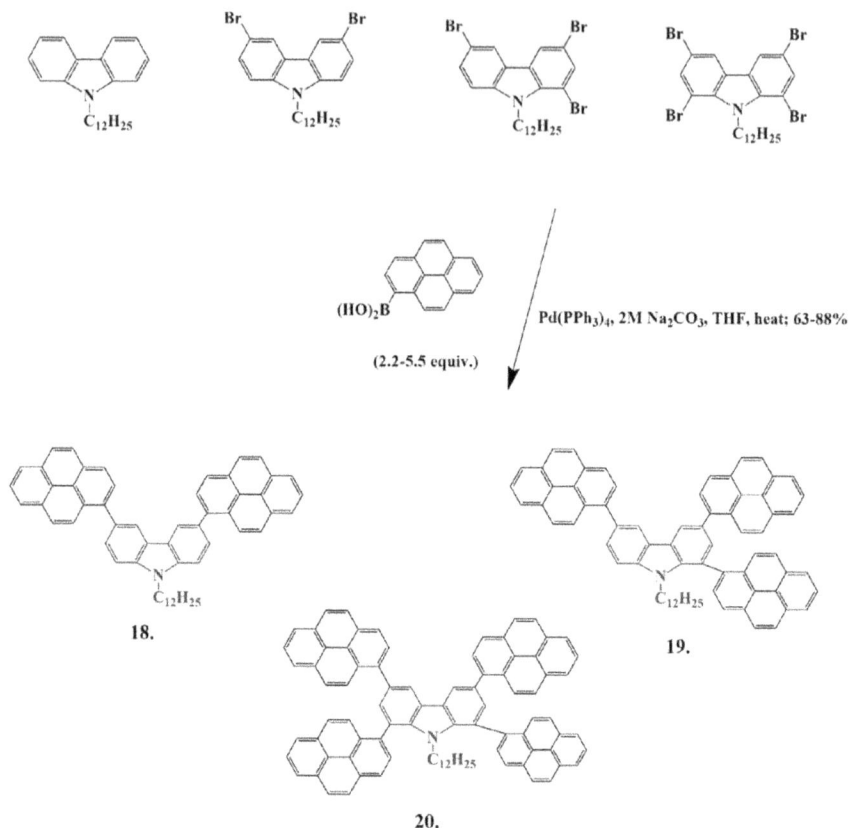

FIGURE 11.8   Synthetic pathway for molecules **18–20**.

11.6   CONCLUSION

In this chapter, we have discussed a number of methodologies for the synthesis of blue fluorescent organic materials used in OLEDs. These materials have enhanced quality values, such as for better efficiency, excellent thermal stability and longer lifetime, using the derivatives of different heterocyclic compounds, such as fluorene, indolocarbazole, pyrimidine, isoxazole, phenanthroimidazole, acridine, dibenzofuran, dibenzothiophene, pyrazolone and pyrene.

ACKNOWLEDGMENTS

AR and PKS thank the Department of Chemistry and L.N. Mithila University. AKS thanks the Department of Science & Technology, New Delhi, India for INSPIRE Research Grant (IFA15/CH-212), and University Grants Commission, New Delhi, India for the UGC-BSR Research Start-Up Grant.

REFERENCES

1. (a) Bernanose, A.; Comte, M.; Vouaux, P. A new method of emission of light by certain organic compounds. *J. Chim. Phys.* **1953**, *50*, 64–68. (b) Bernanose, A.; Vouaux, P. Organic electroluminescence: a study of the emission mode. *J. Chim. Phys.* **1953**, *50*, 261. (c) Bernanose, A. Sur le méchanisme de l'électroluminescence organique. *J. Chim. Phys.* **1955**, *52*, 396–400. (d) Bernanose, A.; Vouaux, P. Relation between organic electroluminescence and concentration of active product. *J. Chim. Phys.* **1955**, *52*, 509–510.

2. (a) Kallmann, H.; Pope, M. Positive hole injection into organic crystals. *J. Chem. Phys.* **1960**, *32*(1), 300–301. (b) Kallmann, H.; Pope, M. Bulk conductivity in organic crystals. *Nature.* **1960**, *186*(4718), 31. (c) Mark, P.; Helfrich, W. Space-charge-limited currents in organic crystals. *J. Appl. Phys.* **1962**, *33*(1), 205–215.

3. Pope, M.; Kallmann, H. P.; Magnante, P. Electroluminescence in organic crystals. *J. Chem. Phys.* **1963**, *38*(8), 2042–2043.

4. Sano, M.; Pope, M.; Kallmann, H. Electroluminescence and band gap in anthracene. *J. Chem. Phys.* **1965**, *43*(8), 2920–2921.

5. Helfrich, W.; Schneider, W. Recombination radiation in anthracene crystals. *Phys. Rev. Lett.* **1965**, *14*(7), 229.

6. Gurnee, E.; Fernandez, R. Organic electroluminescent phosphors. U.S. Patent 3,172,862, 9 March 1965.

7. Roger, P.; Feedback: Friend and rival. *Phys. World*, 14(1).

8. Tang, C. W.; Vanslyke, S. A. Organic electroluminescent diodes. *Appl. Phys. Lett.* **1987**, *51*(12), 913–915.

9. Burroughes, J. H.; Bradley, D. D. C.; Brown, A. R.; Marks, R. N.; MacKay, K.; Friend, R. H.; Burns, P. L.; Holmes, A. B. Light-emitting diodes based on conjugated polymers. *Nature* **1990**, *347*(6293), 539–541.

10. Burroughes, J. H.; Bradley, D. D. C.; Brown, A. R.; Marks, R. N.; Mackay, K. Light-emitting diodes based on conjugated polymers. *Nature* **1990**, *347*, 539B.

11. (a) Hara, Y. Sanyo, Kodak ramp OLED production line. *EE Times.* 6 December 2001. www.eetimes.com/sanyo-kodak-ramp-oled-production-line/. (b) Richard, S. Kodak, Sanyo demo OLED display. *CNET.* Retrieved 6 October2019. www.cnet.com/culture/kodak-sanyo-demo-oled-display/

12. Homer, A. Overview of OLED display technology. IEEE.

13. Sony XEL-1: The world's first OLED TV. Wayback Machine. 17 November 2008. Retrieved 5 February 2016. OLED-Info.com.

14. JOLED begin commercial shipment of world's first printing OLED Panels. *Printed Electronics World.* 12 December 2017. Retrieved 28 November 2019.

15. Kho, M.-J.; Javed, T.; Mark, R.; Maier, E.; David, C. Final report: OLED solid state lighting – Kodak European Research. MOTI (Management of Technology and Innovation) Project, Judge Business School of the University of Cambridge and Kodak European Research, Final Report presented on 4 March 2008, at Kodak European Research at Cambridge Science Park, Cambridge, 2008, pp. 1–12.

16. Organic light-emitting diodes based on graded heterojunction architecture has greater quantum efficiency. University of Minnesota. Archived from the original on 24 March 2012. Retrieved 31 May 2011.

17. Holmes, R.; Erickson, N.; Lüssem, B.; Leo, K. Highly efficient, single-layer organic light-emitting devices based on a graded-composition emissive layer. *Appl. Phys. Lett.* **2010**, *97*(1), 083308.

18. Peng, L. K.; Ramadas, K.; Burden, A.; Soo-Jin, C. *IEEE Trans. Electron Devices.* **2006**, *53*(6), 1483.

19. Carter, S. A.; Angelopoulos, M.; Karg, S.; Brock, P. J.; Scott, J. C. Polymeric anodes for improved polymer light-emitting diode performance. *Appl. Phys. Lett.* **1997**, *70*(16), 2067–2069.
20. Tang, C. W.; Vanslyke, S. A. Organic electroluminescent diodes. *Appl. Phys. Lett.* **1987**, *51*, 913–915.
21. Hosokawa, C.; Higashi, H.; Nakamura, H.; Kusumoto, T. Highly efficient blue electroluminescence from a distyrylarylene emitting layer with a new dopant. *Appl. Phys. Lett.* **1995**, *67*, 3853–3855.
22. Hosokawa, C.; Sakai, H.; Fukuoka, K.; Tokailin, H.; Hironaka, Y.; Ikeda, H.; Funahashi, M.; Kusumoto, T. Organic EL materials based on styryl and amine derivatives. *SID Symp. Digest Tech. Papers* **2001**, *32*, 522–525.
23. Funahashi, M.; Yamamoto, H.; Yabunouchi, N.; Fukuoka, K.; Kuma, H.; Hosokawa, C.; Kambe, E.; Yoshinaga, T.; Fukuda, T.; Kijima, Y. Highly efficient fluorescent deep blue dopant for "super top emission" device. *SID Symp. Digest Tech. Papers* **2008**, *39*, 709–711.
24. (a) Kawamura, M.; Kawamura, Y.; Mizuki, Y.; Funahashi, M.; Kuma, H.; Hosokawa, C. Highly efficient fluorescent blue OLEDs with efficiency-enhancement layer. *SID Symp. Digest Tech. Papers* **2010**, *41*, 560–563. (b) Kawamura, Y. New deep blue fluorescent materials and their application to high performance OLEDs. *SID Symp. Digest Tech. Papers* **2011**, *42*, 829–832.
25. Ogiwara, T.; Ito, H.; Mizuki, Y.; Naraoka, R.; Funahashi, M.; Kuma, H. Efficiency improvement of fluorescent blue device by molecular orientation of blue dopant. *SID Symp. Digest Tech. Papers* **2013**, *44*, 515–518.
26. Hosokawa, C.; Eida, M.; Matsuura, M.; Fukuoka, K.; Nakamura, H.; Kusumoto, T. Full-color organic EL display. *SID Symp. Digest Tech. Papers* **1997**, *28*, 1073.
27. (a) Spindler, J. P.; Begley, W. J.; Hatwer, T. K.; Kondakov, D. Y. High-efficiency fluorescent red- and yellow-emitting OLED devices. *SID Symp. Digest Tech. Papers* **2009**, *41*, 420–423. (b) Young, R. H.; Lenhard, J. R.; Kondakov, D. Y.; Hatwar, T. K. Luminescence quenching in blue fluorescent OLEDs. *SID Symp. Digest Tech. Papers* **2008**, 39, 705–708. (c) Kondakov, D. Y.; Pawlik, T. D.; Hatwar, T. K.; Spindler, J. P. Triplet annihilation exceeding spin statistical limit in highly efficient fluorescent organic light-emitting diodes. *J. Appl. Phys.* **2009**, *106*, 124510.
28. Yokoyama, D. J. Molecular orientation in small-molecule organic light-emitting diodes. *Mater. Chem.* **2011**, *21*, 19187–19202.
29. (a) Frischeisen, J.; Yokoyama, D.; Endo, A.; Adachi, C.; Brütting, W. Increased light outcoupling efficiency in dye-doped small molecule organic light-emitting diodes with horizontally oriented emitters. *Org. Electron.* **2011**, *12*, 809–817. (b) Frischeisen, J.; Yokoyama, D.; Adachi, C.; Brütting, W. Determination of molecular dipole orientation in doped fluorescent organic thin films by photoluminescence measurements. *Appl. Phys. Lett.* **2010**, *96*, 073302.
30. Ogiwara, T.; Takahashi, J.; Kuma, H.; Kawamura, Y.; Iwakuma, T.; Hosokawa, C. Degradation analysis of blue phosphorescent organic light emitting diode by impedance spectroscopy and transient electroluminescence spectroscopy. *IEICE Trans. Electron.* **2009**, *E92-C*, 1334–1339.
31. Patil, V. V.; Lee, K. H.; Lee, J. Y. A novel fluorene–indolocarbazole hybrid chromophore to assemble high efficiency deep-blue fluorescent emitters with extended device lifetime. *J. Mater. Chem. C* **2020**, *8*, 3051–3057.
32. Lv, X.; Sun, S.; Zhang, Q.; Ye, S.; Liu, W.; Wang, Y.; Guo, R.; Wang, L. A strategy to construct multifunctional TADF materials for deep blue and high efficiency yellow fluorescent devices. *J. Mater. Chem. C* 2020, *8*, 4818–4826.

33. Chavez, R.; Cai, M.; Tlach, B. C.; Wheeler, D. L.; Kaudal, R.; Tsyrenova, A.; Shinar, R.; Shinar, J.; Tomlinson, A. L.; Jeffries-El, M. Benzobisoxazole cruciforms: A tunable, cross-conjugated platform for the generation of deep blue OLED materials. *J. Mater. Chem. C* **2016**, *4*, 3765–3773.

34. Chen, S.; Lian, J.; Wang, W.; Jiang, Y.; Wang, X.; Chen, S.; Zeng, P.; Peng, Z. Efficient deep blue electroluminescence with CIEy ∈ (0.05–0.07) from phenanthroimidazole–acridine derivative hybrid fluorophores. *J. Mater. Chem. C* **2018**, *6*, 9363–9373.

35. Chen, S.; Wu, Y.; Zhao, Y.; Fang, D. Deep blue organic light-emitting devices enabled by bipolar phenanthro[9,10-d]imidazole derivatives. *RSC Adv.* **2015**, *5*, 72009–72018.

36. Gusev, A. N.; Kiskin, M. A.; Braga, E. V.; Kryukova, M. A.; Baryshnikov, G. V.; Karaush-Karmazin, N. N.; Minaeva, V. A.; Minaev, B. F.; Ivaniuk, K.; Stakhira, P.; Ågren, H.; Linert, W. Schiff base zinc (II) complexes as promising emitters for blue organic light-emitting diodes. *ACS Appl. Electron. Mater.* **2021**, *3*, 3436–3444.

37. Hu, L.; Liu, S.; Xie, G.; Yang, W.; Zhang, B. Bis(benzothiophene-S, S-dioxide) fused small molecules realize solution-processable, high-performance and non-doped blue organic light-emitting diodes. *J. Mater. Chem. C* **2020**, *8*, 1002–1009.

38. Kotchapradist, P.; Prachumrak, N.; Tarsang, R.; Jungsuttiwong, S.; Keawin, T; Sudyoadsuk, T.; Promarak, V. Pyrene-functionalized carbazole derivatives as non-doped blue emitters for highly efficient blue organic light-emitting diodes. *J. Mater. Chem. C* **2013**, *1*, 4916–4924.

12 Fundamental Perspective of Phosphorescent Organic Materials for OLEDs

Sumit Kumar

12.1 INTRODUCTION

The need for lighting today consumes nearly one fifth of total domestic energy consumption in the world. In this consumption, the contribution due to incandescent as well as fluorescent lights is significant. Soon these light sources will be replaced by other sources such as light emitting diodes (LEDs). A fluorescent light provides better power efficiency in comparison to the incandescent light but this light source contains mercury and therefore cannot meet the benchmarks for a clean energy source.[1-4] Thus, a suitable artificial source of lighting is very much needed and this gap has been filled by the next generation of LEDs, named organic light emitting diodes (OLEDs).[2-8] The OLEDs are found to be more efficient in comparison to LEDs. Interestingly, in comparison to the fluorescent and incandescent light sources, OLEDs do not produce much heat. A general source of lighting produces enormous amounts of heat during the conversion of electrical energy to the light; therefore, the surface temperature of these light sources reaches to approximately 90°C, whereas a fluorescent light source produces relatively less heat and hence the surface temperature is nearly 65°C. This enormous amount of heat is a source of risk of fire. But OLEDs are comparatively cold enough to keep the surface temperature at nearly 35°C. In this way, OLEDs are not only reducing the risk of heating but also increasing the device efficiency by impeding the unnecessary use of operating electrical energy.

OLEDs have gained a lot of importance in recent years due to their use in the development of the display of electronic goods such as televisions, mobiles, etc.[7,9-13] Over the last two decades, extensive studies have been performed to get suitable and efficient methods to obtain light from OLEDs.[10,11,14] There are some alternatives to OLEDs, such as liquid crystal displays (LCDs), available in the market but no other devices supersede OLEDs, due to their flexibility to produce variable colour intensity and colour quality.[15] The element of OLEDs is comparatively much thinner and therefore lighter compared to other competitor devices.[3,15,16] OLEDs are efficient enough to provide a long range of contrast of colours and can be viewed even from wider angles. It has faster response time on applying the suitable voltage and can produce transparent displays, which can be bent to various dimensions to produce tailor-made designs of screens. There have been some technical challenges to select the proper materials that can provide electronically

DOI: 10.1201/9781003260417-12

improved fundamentals to produce efficient light emitting processes. Investigation continues in the field of OLEDs to get suitable red or green emitters but challenges have been faced in the preparation of stable blue emitters. A thermally and morphologically stable material is also needed for the durability of OLEDs.

Interestingly, OLEDs convert electrical energy into light energy through an electroluminescence process.[17,18] The electroluminescence occurs as a result of the recombination of electrons and holes, where excited electrons produce light as photons in response to the application of a strong electric field or by passing the electric current through photo-electro functional materials such as semiconductors. One of the initial applications of the electroluminescence phenomenon was reported for zinc sulphide (ZnS) in early 1936. Another study reported this phenomenon in an inorganic electroluminescence device, where gallium arsenide (GaAs) material showed the electron and hole recombination due to doping of material in order to form a p-n junction.[3] Several inorganic materials, such as zinc sulphide doped with sulphur/copper/manganese, iridium phosphide, etc., are used in the electroluminescence devices in the preparation of LEDs. But recently, organic molecules are more widely used in LEDs, which are abbreviated as OLEDs.[17]

12.2   ELECTRON-HOLE RECOMBINATION PHENOMENON AND SPIN-RELATED STATISTICS IN OLEDs

OLEDs are currently used in the preparation of television screens and display of mobile phones, tablets, and other electronic instruments. OLEDs have mostly replaced LEDs. This is because the radiative rate constant for $S_1 \to S_0$ transition in hydrocarbons is nearly 10^6 times larger than the $T_1 \to S_0$ states because $T_1 \to S_0$ transition is spin forbidden.[19] Therefore, the utilisation of triplet as well as singlet excitons is challenging to increase the internal quantum efficiency (IQE) of LEDs. But the same is possible in the case of phosphorescence via dipolar activity through the perturbation due to spin-orbit coupling.[20] Interestingly the spin orbit coupling is very weak due to low-lying triplet and singlet excited states in the case of conjugated organic systems. The conjugation in the organic systems produces intense transition due to closely placed π bonding and π^* antibonding energy states.[2,6] Thus, these conjugated organic systems or chromophores cannot provide brighter light emission due to weak spin-orbit coupling perturbation and therefore weak phosphorescence due to less contribution of triplet excitons. Therefore, it is harder to compel the barrier of 25% IQE for OLEDs.[2,6,21]

There are two ways to surpass the barrier of 25% IQE. First, the utilisation of phosphorescence can be enhanced in competition to nonradiative relaxation of the population of triplet excitons.[2,21] The second is that the excitons in the triplet state of an organic system can be shared with other species which could produce light.[2,21]

12.3   PRINCIPLES OF ELECTROPHOSPHORESCENCE IN MULTILAYER DEVICES

In the beginning OLEDs more commonly utilised singlet excitons, which means that the materials responsible for fluorescence were used. As the ratio of excitons present in singlet and triplet states is approximately 1:3, in the case of fluorescent materials, only one quarter of the electronically produced excitons are being used, while three quarters of those remaining in the triplet state are not participating.[2,6,22] Eventually, the IQE of these OLEDs cannot be more than 25%. But this barrier was crossed in 1998 with the use of triplet metal to legend charge transfer (MLCT) in the case of OLEDs containing Pt(II) and Os(II). As these phosphorescent OLEDs materials utilise both singlet and triplet excitons, the IQE is better than fluorescent OLEDs. The above study also shows that the heavy transition metal spread over the skeleton of the organic molecule more effectively increases spin-orbit coupling to promote interest system crossing (ISC), which could enhance the population of triplet excitons. Thus, phosphorescent organic LEDs are more efficient and therefore more useful for scientists.

To reach higher IQE of OLEDs, the major challenge is observed as triplet-triplet annihilation.[2,9,23] To get higher intensity of light, once the high current has been applied over the LEDs, there is a possibility of triplet-triplet annihilation due to the long lifetime of triplet excitons. To get rid of this challenge, the phosphorescent material has been doped over the host material at a low concentration. It has the ability to overcome the triplet-triplet annihilation by differing the quenching of concentration of triplet excitons in the OLEDs.[2,6,23]

12.4   GENERAL SCHEME OF MULTILAYER OLED

It is now worth looking at the general scheme of multilayer OLEDs, which is shown in Figure 12.1. The OLEDs are generally made up of thin multilayers containing organic semiconductors stacked linearly. The thickness of the layers varies

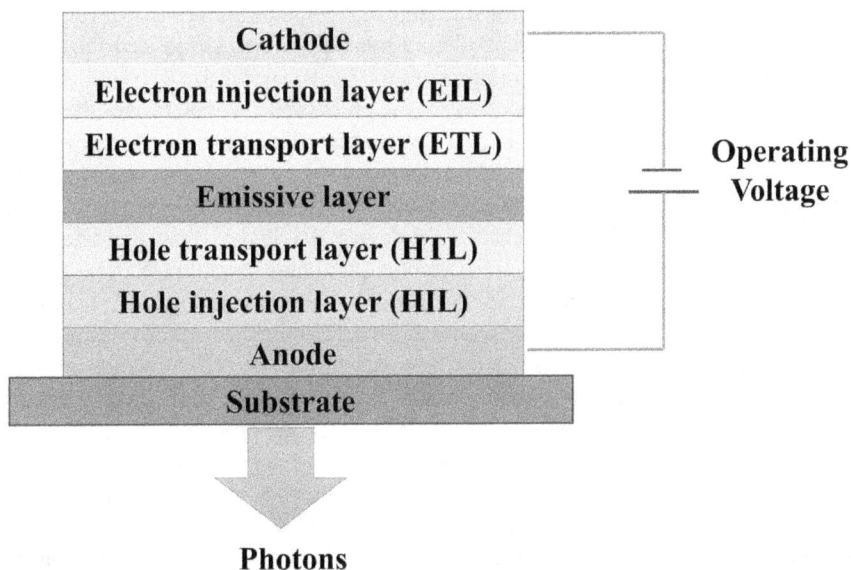

FIGURE 12.1   General scheme of multilayer OLEDs.

depending upon the requirement of the electrical luminescence. The anode is usually placed over the substrate, for example, tin-doped iridium oxide spread on the thin transparent glass substrate. In contrast the cathode is placed at the other corner and is responsible for the generation of electrons. Obviously, it should have a low value of electronic work function in order to efficiently generate electrons. There are some specific classes of metals that fulfil this criterion, including the alkaline earth metals. Interestingly the main idea is to use multilayers to promote electron-hole recombination. The adjacent layer to the anode is the hole injection layer (HIL), which is used along with the hole transport layer (HTL) for the balanced transport of holes into the doped emission layer. On the opposite side, adjacent to the cathode an electron transport layer has been placed, which is covered with a thin layer of lithium or cesium fluoride to reduce the injection barrier as well as to prevent the electron transport layer from chemical reactions.[2,5,6,15] Further for efficient electron-hole recombination of the electrons from the cathode and the holes from the anode after passing the multiple layers, the challenge of ohmic loss has to be surpassed, which is achieved by the use of an electron-blocking layer between the emissive layer and the electron transport layer and a hole-blocking layer between the emissive layer and the hole transport layer. These layers are not shown in the general scheme shown in Figure 12.1, but they are necessary for efficient electron-hole recombination. There is a difficulty arising from these blockings due to an increase of high electron charge density close to the interfaces, which leads to the lowering of the lifetime of OLEDs.

The emission layer in OLEDs is the most important and consists of two parts, host and guest materials. The host material is the solid matrix, and the guest material is embedded over the host. One of the commonly used host materials is 4,4′-N,N′-dicarbazolebiphenyl (CBP). The host material is mainly decided based on the lifetime of the phosphorescence process considering the competing nonradiative relaxation pathway of triplet excitons. The completing relaxation process reduced the IQE of OLEDs; therefore, to get rid of this process the host material is deposited over the guest material to enhance the phosphorescence process. It also determines the colour of the device. As these improvements are worth investigating, Baldo and coworkers have studied various transition metal complexes along with organic ligands, which are dropped over the host materials, to check the efficiency of the device as well as to measure the colour and luminescence of the device.[10,22] Studies show that one example of an efficient host material is Ir(III) cation over phenyl pyridinate (ppy) anions.

In the emissive layer, the nonradiative relaxation of excitons occurs mainly through Forster and Dexter energy transfer processes, whereas there are three possible ways for the emission from the dopants in the host-guest (dopant) system, which is shown in Figure 12.2. The first process is the singlet excitons present in the host-guest system through the process of intersystem crossing (ISC) and producing the triplet excitons. In the second process, the triplet excitons present in the host are transferred to the phosphors though Forster and Dexter energy transfer processes and phosphorescence from phosphors emits the light. The third process is electron hole recombination of the electrons injected from the cathode and the holes injected from the anode. The combination produces the triplet excitons, which are trapped in the guest molecule and relax to the ground state via phosphorescence, producing light energy.

The Forster and Dexter energy transfer processes are also explained in Figure 12.2a and Figure 12.2b. D* is an energy donor, which is an excited molecule generated after the electron-hole recombination. The Forster energy transfer process is a coulombic interaction and therefore it is fast and designated as a long-range interaction in the order of 10 nm. Further the Dexter electronic transfer process is an electron exchange interaction for the host and the dopant; therefore, it is a short-range interaction and lies between 1 nm and 2 nm.

The emission spectrum observed for the host matrix overlap over the absorption spectra measured for the dopant is the most crucial phenomenon of the Forster energy transfer process. Another important factor is the dipole-dipole interaction between the transition dipole moments for host and dopant, which can be viewed from the excitation of electrons from the highest occupied molecular orbital (HOMO) to the lowest unoccupied molecular orbital (LUMO). Nuclear vibration as well as vibronic coupling also play a significant role in the energy transfer.

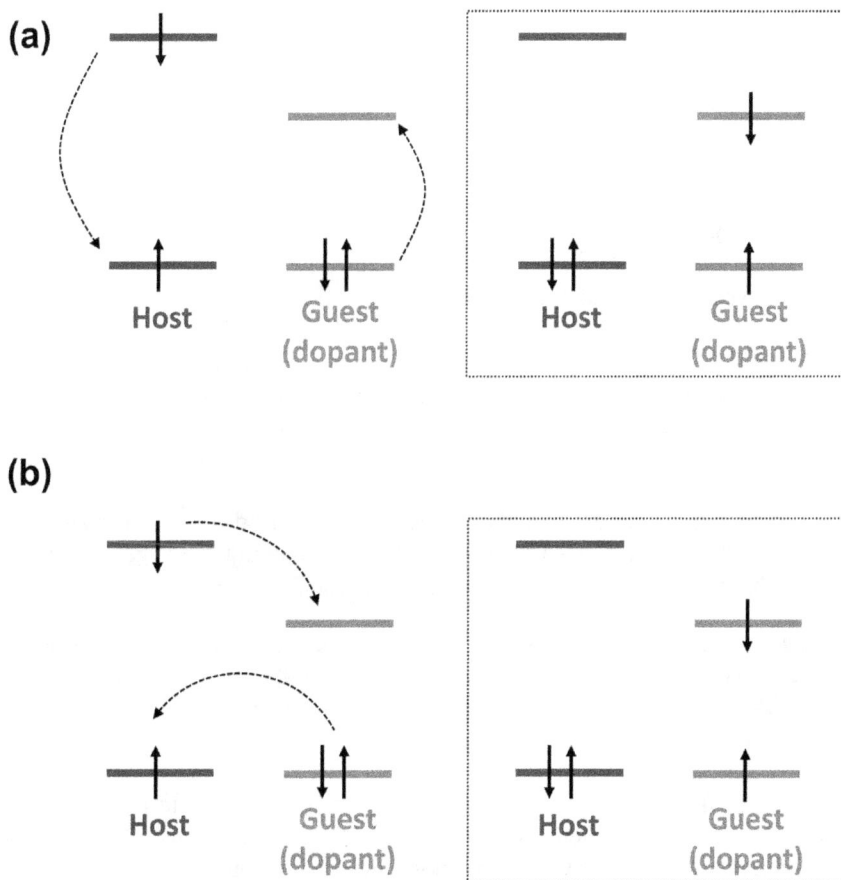

FIGURE 12.2 (a) Forster and (b) Dexter energy transfer processes: electron-hole transport processes in emissive dye doped over host material.

The Dexter energy transfer processes depend significantly upon the electron exchange process and therefore the exchange integral is the important factor, which is determined from the electron exchange rate between two states, that is, the HOMO of the donor and the LUMO of the acceptor molecule. The electron exchange rate depends upon the overlap between the orbitals of these two states and therefore the Dexter energy transfer process is a short-range interaction.

There is a limitation for Forster energy transfer that it cannot be applied for spin-forbidden transitions but the same can occur in the Dexter energy transfer process so that this process is useful for the determination of triplet-triplet animation and various spin-dependent processes/energies. An offset between the states of HOMO and LUMO is required for efficient charge trapping to promote the phosphorescence process for phosphors.

12.5 PROPERTIES TO SELECT A SUITABLE HOST MATERIAL FOR PHOSPHORESCENT OLEDs

The electrophosphorescence process is at the heart of the chemistry involved in selecting efficient host materials to prepare phosphorescent OLEDs. There are a few requirements to select proper host material for the device, which are as follows.[2,5,6,15]

a) The energy states for the triplet excitons for the host have to be higher in comparison to those for guest/dopant molecules so that the reverse electronic process of the decay of the excited electrons from host to guest molecule can be hindered and at the same time, the loss of excitons from the emissive layer can also be prevented.

b) The hole and electron injection barriers have to be low. For this, the HOMO and LUMO for the host material have to be considered with that of the adjacent layer during the preparation of LEDs.

c) For an efficient electroluminescence process, the electron-hole recombination is an important feature, which is modulated through the charge carrier transport property of the host.

d) To achieve the durability of the device for a long period of time, the material has to be thermally and morphologically stable. It is required to get dynamic stability between the interfaces and to prevent phase transformation.

It is also important to use proper glass material for the device to get thermal stability as well as morphological stability. For this, a bulky molecular configuration of the glass material is used in order to maintain the stability of the crystal at high temperature.

Heavy transition metals spread the charge over the skeleton of the organic molecule, more effectively increasing spin-orbit coupling to promote interest system crossing (ISC) and enhancing the population of triplet excitons. To date many phosphorescent emitters have been reported in the literature. Among them, a few are used commonly in the preparation OLEDs, which are shown in Figure 12.3. These phosphorescent Ir[III] complexes produce different primary colours. These can be red, green, and blue colour of light. bis[(4′,6′-difluorophenyl)-pyridinato-$N,C^{2'}$]

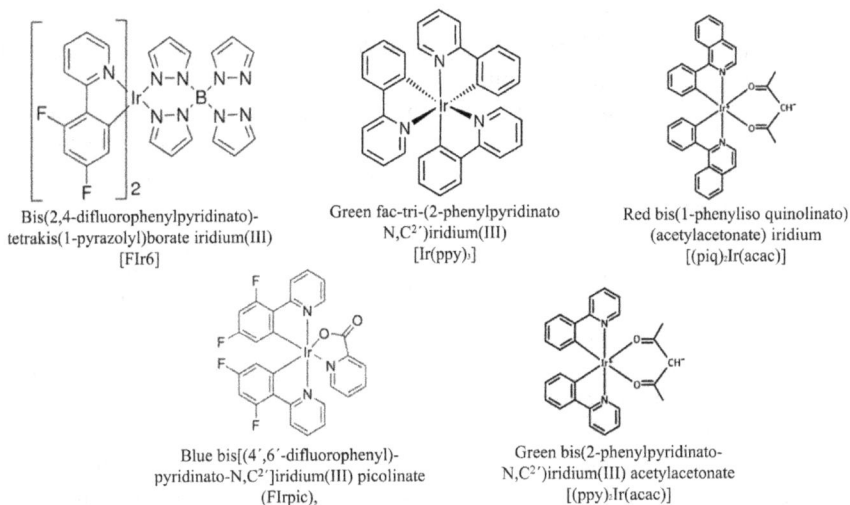

Bis(2,4-difluorophenylpyridinato)-
tetrakis(1-pyrazolyl)borate iridium(III)
[FIr6]

Green fac-tri-(2-phenylpyridinato
N,C²′)iridium(III)
[Ir(ppy)₃]

Red bis(1-phenyliso quinolinato)
(acetylacetonate) iridium
[(piq)₂Ir(acac)]

Blue bis[(4′,6′-difluorophenyl)-
pyridinato-N,C²′]iridium(III) picolinate
(FIrpic),

Green bis(2-phenylpyridinato-
N,C²′)iridium(III) acetylacetonate
[(ppy)₂Ir(acac)]

FIGURE 12.3 The molecular structure of commonly used coloured triplet emitters.

iridium(III) picolinate (FIrpic) produces blue, whereas fac-tri-(2-phenylpyridinato-
N,C²′)iridium(III) [Ir(ppy)₃] is responsible for green light, and at the same time
bis(1-phenylisoquinolinato)(acetylacetonate)iridium [(piq)₂Ir(acac)] is producing
red light of photons.

12.6　TRANSPORT MATERIALS FOR PHOSPHORESCENT OLEDs

Phosphorescent OLEDs should show good electrophosphorescence determined by
the transport abilities of the charge carrier. The molecular structure is mainly responsible for the accumulation of charges, that is, either electron or hole. The important
feature is that the triplet energy of these materials should be higher in comparison
to that of the emitters so that the population of the triplet excitons should not fall
and remain trapped in the emissive layer. The transport materials responsible for
the transmission of holes and electrons are discussed separately, focusing especially
on the triplet energy as it is the most significant value to be noticed. The structural
property performance relationship for the transport material has been measured for
both small molecules and polymers. The performance of OLEDs has been enhanced
by controlling the charge injection, transmission and recombination. Both physical properties, such as molecular architecture, morphology and thermal stability,
and electronic structural properties, such as electronic mobility and triplet excitons
energy, are necessary to enhance the OLED working performance such as brightness,
luminescence, efficiency and driving voltage.

12.6.1　HOLE TRANSPORT MATERIALS

There are some specific properties of OLEDs to be used as hole transport materials.
Mainly electron donating species are used in transport materials. The electron-donating

groups used for this purpose are carbazole, diphenylamine, triarylamine, etc. These materials should contain higher mobility for holes as well as glass material used for its morphological and thermal stability.

For efficient electroluminescence, the energy barrier for the hole injection from the anode has to be lowered by selecting the proper energy level of the HOMO. Simultaneously, the transmission of holes from the emissive layer to the hole transport layer is prevented by selecting the proper LUMO energy level of the transport materials. Examples are 4,4′-bis[N-(p-tolyl)-N-phenylamino]biphenyl (TPD) and N,N′-Di(1-naphthyl)-N,N′-diphenylbenzidine (NPB).[24] These are shown in Figure 12.4. TPAD and NPB have HOMO and LUMO energy levels at identical positions at 5.4 and 2.4 eV, respectively. The triplet energy position is at 2.34 and 2.29 eV for TPAD and NPB, respectively.

The efficiency of the electrophosphorescence process is driven mainly by the triplet excitons and therefore influenced by the energy levels of the triplet state in the hole transport-type host material. A study shows an interesting analysis of the confinement of triplet excitons of Ir(ppy)$_3$ with some common host materials such as TPD

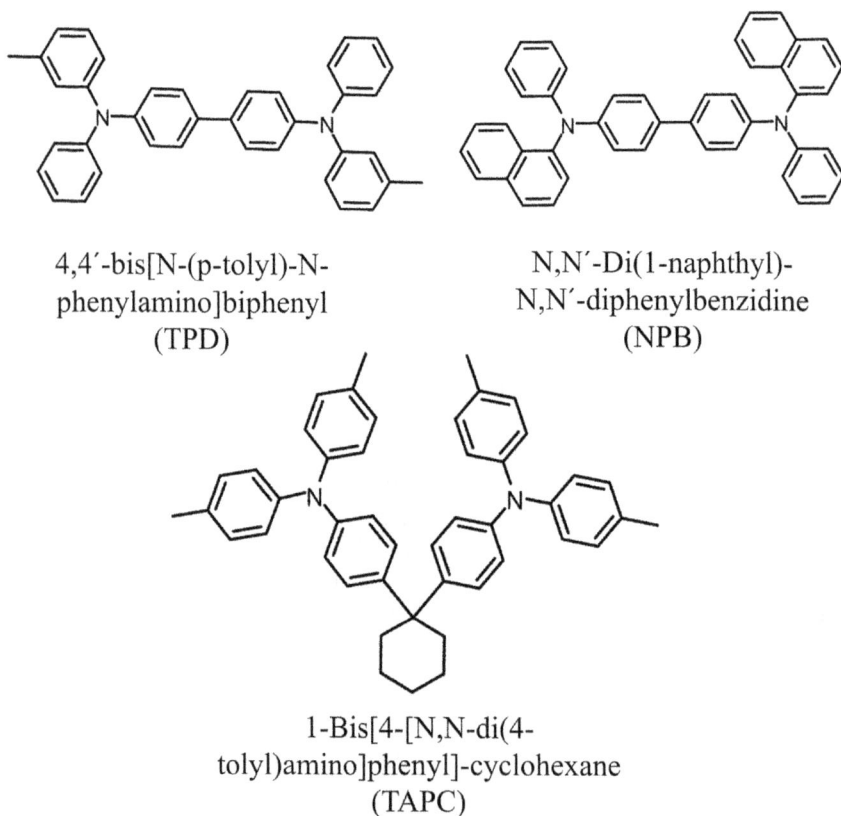

4,4′-bis[N-(p-tolyl)-N-
phenylamino]biphenyl
(TPD)

N,N′-Di(1-naphthyl)-
N,N′-diphenylbenzidine
(NPB)

1-Bis[4-[N,N-di(4-
tolyl)amino]phenyl]-cyclohexane
(TAPC)

FIGURE 12.4 Carbazole and triarylamine derivatives of transport materials for hole transmission.

and TAPC.[25] The energy dispersion pathways of these triplet excitons have also been monitored. The back-and-forth energy transfer of triplet excitons of Ir(ppy)₃ occur to the triplet levels of TPD, whereas 1-Bis[4-[N,N-di(4-tolyl)amino]phenyl]-cyclohexane (TAPC) as host transport material nicely confines the Ir(ppy)₃ triplet excitons.[25]

12.6.2 ELECTRON TRANSPORT MATERIALS

Electron transport materials are mainly electron-withdrawing groups. Opposite to the hole transporting materials the electron transporting materials facilitate the mobility of electrons and prevent mobility of the holes.[8,26] For this, the energy levels of the HOMO and LUMO have been appropriately chosen to promote electrons and block holes injections through the material. The energy of the triplet excitons is suitably selected to be confined to the emissive layer. Some of the common electron transport materials containing electron withdrawing groups are shown in Figure 12.5. One of the examples of electron transport materials is tris(8-hydroxyquinoline)aluminium (Alq3), which has higher value of LUMO to promote electrons and block holes in the transport layer.[27,28] It has good thermal and morphological stability.[27,28] But Alq3 can be used as a hole-blocking layer (HBL)/ETL due to its high HOMO level and low triplet energy. In contrast, 2,9-dimethyl-4,7-diphenyl-1,10-phenanthroline (BCP) is

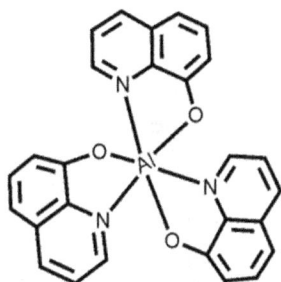

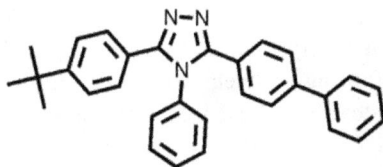

tris(8-hydroxyquinoline) aluminium (Alq3)

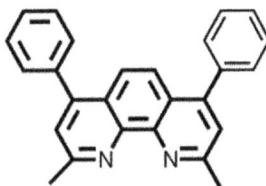

2,9-dimethyl-4,7-diphenyl-1,10-phenanthroline (BCP)

1,2,4-triazole-based 3-(biphenyl-4-yl)-5-(4-tertbutylphenyl)-4-phenyl-4H-1,2,4-triazole (t-Bu-TAZ)

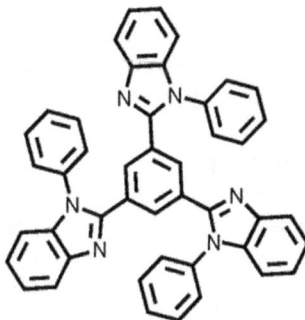

1,3,5-tris(N-phenylbenzimidazol-2-yl) benzene (TPBI)

FIGURE 12.5 Electron transport material containing electron-withdrawing groups.

easily used as HBL/ETL along with Alq3 because it has a deep HOMO level (6.5–6.7 eV) and high triplet energy (2.5 eV).[29,30]

The triplet energy of benzimidazole-based 1,3,5-tris(N-phenylbenzimidazol-2-yl) benzene (TPBI) is 2.74 eV and the HOMO and LUMO of the material are 6.2 eV and 2.7 eV, respectively.[30] The triplet energy (2.7 eV) and HOMO/LUMO (HOMO/ LUMO: 6.3/2.7 eV) of 1,2,4-triazole-based 3-(biphenyl-4-yl)-5-(4-tertbutylphenyl)- 4-phenyl-4H-1,2,4-triazole (t-Bu-TAZ) has been found to be relatively close to that of TPBI.[31] These host materials are commonly used in the hole-blocking layer or electron transport layer in blue phosphorescent OLEDs. This is due to high triplet energies as well as the favourable energy levels of TPBI and t-Bu-TAZ.[29,30]

12.7 HOST MATERIAL FOR PHOSPHORESCENT OLEDs

The host material of the device also plays a pivotal role to determine the efficiency of the electrophosphorescence in the OLEDs.[5] The host material can be of the hole transport type, electron transport type and bipolar transport host type.[2]

12.7.1 HOLE-TRANSPORT-TYPE HOST MATERIALS

The hole-transport-type host material mostly contains the electron-donating functional groups in the molecule structure to promote the transmission of holes in comparison to the electrons.

Either carbazole derivatives or triphenylamine derivatives are the two types of host materials prefered. Some of these hole-transport-type host materials are shown in Figure 12.6. 4,4′,4″-tris(N-carbazolyl) triphenylamine (CBP) is an example having both electron transport type and hole transport type in its emissive layer. CBP-containing OLEDs generally have high peak efficiency for blue-green host material. Tokitoa et al. have reported confinement of the triplet energy in the phosphorescent molecule to enhance the efficiency of OLEDs based on iridium(III)bis[4,6-di-fluoropheny)-pyridinato-$N,C^{2'}$]picolinate (FIrpic).[32] It has been reported that 4,4′-bis(9-carbazolyl)-2,2′-dimethyl-biphenyl (CDBP), which is a derivative of CBP, has the higher triplet energy and is used as a charge carrier for the transport layer in FIrpic-based OLEDs.[32] It was found that the application of CDBP enhances the external quantum efficiency to 10.4% by the application of 20.4 Cd/A and power efficiency of 10.5 lm/W due to confinement of triplet energy in FIrpic-based OLEDs.[32]

10-(4-(9H-carbazol-9-yl)phenyl)-8,8-di-p-tolyl-8Hindolo[3,2,1-de]acridine (BCBP) is prepared by modifying the CBP using arylmethylene bridge linkage.[33] The bridge linkage makes the molecule system bulky and BCBP has higher T_g of 173°C compared to CBP, which has T_g of 62°C. Interestingly the bulkiness of BCBP does not decrease the triplet energy with respect to CBP and therefore BCBP is used as host material for Ir(ppy)$_3$-based green OLEDs as emitter, which show a low turn-on voltage (2.9 V) and high power efficiency of 41.1 lm/W. BCBP as host material with blue FIrpic-based phosphorescent OLEDs show maximum current and power efficiencies of 9.0 Cd/A and 6.3 lm/W, respectively. The study confirms that the aryl bridge increases the molecular size without changing the ability of molecular conjugations and therefore increases both current and power efficiencies of OLEDs.

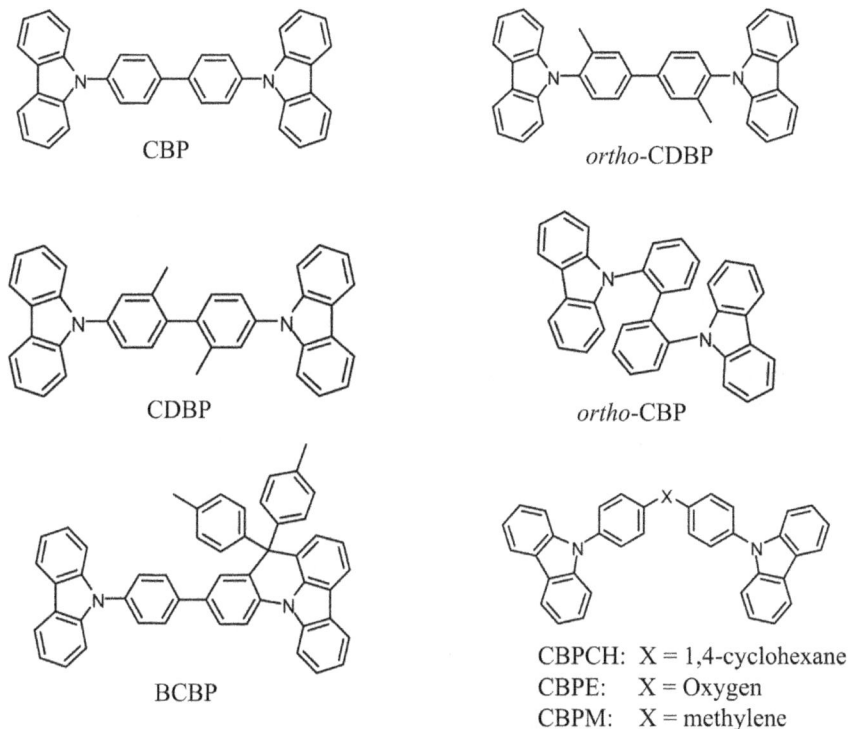

CBP

ortho-CDBP

CDBP

ortho-CBP

BCBP

CBPCH: X = 1,4-cyclohexane
CBPE: X = Oxygen
CBPM: X = methylene

FIGURE 12.6 Hole-transport-type hole material consisting of carbazole derivatives.

Jian et al. have synthesized a series of CBP derivatives through changing the linkage group between two 9-phenyl carbazole groups.[34] These molecules are prepared by methyl substitution (o-CDBP), by attaching the one over the ortho position of another 9-phenyl carbazole group (o-CBP).[34] Further derivatives are prepared by separating the biphenyl by the help of a nonconjugated bridge of methylene (CBPM), ether bond (CBPE), or by using a cyclohexane group (CBPCH).[34] These CBP derivatives have triplet energy between 2.73 and 3.01 eV and systematic increase of bridging groups increases the T_g from 78°C to 115°C. Many other carbazole-based hole transport host materials are prepared due to its individual advantage over others.

Similar to carbazole derivatives, derivatives of triphenylamine (TPA) such as TPD, TAPC and TCTA are used as common hole transport materials due to high triplet energy of 3.04 eV.

12.7.2 Electron-Transport-Type Host Materials

In contrast to the hole-transport-type host materials, electron-withdrawing materials have been chosen for the preparation of OLEDs to facilitate the injection/transmission of electrons and prevent hole transmission/injection.

Electron mobility is found to be lower than hole mobility; therefore, the hole-blocking layer is generally applied before electron-transport-type host materials or sometimes host material having high electron mobility of electron has been chosen as these have higher triplet energy and this triplet energy is prevented from returning to the host material. The low injection barrier also is one of the parameters to be checked to reduce the driving voltage for the OLEDs.

For more insight, some of the commonly used electron-transport-type host materials are depicted in Figure 12.7. One of the commonly used electron-transport-type host materials is tris(8-hydroxyquinoline)aluminium (Alq3), which is selected for its good electroluminescence as it has low triplet energy of 2.0 eV. Leung et al. have reported the fluorescence and phosphorescence of 2,2′-bis(1,3,4-oxadiazol-2-yl) biphenyls (BOBPs) and found the phosphorescence of BOBPs is an unusual red shift due to 1,3,4-Oxadiazoles (OXD), which properly matched with the spectral properties of the BOBPs and Ir(ppy)₃ based OLEDs and therefore the maximum brightness and current efficiency are reported as 45 000 cd m^{-2} and 26 cd/A, respectively, for the reported device (ITO/NPB/BOBP3: Ir(ppy)₃/BCP/BeBq2/LiF/Al).[35] Chen et al. have synthesized a long series of oxadiazole derivatives (TPO, PPO, MTPO, MPPO, BTPO and BPPO). These materials are used as host materials along with the blue phosphorescent dopant emitter iridium(III)bis[4,6-di-fluorophenylpyridinato-$N,C^{2'}$] picolinate and found maximum luminance of 4484 cd/m^2 and also external quantum efficiency increases to 6.20%.[35]

12.7.3 Bipolar Transport Host Materials

The charge carrier injection as well as transporting ability is the major factor in selecting proper host material for the preparation of these devices. Materials with both hole- and electron-transporting abilities are used as emitters of triplet excitons. For efficient electroluminescence efficiency only, these materials are preferably

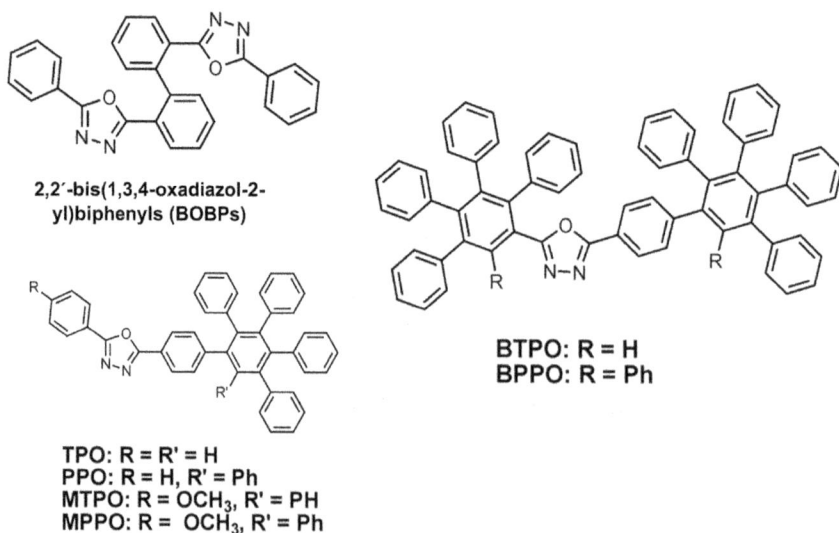

2,2′-bis(1,3,4-oxadiazol-2-yl)biphenyls (BOBPs)

BTPO: R = H
BPPO: R = Ph

TPO: R = R' = H
PPO: R = H, R' = Ph
MTPO: R = OCH₃, R' = PH
MPPO: R = OCH₃, R' = Ph

FIGURE 12.7 Commonly used electron-transport-type host materials.

chosen for the preparation of these devices, which have balanced abilities to inject as well as transport both types of charge careers, that is, holes or electrons. Exceptionally in some of the cases the higher value of electroluminescence has been achieved even at lower current densities but in those cases the efficiency roll-off is the major factor to be considered at higher values of current densities. The efficiency roll-off is one of the major challenges in the preparational technology of these devices where the efficiency decreases as the brightness increases. For these cases single triplet annihilation is found to be one of the major reasons.

The unipolar host materials mainly show lower values of electroluminescence due to unbalanced charge career injection as well as transporting abilities and therefore the performance of the device is degraded due to efficiency roll-off.[36] The unbalanced injection and transportation of charge carriers significantly narrow the hole-electron recombination zone and eventually shift this zone closer to the electron-transporting layer or the hole-transporting layer from the middle of the emissive layer, which is clearly shown in Figure 12.8.[36]

The shift of the hole-electron recombination zone enhances the triplet-triplet annihilation and also increases the possibility of decrease of emission colour purity, which is responsible for the degree of resemblance of colour with its hue. If the charge transport is very much unbalanced then there is a chance of charge leakage and therefore the operational lifetime of the device decreases a lot. The challenge of unbalanced charge transport has been overcome by using a hole-blocking layer and an electron-blocking layer in the recombination zone or by applying the electron-type and hole-type host systems in a single emissive layer. The hole-blocking layer is applied between the emissive layer and the electron-transporting layer whereas the electron transport layer is placed between the emissive layer and the hole transport layer. Therefore, the preparation of the device is much more complicated and results in decreased lifetime of the device and the phase separation of host systems. To overcome this challenge, the bipolar host materials have been preferably chosen to consider balanced abilities of both electron and hole injection and transportation through the material. This type of host materials enhances the performance of the device.

The bipolar host materials have electron-donating groups to promote the transmission/injection of the holes and electron-deficient groups to enhance the transmission/injection of electrons.[37] When both electron-donating and electrondeficient groups occur in the same material there is a chance of intramolecular charge transfer to

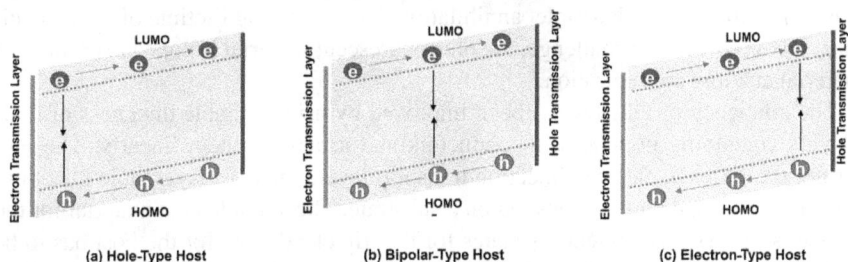

(a) Hole-Type Host   (b) Bipolar-Type Host   (c) Electron-Type Host

FIGURE 12.8 Electron-hole recombination zone in the case of (a) hole-type, (b) bipolar (c) electron-type host materials.

decrease the band gap. Thus, the electron donor-acceptor interaction is minimized by selecting the proper host material having higher triplet energy. The decrease of electron-hole interaction as well as the corresponding increase of triplet energy can be possible by using multiple sizes of π-conjugated molecular system as spacer to add electron-donating or -withdrawing groups at ortho- or meta-positions rather than at the para-position of the molecule to increase the branching of the chain. The branching and application of a spacer disperse the molecular system to reduce electron-hole interaction in the chain. The same can be achieved by inserting the sp^3 hybridized bond between electron-withdrawing or -donating groups and the conjugated molecular system is used as a host material to reduce the effect of the substituent groups. There is a requirement of efficient host materials to prevent the reverse transfer processes of energy and confining the excitons in the guest material. For the suitable bipolar host materials, the energy levels (HOMO and LUMO) should coincide with that of the neighbouring layer to increase the chance of injection/transportation of electrons/holes.

There are a number of combinations of electron donor and acceptor moieties reported in the literature for the preparation of suitable bipolar host materials. The common electron donor is carbazole, di- or triphenylamine, etc. In contrast, the molecule used as electron acceptor has been more widely investigated. The examples reported are phosphine oxide, phenylsulfonyl, N-heterocycles groups, etc.

12.8 CONCLUSION

The consumption of electrical energy is one of the major concerns in today's scenario, where 20% of all domestic usage is for lighting. OLEDs are the most suitable electrical device to replace other lighting elements, that is, incandescent or fluorescent light sources, as they consume less electric energy and are not misused as heat.

OLEDs have a bright future due to their use in displays for electronic goods such as televisions, mobiles, etc. OLEDs are an important source of light converting from electrical energy through the electroluminescence process. OLEDs are comparatively more efficient than the traditional fluorescent LEDs as they utilise more of the population of excited electrons, including triplet excitons, whereas the traditional fluorescent LEDs only utilize the singlet excitons. As the ratio of excitons present in singlet and triplet states is approximately 1:3, in the case of fluorescent materials, only one quarter of the electronically produced excitons is being used; therefore, the IQE of OLEDs is only 25%. To reach higher IQE of OLEDs, the major challenge was observed to be triplet-triplet annihilation due to the long lifetime of triplet excitons. To overcome this challenge, the phosphorescent material is doped over the host material at a low concentration.

The efficiency of OLEDs has been improved by using suitable thickness of multilayers containing organic semiconductors and arranging them linearly. The layers are the cathode, electron injection/transport layer, emissive layer, hole injection/transport layer and anode deposited on a substrate. The suitable host material for the device is selected having energy states for the triplet excitons for the host has to be higher in comparison to those for guest/dopant molecules and the hole as well as electron injection barriers has to be low. To achieve the durability of the device for long periods of time, the material has to be thermally and morphologically stable.

The performance of phosphorescent OLEDs is determined by the electrophosphorescence process and therefore the transport abilities of charge carrier. The triplet energy of all the materials used should be higher than that of phosphorescent emitter so that the triplet must be confined inside the emissive layer. Both hole and electron transport layers have been prepared based on this fundamental.

The host material is significant to determine the efficiency of the electrophosphorescence in OLEDs. Hole transport type, electron transport type and bipolar transport host type are the main categories. Bipolar-transport-host-type material is most useful in comparison to the other two types. The efficiency roll-off and triplet-triplet annihilation are found to be prominent phenomena here. The unipolar host materials mainly show the lower values of electroluminescence due to unbalanced charge career injection as well as transporting abilities and therefore the performance of the device is degraded due to efficiency roll-off. Many other parameters of note are discussed in the chapter.

ACKNOWLEDGMENT

I would like to thank Magadh University, Bodh Gaya, India, for providing lab facilities. The Department of Science and Technology (DST), India (Grant No. SRG/2019/002284), is acknowledged for financial support.

REFERENCES

1. Khanna, V. K., Fundamentals of solid-state lighting: LEDs, OLEDs, and their applications in illumination and displays. CRC Press: 2014.
2. Yersin, H., Highly efficient OLEDs with phosphorescent materials. John Wiley & Sons: 2008.
3. Brütting, W.; Berleb, S.; Mückl, A. G., Device physics of organic light-emitting diodes based on molecular materials. Org. Electron. **2001,** 2 (1), 1–36.
4. Brütting, W.; Frischeisen, J.; Schmidt, T. D.; Scholz, B. J.; Mayr, C., Device efficiency of organic light-emitting diodes: Progress by improved light outcoupling. Phys. Status Solidi A. **2013,** 210 (1), 44–65.
5. Fangfang, W.; Youtian, T.; Wei, H., Recent progress of host materials for highly efficient blue phosphorescent OLEDs. Acta Chim. Sin. **2015,** 73 (1), 9–22.
6. Minaev, B.; Baryshnikov, G.; Agren, H., Principles of phosphorescent organic light emitting devices. Phys. Chem. Chem. Phys. **2014,** 16 (5), 1719–1758.
7. Tang, C. W.; VanSlyke, S. A., Organic electroluminescent diodes. Appl. Phys. Lett. **1987,** 51 (12), 913–915.
8. Kulkarni, A. P.; Tonzola, C. J.; Babel, A.; Jenekhe, S. A., Electron transport materials for organic light-emitting diodes. Chem. Mater. **2004,** 16 (23), 4556–4573.
9. Tao, Y.; Yang, C.; Qin, J., Organic host materials for phosphorescent organic light-emitting diodes. Chem. Soc. Rev. **2011,** 40 (5), 2943–2970.
10. Baldo, M. A.; O'Brien, D. F.; You, Y.; Shoustikov, A.; Sibley, S.; Thompson, M. E.; Forrest, S. R., Highly efficient phosphorescent emission from organic electroluminescent devices. Nature **1998,** 395 (6698), 151–154.
11. Lamansky, S.; Djurovich, P.; Murphy, D.; Abdel-Razzaq, F.; Lee, H.-E.; Adachi, C.; Burrows, P. E.; Forrest, S. R.; Thompson, M. E., Highly phosphorescent bis-cyclometalated iridium complexes: synthesis, photophysical characterization, and use in organic light emitting diodes. J. Am. Chem. Soc. **2001,** 123 (18), 4304–4312.

12. Chi, Y.; Chou, P.-T., Transition-metal phosphors with cyclometalating ligands: fundamentals and applications. Chem. Soc. Rev. **2010,** 39 (2), 638–655.
13. Chou, P. T.; Chi, Y., Phosphorescent dyes for organic light-emitting diodes. Chem. A Eur. J. **2007,** 13 (2), 380–395.
14. Wong, W.-Y.; Ho, C.-L., Functional metallophosphors for effective charge carrier injection/transport: new robust OLED materials with emerging applications. J. Mater. Chem. **2009,** 19 (26), 4457–4482.
15. Geffroy, B.; Le Roy, P.; Prat, C., Organic light-emitting diode (OLED) technology: materials, devices and display technologies. Polym. Int. **2006,** 55 (6), 572–582.
16. Wong, W.-Y.; Ho, C.-L., Heavy metal organometallic electrophosphors derived from multi-component chromophores. Coord. Chem. Rev. **2009,** 253 (13–14), 1709–1758.
17. Gather, M. C.; Köhnen, A.; Meerholz, K., White organic light-emitting diodes. Adv. Mater. (Weinheim, Ger.) **2011,** 23 (2), 233–248.
18. Holder, E.; Langeveld, B. M.; Schubert, U. S., New trends in the use of transition metal–ligand complexes for applications in electroluminescent devices. Adv. Mater. (Weinheim, Ger.) **2005,** 17 (9), 1109–1121.
19. Yersin, H., Triplet emitters for OLED applications. Mechanisms of exciton trapping and control of emission properties. In Transition metal and rare earth compounds: Excited States, Transitions, Interactions III, Yersin, H., Ed. Springer: **2004,** 1–26.
20. Rausch, A. F.; Homeier, H. H.; Yersin, H., Organometallic Pt (II) and Ir (III) triplet emitters for OLED applications and the role of spin–orbit coupling: A study based on high-resolution optical spectroscopy. Photophys. Organometallics **2010,** 193–235.
21. Adachi, C.; Baldo, M. A.; Thompson, M. E.; Forrest, S. R., Nearly 100% internal phosphorescence efficiency in an organic light-emitting device. J. Appl. Phys. **2001,** 90 (10), 5048–5051.
22. Baldo, M. A.; O'Brien, D. F.; Thompson, M. E.; Forrest, S. R., Excitonic singlet-triplet ratio in a semiconducting organic thin film. Phys. Rev. B. **1999,** 60 (20), 14422–14428.
23. Kondakov, D. Y., Triplet–triplet annihilation in highly efficient fluorescent organic light-emitting diodes: current state and future outlook. Philos. Trans. R. Soc. A. **2015,** 373 (2044), 20140321.
24. Lee, D.-H.; Liu, Y.-P.; Lee, K.-H.; Chae, H.; Cho, S. M., Effect of hole transporting materials in phosphorescent white polymer light-emitting diodes. Org. Electron. **2010,** 11 (3), 427–433.
25. Goushi, K.; Kwong, R.; Brown, J. J.; Sasabe, H.; Adachi, C., Triplet exciton confinement and unconfinement by adjacent hole-transport layers. J. Appl. Phys. **2004,** 95 (12), 7798–7802.
26. Hughes, G.; Bryce, M. R., Electron-transporting materials for organic electroluminescent and electrophosphorescent devices. J. Mater. Chem. **2005,** 15 (1), 94–107.
27. Higginson, K. A.; Zhang, X.-M.; Papadimitrakopoulos, F., Thermal and morphological effects on the hydrolytic stability of aluminum tris (8-hydroxyquinoline)(Alq3). Chem. Mater. **1998,** 10 (4), 1017–1020.
28. Ishii, H.; Sugiyama, K.; Ito, E.; Seki, K., Energy level alignment and interfacial electronic structures at organic/metal and organic/organic interfaces. Adv. Mater. (Weinheim, Ger.) **1999,** 11 (8), 605–625.
29. Adamovich, V. I.; Cordero, S. R.; Djurovich, P. I.; Tamayo, A.; Thompson, M. E.; D'Andrade, B. W.; Forrest, S. R., New charge-carrier blocking materials for high efficiency OLEDs. Org. Electron. **2003,** 4 (2–3), 77–87.
30. Kondakova, M. E.; Pawlik, T. D.; Young, R. H.; Giesen, D. J.; Kondakov, D. Y.; Brown, C. T.; Deaton, J. C.; Lenhard, J. R.; Klubek, K. P., High-efficiency, low-voltage phosphorescent organic light-emitting diode devices with mixed host. J. Appl. Phys. **2008,** 104 (9), 094501.

31. Wang, Q.; Ding, J.; Ma, D.; Cheng, Y.; Wang, L.; Jing, X.; Wang, F., Harvesting excitons via two parallel channels for efficient white organic LEDs with nearly 100% internal quantum efficiency: fabrication and emission-mechanism analysis. Adv. Funct. Mater. **2009,** 19 (1), 84–95.

32. Tokito, S.; Iijima, T.; Suzuri, Y.; Kita, H.; Tsuzuki, T.; Sato, F., Confinement of triplet energy on phosphorescent molecules for highly-efficient organic blue-light-emitting devices. Appl. Phys. Lett. **2003,** 83 (3), 569–571.

33. Jiang, Z.; Xu, X.; Zhang, Z.; Yang, C.; Liu, Z.; Tao, Y.; Qin, J.; Ma, D., Diarylmethylene-bridged 4, 4′-(bis (9-carbazolyl)) biphenyl: morphological stable host material for highly efficient electrophosphorescence. J. Mater. Chem. **2009,** 19 (41), 7661–7665.

34. He, J.; Liu, H.; Dai, Y.; Ou, X.; Wang, J.; Tao, S.; Zhang, X.; Wang, P.; Ma, D., Nonconjugated carbazoles: a series of novel host materials for highly efficient blue electrophosphorescent OLEDs. J. Phys. Chem. C. **2009,** 113 (16), 6761–6767.

35. Leung, M.-K.; Yang, C.-C.; Lee, J.-H.; Tsai, H.-H.; Lin, C.-F.; Huang, C.-Y.; Su, Y. O.; Chiu, C.-F., The unusual electrochemical and photophysical behavior of 2, 2 ‘-bis (1, 3, 4-oxadiazol-2-yl) biphenyls, effective electron transport hosts for phosphorescent organic light emitting diodes. Org. Lett. **2007,** 9 (2), 235–238.

36. Murawski, C.; Leo, K.; Gather, M. C., Efficiency roll-off in organic light-emitting diodes. Adv. Mater. (Weinheim, Ger.) **2013,** 25 (47), 6801–6827.

37. Yook, K. S.; Lee, J. Y., Bipolar host materials for organic light-emitting diodes. *Chem. Rec.* **2016,** *16* (1), 159–172.

Index

For Product Safety Concerns and Information please contact our EU
representative GPSR@taylorandfrancis.com
Taylor & Francis Verlag GmbH, Kaufingerstraße 24, 80331 München, Germany